BUCKMINSTER FULLER'S

UNIVERSE

BUCKMINSTER FULLER'S
UNIVERSE

LLOYD STEVEN SIEDEN

Foreword by

NORMAN COUSINS

PERSEUS PUBLISHING

Cambridge, Massachusetts

Library of Congress Catalog Card Number: 00-105205

ISBN 0-7382-0379-3

The Foreword appears courtesy of Omni International Publications Ltd. It originally appeared in the September 10, 1983, issue of the *Saturday Review*.

Perseus Publishing is a member of the Perseus Books Group

First paperback printing July, 2000

1   2   3   4   5   6   7   8   9—02   01   00

Find Perseus Publishing on the World Wide Web at
http://www.perseusbooks.com

To Clarice:
Whose love is the stuff of which this book is bound,
and without whose unswerving confidence, support,
and courage this project would not have been possible.

To Jesse, Daniel, Karley, David, and all the children of
Spaceship Earth—trim tabs of the future.

I wish to thank all those who contributed their expertise and resources to the assembling of the photographs presented in this volume. Although my research uncovered the fact that the majority of these photographs have been reproduced so many times in various books, magazines, and other publications that they are now considered to be in the public domain, reproducible facsimiles were provided by various individuals and institutions. Those who wished to be credited are listed below.

Unfortunately, the names of most of the actual photographers have been lost to history over time, and I have no way of crediting them in these pages. Fuller and the institutions with which he worked constantly received photographs of his projects from grateful admirers and associates and, because of his commitment to providing the best possible information to the public, he often published many photographs without crediting unknown sources. As Fuller and others have done in the past, I apologize to any original sources I could not trace.

Beech Aircraft Corporation: Figures 12–1, 12–2, 12–3, 12–4, 12–5, and 12–7

Bridgeport Public Library: Figures 7–1, 7–4, and 8–1 (photograph by F.S. Lincoln, New York)

Buckminster Fuller Archives: Figures 6–1, 6–3, 7–5, 9–2, 9–3, 11–1, 11–3, 13–1, 13–4, 14–1, 14–2, 14–13, 15–1, 15–4, 15–5, 15–12, 15–13, 15–14, 15–15, 15–16, 16–1, 17–3, and 17–4

Butler Manufacturing Company: Figures 9–4, 9–5, 9–6, and 9–7

Chicago Museum of Science and Industry: Figures 5–2, 6–2, 11–2, 13–1, 14–12, and 17–1

Chrysler Corporation: Figure 8–2

Mary Anne Fackelman-Miner, The White House: Figure 17–7

Ford Motor Company: Figures 14–5, 14–6, 14–7, and 14–8

Honeywell Inc.: Figure 10–3

Kaiser Corporation: Figures 15–2 and 15–3

Lyndon Baines Johnson Library: Figures 15–11 and 17–2

Milton Academy: Figures 1–4, 1–6, 1–7, and 3–2

North Carolina Department of Cultural Resources: Figures 14–3 and 14–4

Warren Schepp: Figures 2–1, 2–2, 2–3, 2–4, 5–3, 7–2, 7–3, 9–1, 10–1, 10–4, 10–5, and 12–6

Southern Illinois University: Figures 11–4 and 11–5 (photograph by Chris Bloomquist)

U.S. Marine Corps: Figures 14–9, 14–10, and 14–11

Windstar Foundation: Figures 17–5 and 17–6

Wrather Port Properties: Figures 15–6, 15–7, 15–8, 15–9, and 15–10

Doubleday: "Roam Home to a Dome," pages 333 to 334 (From *The Dymaxion World of Buckminster Fuller* by Buckminster Fuller and Robert Mark. Copyright © 1960 by R. Buckminster Fuller. Reprinted by permission of Bantam, Doubleday, Dell Publishing Group, Inc.)

Allegra Snyder: Poetry on page 418

Foreword

Memories of Bucky
Norman Cousins

 Once, when we were members of an American delegation in Moscow for the purpose of exploring outstanding issues between the two countries, Bucky Fuller gave a talk on the human future that no one who heard him soon forgot.

What happened was that one of the Russians had suggested that, as a relief from our regular conference sessions, we might stage a debate on what the world would be like in the year 2000. We agreed. Their debater would be Eugene Fyodorov, the famous meteorologist and futurist. Invariably, for our forensic gladiator we picked Buckminster Fuller, architect, inventor, cosmic chronicler, philosopher, and poet.

I groaned when I learned the ground rules set up by the Russians. Each debater would have fifteen minutes, precisely. I had never heard Bucky speak publicly for less than two hours. When the rigid time limitation was explained to Bucky, he merely shrugged, giving us the impression that fifteen-minute talks were a matter of casual routine.

Professor Fyodorov spoke first. Systematically, and methodically, he presented a checklist of all the factors that he believed would have a bearing on the world economic situation in 2000. He extrapolated figures with respect to world population, world food supply, world supply of vital resources, and other parameters. I looked around at the Russians as their champion spoke. It was obvious that they were pleased with the coldly scientific and comprehensive nature of the presentation.

After fifteen minutes, give or take ten seconds, Professor Fyodorov completed his talk and sat down. Substantial applause from all present.

Bucky started to speak. Within three minutes, he cast a spell over the entire group. The Russians sat forward in their seats. The world's greatest resources, he said, were to be found in human intelligence, ingenuity, and imagination. He then identified the principal problems of the riders on Spaceship Earth and gave the basis for his belief that these problems were well within human capacity. His earnestness, enthusiasm, creativity, and knowledge were beautifully blended.

Bucky sailed through the fifteen-minute barrier with the ease and confidence of Roger Bannister going through the four-minute mile.

As chairman of the evening session, I started to rise for the purpose of informing Bucky his time had expired. I felt a restraining hand on my arm. "Please let Mr. Fuller continue," Professor Fyodorov said. "He is magnificent, absolutely magnificent. You must not stop him."

I settled back in my seat. Bucky continued for almost an hour. The Russians were mesmerized. In the midst of the applause following his talk, Professor Fyodorov whispered in my ear.

"It was no contest," he said. "Mr. Fuller is the winner. Never in my life have I heard anything so wonderful. I am sorry he stopped so soon. Tell me, what did he say?"

The professor was not being sarcastic. Audiences all over the world have had the same experience. They may not have known or understood quite what Bucky was saying, but they felt better for his having said it. He gave people pride in belonging to the human species. He gave them confidence in their innate abilities to overcome the most complex problems. He made them feel at home in the cosmos.

Bucky made love to his audiences. And they loved him back. Well into his eighties, he maintained a travel schedule that seemed well beyond human capacity. He bounced back and forth through the time zones with the frequency of an airline pilot. I doubt that he gave fewer than an average of 125 lectures each year during the last quarter century of his life. Even after his 1981 hip operation, he continued his global one-man chautauqua.

Very early in his life he discovered the secret of perpetual curiosity and spent the rest of his life trying to give the secret away. Of all his attributes, none was more compelling than his ability to transmit to others his kinship with the universe. The goodwill that radiated from him, his enthusiasm for new directions and options, his capacity for liberating the human imagination from earthbound concepts and for propelling human beings into a new relationship with the world around them—all this lit up the minds of his listeners.

His hold on human beings was the product of many things. He was felt, and not merely heard. He created new energies in people by connecting them to the finite and the infinite. His uniqueness as a teacher in these respects was that he saw poetry in everything. Fuller viewed physics, astronomy, chemistry, and other sciences as much through the creative imagination as through equations and formulas.

In so doing, he refuted C. P. Snow's notion of the gulf between the "two cultures." Bucky Fuller regarded science and the fine arts as extensions of each other, as manifestations of an integrated reality.

The affection of students for Bucky Fuller offered the strongest possible evidence that young people are responsive to the values that give affirmative energy to a society. What they understood through him was that the main end of science is not to answer questions but to generate new ones; not to relieve curiosity but to enlarge it and ignite it; not to build better machines but to enable people to run them and control them. I have known very few people who came away from Bucky who did not forever after have an enhanced sense of sublime wonder when looking at a starlit sky.

If we turn to Buckminster Fuller solely for information, we will obtain information, but we will be cheating ourselves. We should study him for the increased respect he gives us for human potentiality; for helping us to learn that there are no boundaries to the human mind, which he celebrates above all else. The great poets have attempted to describe the human mind and spirit, but I doubt that any of them have been able to do so more provocatively than Bucky. The reason perhaps is that Bucky was not only inspired and nourished by the weightless and all-embracing entity called the human mind; he had a way of opening our minds to the phenomena within it. In this way, he introduced us to ourselves.

Preface

Exactly five years ago, R. Buckminster Fuller passed on. At that moment, I sat watching the movie *Wargames*, in which two young people saved our Planet from military annihilation. Although that motion picture was fictitious, it provides a striking analogy for Fuller's fundamental message. During the majority of his life, the man who preferred to be called Bucky explained that humanity's primary hope rested not with governments, corporations, religions, or other formalized organizations, but rather with individual human beings. He was particularly resolute in his belief that the single category of individuals which could best create and maintain the success of all humanity was the younger generations.

His belief in youth rested primarily upon the fact that young people tend to remain aloof from the traditions of the past, particularly the scarcity mentality which, according to Fuller while justified only decades ago, became obsolete in 1970. He felt that young people who had not yet been indoctrinated into the "you *or* me" system which has dominated human societies since the dawn of recorded history could facilitate humanity's entrance into an entirely new era predicated upon a "you *and* me" context.

Fuller himself was born into a generation in which there were not sufficient resources available to support everyone, and consequently, fighting for one's own share could be argued as acceptable. Even as a young man, however, Bucky was anticipating future trends and realized that in his lifetime, humanity might reach a point when there would be enough material goods to support all life on Earth. He also understood that if and when such a moment occurred, a new era—unlike what most people could even imagine—would instantly come to fruition. Displaying his characteristic love of unique possibilities and being of service to others, Fuller soon embarked on a lifelong mission of doing whatever he perceived was necessary to support humanity's attainment of that transformational moment as well as its entrance into the new era of prosperity which would follow.

While working at that mission, Fuller also earned the longest listing in the history of *Who's Who in America* and at various times during his life was classified as an architect, teacher, poet, cartographer, philosopher, scientist, writer, lecturer, businessman, mathematician, guru, futurist, visionary, pilot, inventor, administrator, entrepreneur, designer, cosmogonist, engineer, artist, and prophet—among other things. Yet, when questioned on his calling, Bucky responded in one of two ways. Either he would describe himself as a "comprehensive, anticipatory design scientist" ("comprehensivist" for short), or he would portray himself as being akin to part of a ship, the "trim tab." As he was raised in the seaside Boston suburb of Milton during a period when the majority of global trade depended on maritime transportation, it is not surprising that Fuller often likened his life's operation to that of a sailing ship.

He would explain that the trim tab was the tiny flap on the trailing portion of a ship's rudder or of an airplane's wing. Most people who are not familiar with aeronautical operation believe that a ship's steering is accomplished by the rudder; however, even massive ocean liners are, in fact, maneuvered by the slightest of pressures on the tiny trim tab, which, in turn, moves the rudder. The shift in the rudder then changes the ship's course.

During his twenties, Bucky came to believe that any individual could, by simply exerting the subtlest of pressure in the proper place, become a trim tab for humanity, and he set out to live his life in that manner. Fuller became so adamant about the contribution he could make that in 1927 he actually created an experiment using himself as "Guinea Pig B" (for "Bucky"). The specific purpose of his experiment was to determine and document what one individual could accomplish on behalf of all humanity which could not be achieved by any organization, government, or business, regardless of its size or power. That experiment remained a critical element of Fuller's daily life until his death fifty-six years later. The following pages recount much of Fuller's experiment, both strengths and weaknesses, critically scrutinizing it as any experiment should be examined.

Throughout his life, Buckminster Fuller shared his vision of the impending new era. He spoke of and worked to support a period of global abundance in which every human being living on the Planet he named Spaceship Earth would be provided with the necessities of life. Yet his claim that such an option became a reality in 1970 was no idle boast. Rather, it was the documented contention of a pragmatic engineer and inventor who had inventoried and studied global resources for decades and was certain that there were enough resources to support everyone. How Fuller came to that conclusion is essential to his life, and although that understanding of resources provided the context for his operation, it is only one facet of a man who refused to be categorized and who maintained a childlike interest in nearly everything he encountered.

But this book is not so much about Fuller's life as it is about the principles which guided him and which can provide inspiration and insight to those who attain an understanding of them. In the future, someone may well write a complete biography of Buckminster Fuller, filled with all the quirks of his personal life. This is not that book. Rather, it is a translation of Fuller's own ideas and the principles he discovered from the generally convoluted style dubbed *Fullerese* into a language which can be understood by almost anyone. Readers seeking additional biographical and historical information may refer to the chronology in the back of this book.

I have endeavored to intersperse those ideas and principles within a biographical context to provide the reader with a sense of how an undersized, nearsighted boy born prior to the invention of most of the conveniences we now take for granted could, like Leonardo da Vinci, develop innovations and ideas which may not be practically employed for decades. Using Fuller's life as the continuity for presenting the principles he discovered and lived by also provides some sense of how a single committed individual can make such an enormous contribution to so many others.

I do not want to imply that Fuller was, by any stretch of the imagination, a saint. On the contrary, he was, as he frequently explained to his audiences, simply a normal human being who responded to circumstances. The magnitude of the circumstances he noticed and responded to was, however, much larger and more encompassing than most of us ever notice, much less consider doing something about. It was his vision of the comprehensive nature of human beings and all Universe to which Fuller responded and which he discussed with audiences throughout his life.

By speaking about his vision rather than himself, he created a persona far different from the personality who loved fast cars and sailing ships as well as the companionship he found in drinking and partying for so many years. In attempting to provide the most significant elements of Fuller's work, I have chosen, as Bucky did, not to focus on his personality. Like Fuller, I have not attempted to hide the truly human characteristics which establish his human frailty, but rather, to concentrate on the elements of his philosophy which supported his successful contributions to so many others.

When this project began over four years ago, I had no intention of writing a book. At the time, I was presenting educational programs about Fuller and his work, and I was constantly asked a question which stumped me: What book would I recommend for those who wanted to learn more about Fuller?

Having spent years reading and studying the books by and about Fuller, I could not, in good conscience, recommend any of them. It was not that other books could not provide the information people sought, but that, to some extent, each of them reflected Bucky's style of communication. Thus, books by and about Fuller tend to be filled with sentences akin to paragraphs and contrived words which Bucky would create when he could not find a precise word which expressed his thought.

After years of reading and listening to his style of communication, I had become accustomed to it and understood it fairly well. However, I also realized that most people picking up *Synergetics, Critical Path,* or any of Fuller's other books are generally confused and frustrated by his style within a few pages. Fuller's genius was as a master thinker, not a communicator. In this book, I have endeavored to break his code and translate his most significant thoughts into accessible language.

To remain in accord with Fuller's essential ideas and intent, I have capitalized certain key words which are not usually written in this manner. Items such as Nature, Greater Intellect, Planet, Earth, and World are proper names and should, in my opinion, be capitalized. As Fuller stated in a 1977 preface for the book *American Space Photography,*

> As I write the introduction to this truly fascinating book, I find its author
> and its publishers spelling our planet's name with a small "e," even

though its 7,926-mile diameter is larger than the mile diameters of
Venus's 7,620, Mars's 4,220, Pluto's 3,600 or Mercury's 3,010. The
first letter of all the other planets' names we honor with capital letters. I
am confident that those who spell Earth with a small "e" are not yet
"seeing ourselves as others can see us" from elsewhere than aboard our
cosmically minuscule planet.

I have also chosen to omit, as Fuller did, the commonly used article
the in front of the word *Universe*. Fuller felt that there was but a single
Universe, of which he and everyone else was a part, and he was not
willing to separate himself from Universe or relegate it to a less significant
status by the article *the,* which implies that others also exist.

He defined Universe as "the aggregate of all humanity's consciously
apprehended and communicated nonsimultaneous and only partially over-
lapping experiences." Because Fuller perceived everything as being composed of energy experiences, his definition of Universe is all inclusive.

This is but one example of Fuller's comprehensive character. The
difficulty for both scholars and individuals from all walks of life has
always been categorizing Fuller's inclusiveness for easy comprehension.
This book does not attempt any such classification. Instead, it presents the
significant underlying principles which guided Bucky in being able to
successfully operate in so many areas. By documenting those principles as
well as presenting examples of how he utilized them, I hope to share some
portion of the understanding and the essence of his wisdom which Bucky
has so graciously contributed to my life.

Lloyd Steven Sieden

July 1, 1988

Contents

BUCKMINSTER FULLER'S
UNIVERSE

CHAPTER 1

Comprehensive Heritage

 Bucky Fuller was born on July 12, 1895 in Milton, Massachusetts, just outside Boston, to Richard Buckminster Fuller, Sr., and Caroline Wolcott (Andrews) Fuller. He was christened Richard Buckminster Fuller, Jr., and given the unusual but ancestral family name Buckminster.[1] That name dated back to his celebrated forebear Colonel Joseph Buckminster, who left the British Army to fight with George Washington in the Revolutionary War.[2]

Before he was ten years old, young Bucky, as he had come to be called by both family and friends, was already developing the classic Fuller flair for selecting his own pathway, and he demonstrated that resolve when deciding how he would inscribe his name. Tradition at the Fuller family's summer retreat on Bear Island in Maine's Penobscot Bay required all visitors, be they family, friend, or stranger, to sign the guest register. As a child, each summer Bucky would experiment in penning his yearly contribution to that register. One year he signed in as "Richard B. Fuller," another year as "Richard (Bucky) Fuller" and another as "R. B. Fuller." Finally, after years of experimentation, Bucky settled on the combination he maintained throughout the balance of his life. He felt that using his first initial with his uncommon middle name was the most sophisticated combination; thus, Bucky became R. Buckminster Fuller.[3]

In many respects, he was the culmination of several generations of New England nonconformity. Not adhering to traditional beliefs was so prevalent in the Fuller family that it had become the status quo.

Fuller's great, great, great, great-grandfather, Lt. Thomas Fuller of the British Navy, traveled to the American Colonies on a furlough in 1630 and became so exhilarated by the revolutionary demands for greater freedom being openly expressed that he settled in New England. Lt. Fuller's offspring shared his love of new ideas and possibilities, and his grandson the Reverend Timothy Fuller graduated from Harvard in 1760 and became a Massachusetts delegate to the Federal Constitutional Assembly. The nonconformist, freethinking Reverend Fuller was so incensed about the fact that the draft Constitution did not prohibit slavery, he refused to vote for its ratification.[4]

His son, Timothy Fuller, Jr., was born in 1778 and gained notoriety as a founding member of Harvard's Hasty Pudding Club. As a penalty for taking part in a student revolt, Timothy, Jr., was graduated second in the Harvard class of 1801 rather than in his rightful place at the head of that class.[5]

Fuller's grandfather the Reverend Arthur Buckminster Fuller graduated from Harvard in 1840 and maintained the family tradition of supporting greater freedom for all people. He was a staunch abolitionist, and although he held positions as a minister in Boston and chaplain of the Fifth Massachusetts Regiment, he was not opposed to fighting for his beliefs. His dedication to the cause of freedom eventually led to the Reverend Fuller's untimely death while leading a successful charge of Union troops during the Civil War.[6]

From records of family history and accomplishments, it is clear that the external achievements of the Fuller family, as in most families prior to the turn of the century, were fulfilled by men. Young Bucky was, however, influenced by one strong female ancestor, his great-aunt Margaret Fuller. Margaret Fuller was the singular, prominent nonconformist female of the Fuller clan, and in her short life from 1810 to 1850, she became the most famous of Bucky's ancestors.[7]

Margaret was a feminist long before the term was coined, and she paved the way for future generations of women to fight for greater freedom and equality.[8] Her platform for change was the written word, and in conjunction with her good friend the poet Ralph Waldo Emerson, Margaret Fuller founded the Transcendentalist literary magazine *The Dial*. She also became the first female foreign correspondent employed by a major newspaper, *The New York Times*.[9]

Although she died decades before Bucky's birth, Margaret Fuller's influence endured within the Fuller family, and young Bucky was not exempt from it. He was told stories of his famous aunt, her distinguished cohorts, and their Nature-based philosophy which stressed the divine orderliness of the Universe.[10]

Years later, when Bucky himself began to establish his comprehensivist philosophy, he could not help but be influenced by the connection between human beings and all Nature which Margaret had championed. His philosophy would, however, also include the mechanical technology which was just beginning to seriously influence lifestyles when Transcendentalism was flourishing and which was overlooked or viewed as a negative influence by most Transcendentalists. Bucky felt that natural creations such as the human hand represented technology at its finest and believed that the best human-created technology was that which most closely mirrored Nature's creations.[11]

Although she significantly influenced Bucky's ideas, Margaret was an exception to the male-dominated society into which he was born. His family was no different than most, in that men were expected to go out into the world while women were supposed to fulfill a role regarded by the men as far less important: maintaining the home.[12] Although Bucky sought to overcome prejudice at all levels, even in the later years of his life, the influence of the male-dominated environment into which he had been born continued to influence him. Still, the strong women of the Fuller family, including his mother and Aunt Margaret, did somewhat balance Bucky's exposure to the traditional male-dominated perspective. Their influence nurtured what would later blossom into a passionate desire to eliminate discrimination wherever he encountered it and to work at projects which benefited all humanity.

Fuller, however, could not completely break free of his early upbringing. Even though, during the final years of his life, he proclaimed that the Equal Rights Amendment to the Constitution was the most significant piece of legislation ever written, women tended to be relegated to support positions rather than partners in his life.[13] He was always quick to acknowledge the contributions of his wife, his mother, and other women, but he rarely spoke or wrote about the accomplishments of women in the same glowing terms as he did those of his male ancestors or men like Henry Ford and Albert Einstein. Special adulation was, however, reserved for his father.

Like all the American Fuller men before him, Bucky's father, Richard Fuller, attended Harvard. Following his graduation in 1883, however, the rebellious Richard became the first male in eight generations who did not observe the family heritage of service and prominence as a minister or a lawyer.[14] Instead, Richard Fuller indulged the family penchant for adventure to become an importer of leather hides and exotic teas. In that position Richard Fuller became one of the few Milton citizens who journeyed farther than could be traveled in one day on foot or by horseback.[15]

The Fullers were one of many families who moved from Boston to Milton, a town which had for generations served primarily as a summer retreat from the Boston heat. In Milton, Richard and Caroline Fuller contracted for the construction of a large house, and in that house their second child and first son, Richard Buckminster, Jr., was born on a cloudless summer day in 1895. At the time, the Fullers' immediate family

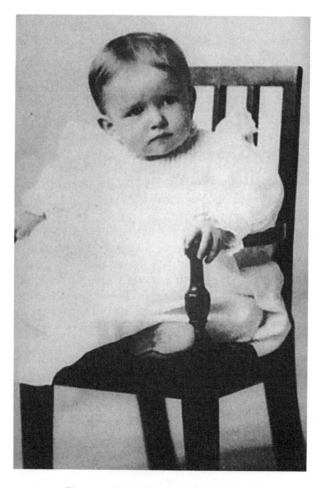

Fig. 1–1   Early baby photograph of Bucky.

Fig. 1-2 Young Bucky (upper left, sitting astride stone ledge) with some of the Fuller clan, 1899. Brother Wolcott and sister Leslie are seated second and third from the left.

consisted of only one other child, Bucky's older sister Leslie, but within a few years two more children were born, brother Wolcott (Wolly) and sister Rosamond (Rosy).[16]

Even as a youngster Bucky was unique within the Fuller clan, which included numerous nearby aunts, uncles, and cousins.[17] Because he had been born with an extreme undiagnosed nearsightedness, his visual perception of reality was limited to fuzzy shapes, and he compensated for that disability by relying on his other senses as well as his independent thinking for an understanding of his environment.[18]

By the time he had reached the age of four, Bucky, as he had been nicknamed by friends and family, had developed an acute sense of hearing, touch, and especially smell. In fact, his sense of smell was so sensitive that later in life he would assert that he was usually able to recognize and make inferences about people only from their scent. And he retained those acute capacities throughout his life. Even in his eighties, he would amaze people by demonstrating an uncanny ability to determine a great deal about a person just from a scent which went undetected by others.[19]

Because of his handicap, during the last of those nearly sightless initial four years of life, young Bucky fabricated what would be the first of many futuristic structures. His kindergarten teacher provided her students with the traditional toy construction materials of the day, dried peas and toothpicks, and the other children immediately began reproducing structures which mirrored the buildings they had observed in daily life. Naturally, their frameworks were rectangular, supported by right angle corners held firm with dried peas. However, with no accurate visual experience to rely upon, young Bucky began creating a structure which satisfied his sense of touch rather than imitated adult construction.[20] Accordingly, his framework was composed of stable triangles and was, in fact, a rudimentary model of the octet truss, which he would invent and patent in 1961.[21]

Although his teacher was fascinated enough to call in the other teachers to see what little Bucky had built, no one had any idea just how important that crude structure would become.[22] Even today, few people who are not involved in the construction industry appreciate the importance of the octet truss. Yet, without that light, incredibly strong structure, a great deal of construction would not be possible. One need only look up during a concert or stage performance to see Fuller's invention. The

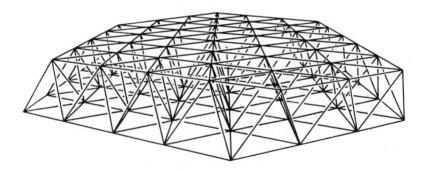

Fig. 1–3   The modern octet truss.

Fig. 1–4   Milton Academy, 1904. Bucky is fifth from the right standing behind his bicycle.

triangulated scaffolding which provides the strength for staging many performances is simply a series of connected octet truss structures.

The year 1900 ushered in a new century as well as an entirely new life for Bucky. Late in 1899, his nearsightedness was diagnosed, and eyeglasses were prescribed. Following that momentous transformation, the five-year-old was able to view a new world beyond his wildest dreams. Prior to receiving glasses, Bucky had believed everyone was confined to the same blurred environment he experienced. So when his sister Leslie described the world as she saw it, he thought he was listening to the output of a fanciful imagination. With his newly discovered, normal vision Bucky embarked upon a journey of outward exploration, questioning, and experimental learning which would continue throughout his life.[23]

The moment he obtained eyeglasses, Bucky felt reborn, excited to discover an endlessly fascinating environment.[24] He was especially captivated by eyes, both human and animal, and quickly became aware of the communication occurring between people's eyes.[25] Because he had functioned without true visual contact for the first years of his life, Fuller valued the human capability of conscious eye-to-eye communication more than most people, and he knowingly relied upon it for information and insight.

In typical Fuller fashion, he turned his apparent visual problem into a benefit over the years. Rather than regarding his myopia as a handicap, Bucky considered himself fortunate to be able to remove his glasses at times and consciously shift his attention inward, using his imagination and intuition, rather than his sight, for guidance. At those times, he could effortlessly recall the inner, imaginative environment he had lived within and explored prior to receiving glasses, and that "in-sight" provided him with many of his extraordinary ideas.[26]

Fuller's formal education was typical for an upper-middle-class New England boy at the turn of the century, but he was perpetually hampered by his small physique and dense eyeglasses. Still, he was an athletic, adventurous lad who perpetually tested limits as well as the patience of his parents and relatives. Along with his best friend, Lincoln Pierce, Bucky constantly attempted to climb and run farther than he had before as well as to build new types of "structures." He also spent a great deal of his time emulating the folk hero Robin Hood. Richard Fuller read stories of Robin Hood to his son at a very young age, and they assumed massive proportions within Bucky's vivid imagination. Once he had found a role model, it was not long before Bucky was patterning his adventures after Robin.[27]

Rather than cowboys and Indians, Fuller's play centered on roaming nearby woods and effecting swift moral and romantic justice for creatures, both human and animal, he found in distress. In time, his unmistakable Robin Hood escapades ceased, but the image of rushing in on his "white horse" to save others from peril remained with Fuller. Thus, he often found himself charging into situations and assuming responsibilities well beyond his legal and physical capabilities in a rash effort to help people in trouble.[28]

Frequently those surprising acts of confidence were rewarded when other people and resources joined his crusade or he succeeded on sheer audacious courage alone, but during the early years of his adult life, Fuller frequently found himself overmatched against established "giants," be they from government bureaucracies or construction trade unions. In those futile challenges, Fuller usually came out the vanquished hero rather than the victorious Robin Hood, and he soon mastered the art of thoroughly considering the competition before jumping into a situation.

The early exploits of Bucky–Robin Hood were most prevalent during his favorite time of year, summer. He loved the family encampment, which was held at their privately owned, primitive complex on Bear Island in Maine. The Fuller family had purchased the entire island as well as a few smaller surrounding islands, and each summer those islands served as a gathering site. Even during the later years of his life when he was traveling around the World several times each year, Bucky always set aside the month of August for his Bear Island retreat.[29]

As a young boy on the island, Bucky learned the skills of the sea from neighboring sailors and fishermen, and there he began to develop a reverence for and appreciation of Nature. And it was in that Maine wilderness that he also came to appreciate Nature as something more than an entity to be tamed and dominated by human beings. Several months on the island each year provided Fuller with valuable experience living in harmony with Nature and observing the elegant, simple technology of Nature itself as demonstrated throughout the island's wild environment. Prone to investigation and inquiry, Bucky quickly initiated a program of engineering inventions which provided a greater degree of creature comforts for the inhabitants of rugged Bear Island.[30] Without a great deal of conscious thought, those inventions mirrored what he would later designate as the "natural technology" of his observations throughout the island.

Natural technology simply referred to employing the extremely efficient techniques he perceived within Nature throughout his own designs. From that study of Nature, Fuller learned to mirror Nature's ingenious, yet uncomplicated principles and designs in increasingly more sophisticated inventions, such as the geodesic dome and the Dymaxion Map.

One of Bucky's earliest Bear Island inventions modeling Nature was a mechanical oar patterned after the motion of jellyfish. It consisted of a tepee-like cone mounted on the end of a pole and resembled an inside-out umbrella. Standing at the rear of his boat, Fuller would pull the pole toward him through a large iron ring attached to the rear of the boat. As the peak of the umbrella-like invention pointed toward the boat, it displayed little resistance. However, when it was pushed backward into the water, the cone opened, propelling the boat forward.[31]

That simple device allowed Bucky to push his boat with almost the same force as if the pole were touching the bottom of the waterway. He employed his "mechanical jellyfish" to propel his boat far more rapidly and with less effort than was possible using conventional rowing techniques.[32] Thus, he found that the time required to complete his daily chore of rowing the four-mile round-trip mail run to Eagle Island was cut in half

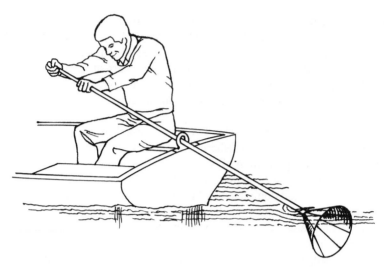

Fig. 1–5   The mechanical jellyfish young Bucky invented to propel his small boat.

by employing his mind rather than mere muscle. Bucky also discovered that his boat trips were much safer and more enjoyable because he could perform the pushing maneuver while facing forward so that he could see any hazardous or interesting events developing in front of him much sooner than if he were rowing facing the rear.[33]

By designing, building, and maintaining boats under the watchful guidance of neighboring seamen, Fuller learned a great deal about basic construction and the special requirements of moving vehicles which would serve him well in later designs.[34] Those same men also taught him to be an accomplished sailor, an avocation which he continued to love throughout his life.

The discipline of sailing provided Bucky with a pragmatic understanding of the importance of designing a precise location for each element of any structure or device, stowing everything for easy access while keeping things out of the way when not in use. The sailing credo of "everything in its place, and a place for everything" thus became an essential component of Fuller's consciousness while he was still a boy, and it served him well throughout the ensuing years.[35]

His nautical training also supported Bucky's understanding and appreciation of natural resources and their efficient deployment. In sailing, he found that each crew member and component of a ship was most valuable when situated in its proper location and could become a serious problem when out of place. He came to understand that, similarly, resources can be employed even if they are out of place, but that they are most beneficial within the confines of a single task.[36] For example, a wrench can be used as a hammer when a hammer is not available, but it is most effective in accomplishing the tasks for which it was designed.

Fuller soon realized that a similar axiom of maximum efficiency could be applied to human beings, and later in life, he employed that insight to consciously determine his niche of maximum efficiency so that he could provide a maximum contribution to humanity. In later years, the cultivation of his insight into the importance of the placement of elements helped him to formulate his perception of pollution as much more than simply a negative component of the environment.[37] Instead, Fuller described pollution as resources not positioned at their maximally effective location. In the most extreme cases, such dislocation is not only less beneficial, but harmful.[38]

Bucky often illustrated that perspective when he discussed sulfur polluting the atmosphere. Almost everyone has observed huge plumes of sulfur pouring out of industrial smokestacks to dirty the air. Yet, at the same moment one industrial giant is releasing large quantities of sulfur into the air, other production facilities are mining it from the ground to be used in manufacturing processes.[39]

Clearly, when examined from a comprehensive perspective, this is a case of resources requiring positive relocation. The polluters need only reclaim the waste sulfur from their processes and relocate it to those industries which require the material. By operating thusly, polluters not only become good citizens, they may also profit from the sale of a substance they once considered waste material.[40]

With his "Robin Hood consciousness," Fuller himself was always searching for such relocation situations. Whenever he uncovered a significant example, he would attempt to have the errant resources transferred to locations and applications in which they would be of benefit rather than a problem. He also assisted others in attaining a perspective which aided them in understanding, discovering, and supporting similar resource reallocation.

Another major influence on Bucky was his father. Although Richard Fuller, Sr., spent much of his time traveling around the World and died early in Bucky's life, he was respected and remembered by his young son as well as the Milton community. His work as an importer caused Richard

Fuller to take many extended voyages to South America and India in pursuit of merchandise. Even decades later, Bucky could vividly recall his father being away for several months at a time and coming home with stories and pictures of exotic destinations.

Upon returning to Milton, the senior Fuller was welcomed as a local hero. Much of the small community would gather at the Fuller home to listen intently as he spun intriguing tales of the foreign lands he had visited, and Bucky, like most young children, was particularly captivated by his father's heroic travels. He also dreamed of the day when he too might undertake such exciting journeys with his father.

Although Bucky would eventually travel to places his father never knew, those journeys would not be in the company of the elder Fuller. Richard Fuller suffered the first of a series of debilitating strokes in 1907, and they progressively limited his activity until his death three years later.[41] During that difficult period, Bucky, who was entering his teenage years, earnestly helped with his father's care whenever possible. In that effort, he was also witness to the rapid deterioration of not only his father, but his hero. That traumatic experience may well have left a lifelong emotional scar which was evidenced in a rigid tenacity that belied his diminutive frame. Although even as an adult he was only five foot six inches tall, Fuller consistently demonstrated an inner tenacity which might have been his way of proving that he did not suffer from the same weakness he had witnessed in his father.

Bucky also developed and exhibited an innate compassion and respect for human life during the three years he helped with his father's care, and those characteristics would eventually become his hallmark as millions of people experienced his caring nature over the course of his life. He would spend hours patiently fanning his father during the hot afternoons or reading aloud to the man who had only years earlier been the one reading aloud to his young son.[42] Still, as his father's condition continued to degenerate, Bucky began to wonder about the force which was abandoning a man who had been so vital only a few years earlier. As with many of his questions, that issue was so complex that many years transpired before Bucky began to appreciate and understand that essence of life. Esoteric questions were abruptly put aside when Richard Fuller succumbed to his illness and died on Bucky's fifteenth birthday in 1910.

His father's death was a tremendous blow to the family and especially to Bucky. In his short lifetime, Richard Fuller had had a significant impact on his children. He had opened them to a World seldom seen by children or adults and had implanted within them a myriad of possibilities. He had also supported them in a luxurious lifestyle which could not continue without his income.

The end of his father's life brought about drastic changes for the household and Bucky. Not only had he lost the strong masculine guidance of a man he loved and respected, but, as the eldest son, he was expected to assume some responsibility as "man of the house."[43] Thus, Richard Buckminster Fuller, Jr., initiated an attempt to become the next Fuller man to maintain the family tradition of education and a distinguished career which had been carried on for centuries. Unfortunately, those family traditions and expectations did not fit into the changing environment Bucky was beginning to experience and in which he was starting to experiment.

The death of Richard Fuller was the first of several emotional shocks which molded Bucky into a rambunctious teenager who frequently overreacted and eventually into an adult who sometimes succumbed to that same characteristic. He needed love and approval during those difficult teenage years, but he was often confronted with austere discipline as well as lack of support and respect from a family strained by an enormous trauma.[44]

Being a strong, proud woman from a family steeped in an American ancestry as deep as that of Bucky's father, his mother attempted to maintain the family's affluent lifestyle; however, the tremendous changes which began with Richard Fuller's first stroke only escalated following his death. As the eldest male, Bucky was expected to take on additional responsibilities while serving as a role model for his brother and his two sisters. Out of financial necessity, the servants who had supported the Fullers' comfortable lifestyle were dismissed, and Bucky assumed the physically demanding chores such as stoking the furnace through long New England winters, cleaning heavy rugs, and all the yardwork required for a large house. It was while toiling at those arduous, yet necessary, chores that Fuller began to develop a respect for housework and appreciate the enormous amount of time and energy it demanded.[45]

At that time, even the simplest cleaning chores required strength and endurance. For instance, heavy rugs were carried outside and hung over clotheslines, where they were beaten and brushed by hand.[46] Years later, Fuller would recall his early house-cleaning experiences when developing the ultramodern Dymaxion House and other dwellings, and because of those recollections, he would mindfully design built-in, labor-saving devices, such as suction ducts which automatically removed loose dirt and dust from a home.

While helping care for his ailing father, Bucky continued his formal education at one of New England's finest and most expensive preparatory schools, nearby Milton Academy. Milton was predominantly a boys' boarding school. However, because of his family's financial problems, Bucky could attend only as a day student from 1904 through his graduation in 1913.

Although his principle interests at Milton Academy were not what his mother preferred, they were no different from those of most boys. He was far more concerned about being accepted by his peers than receiving the best possible education, and he would have been just as happy attending a less prestigious school which would not have taxed the family finances. Mrs. Fuller was, however, determined her eldest son would receive the best possible education, and she scrimped and saved so he could study at Milton, even though studying was not at the top of Bucky's priority list during those years.

Fitting in with the other boys while attending Milton on a limited budget, as well as his budding sense of independence, placed Bucky in a precarious position, and problems quickly arose. Without funds, Bucky was unable to have the clothes and other items which would have helped conceal his ungainly appearance. With large, thick glasses magnifying the size of his eyes and a head too large for his body, Bucky looked to be a scholarly owl.

In addition, he had already begun to develop his insatiable curiosity about everything, and his attention tended to regularly stray far from his own personality and appearance.[47] That wandering curiosity, which would serve as a wellspring of new ideas in later life, was a problem for the teenager. It resulted in his dressing carelessly and not paying attention to the social norms followed by the rich boys, making him even more of an outsider at Milton.

His enormous curiosity and stubbornly independent thinking did bolster him scholastically even though they eventually impeded his progress within the educational system. Bucky was especially adept in science and mathematics, where he generally received A's. He did not fare so well in the subjects which were of little interest, such as classics, Latin, English literature, and history.[48] At Milton, his curiosity and constant questioning led him into some of his first serious conflicts with "authority." The dimensions of his student–teacher confrontations as well as the right-eousness of his position are evident in the following illustration, which Fuller recounted thousands of times during his own lectures.

When Euclidian geometry was introduced to his class, the teacher made a dot on the blackboard and explained that it did not have substance but was merely an imaginary point in space. She continued by declaring that a series of points produced lines and a series of lines produced planes, all of which were nonexistent. The teacher then told the class that planes could be assembled to create three-dimensional shapes, such as cubes and pyramids, which did have substance and actually existed in the physical environment.[49]

It was during such moments of obviously illogical justification that Bucky could no longer restrain himself and entered into incessant questioning. First, he wanted to know how something of substance could be created from nothing. Then, he wanted to know about the cube that had been created. How much did it weigh? What was its size? And where was it located?[50]

Bucky's geometry teacher quickly diagnosed his inquiries as impertinence and "suggested" he suppress his wisecracks before he found himself deeply immersed in trouble. Realizing he had confronted (and threatened) the teacher's authority and that she was either unwilling or unable to answer his questions, Bucky backed off.[51] And after a few such incidents, he also realized that the system would not tolerate his outbursts of curiosity. Overwhelmed by an intense desire to please his family and fit in with the other boys, Bucky then resolved to do his best to keep quiet and "play the game" of school according to its illogical rules.[52]

He did, however, continue thinking independently and noting the differences between his personal experiences and the lessons he was being taught in school.[53] Years later, those obvious differences would become a primary motivation for the creation of his unique philosophy, operating strategy, and search for a truth which conformed to his experience rather than tradition.

Although he truly wanted to understand what he was being taught, like most children, Bucky sought acceptance by his peers above all else. And since he knew that excelling scholastically would not endear him as "one of the fellows," he sought to demonstrate his masculinity on the field of athletics. He exhibited his love of challenges by not selecting a sport suited to his physical attributes; rather, he chose to try out for the school's championship football team.[54] Summers on rugged Bear Island

Fig. 1–6   The Milton Academy football team. Bucky is seated in the front row, far left.

had helped Bucky develop a surprising strength for his small size, but his poor vision and glasses seemed to rule out rugged contact sports.[55]

Still, Bucky was resolute in his intent, and although day students were relegated to second-class status at Milton, the rigid determination Bucky seemed to have acquired from his father's death served him well in the trenches of football scrimmages. The shadow of that incident also continued to haunt him in less positive ways, as was evidenced in a football-related episode. During the first days of practice, he learned that the school did not furnish all the required equipment; each member was expected to provide his own helmet. When he told his mother that he needed a helmet, the frugal Mrs. Fuller, who was attempting to manage without her husband's income, attempted to satisfy her son and preserve her budget by purchasing what she felt was adequate. Consequently, Bucky received a cheap imitation "helmet" made of quilting rather than the genuine, sturdy leather variety used by the other boys.[56]

Although unhappy about his mother's judgment, Bucky politely thanked her for the helmet. He did, however, feel a great deal of shame and embarrassment when he was forced to wear it in order to play, and in conjunction with other experiences, that incident may well have contributed to his attitude toward both women and money. In fact, the feeling of embarrassment was so powerful that even during the last years of his life, Fuller vividly recalled the incident and the accompanying sense of inferiority, as well as his mother's lack of understanding.[57]

Having an intimate familiarity with frequent embarrassment and rejection resulting from lack of material possessions following his father's death provided Bucky with great empathy for less fortunate people everywhere. Accordingly, he later sought, with tremendous fervor, to relieve the suffering of large segments of dispossessed humanity. Most conspicuous was his passion for finding ways to feed and shelter all humanity, and a great deal of his energy focused on reallocating military resources toward such humanitarian endeavors.

Even as a youth, Bucky's enthusiasm was something to behold, and football was no exception. Through sheer zeal he overcame the obstacles of his small size, poor eyesight, and lack of funds to become a star athlete. In fact, he was so successful that during his senior year, Fuller won the coveted position of varsity quarterback. The previous three Milton Academy senior quarterbacks had become quarterbacks at Harvard, and Bucky desperately wanted to continue that tradition when his application to Harvard was accepted in 1913.[58]

He was the fifth consecutive generation of the Fuller family's men to attend Harvard. His father, grandfather, great-grandfather, and great-great-grandfather had all attended that university and had kept detailed diaries and records of their college experiences. Those documents were given to Bucky prior to his departure for college, and he studied them intently.[59] In fact, he may have learned more from his ancestors' records than he did from his formal Harvard education.

Through the diaries, Fuller was exposed to vivid recollections of the living conditions experienced by the Harvard classes of 1883, 1843, 1801, and 1760.[60] Studying those accounts provided him with a unique historical perspective available to few individuals of that, or any, era. The diaries presented a broad portrait of human development in America from the time of George Washington through the industrial revolution. From them, Fuller gained an unparalleled understanding of the changes which had occurred since America had declared its independence and how many of those innovations had enhanced the quality of everyday life for each successive generation of his family.

For example, each of his ancestors recounted two-day round-trips between the Cambridge campus of Harvard and Boston proper. That short journey was time consuming because travel was limited to two possibilities: on foot or by horse. However, in 1913, when Fuller entered Harvard, an entirely new era of transportation had begun, marking the onset of a dramatic transformation in the relationship between human beings and their environment. That shifting relationship was conspicuously demonstrated by the 1913 opening of the subway system, which carried people between Cambridge and Boston in only seven minutes.[61]

Adhering to tradition was not a quality Bucky Fuller ever mastered, and nowhere was his stubborn independence more strikingly evident than

Fig. 1–7   A 1913 formal portrait taken just prior to Fuller's entering Harvard.

in his ideas on education. Although he had kept relatively quiet during his years at Milton Academy, Bucky's attempt to follow in his ancestors' educational footsteps was another matter. He entered Harvard in 1913 with a set of beliefs very removed from the reality which existed at that institution.[62]

Throughout his boyhood, Bucky had been exposed to the idealistic Harvard reminiscences of his father and his father's friends, and as is true of most remembrances, those men had spoken in glowing terms about their most positive experiences. Thus, Bucky became intoxicated with stories extolling Harvard athletics, scholarship, and gentlemanly fellowship and he entered Harvard believing it to be an essentially mythical kingdom of superathletes and heroic individuals guided by the highest of ideals.[63]

Young and impetuous, Bucky was not, however, extremely interested in the superb formal education being provided for him. Like most eighteen-year-olds, he was far more concerned with socializing and the singular issue which continued to haunt him: being accepted by his peers.[64]

At that time, the Harvard social scene was dominated by clubs similar to modern fraternities, and all his Milton Academy friends were being invited to join those clubs. No such invitations came Bucky's way. Years later, he would explain that peer rejection as a direct consequence of his lack of family wealth and a father to arrange an entry, but it also had a great deal to do with his tenacious individuality.[65]

Rather than conform to the standards of the young aristocrats who dominated Harvard, consciously and unconsciously Bucky tended to overlook his appearance, while continuing to pay a great deal of attention to his experiences and environment. It was as if some part of him realized that the system was unjust and sought to keep him apart from it, while another aspect of his character battled for peer acceptance. In the midst of such internal contradictions, something had to give way, and it was not long before a troublesome incident occurred.

His first semester at Harvard was filled with frustration and dejection. To compensate for the fact that all his Milton friends were becoming members of fraternal clubs and would no longer associate with him, Bucky again focused on athletics. Football had gained him acceptance at Milton, and he felt that sport would again serve his purpose. Having won the position of varsity quarterback at Milton, Bucky was a prime candidate to attain that position on Harvard's team, but his dream was abruptly shattered when he broke his knee during a practice session.[66] To aggravate matters, Bucky fell in love for the first time but was left heartbroken when the girl dropped him for another young man who belonged to one of the prestigious social clubs.[67]

Desperately he concentrated even more on impressing his former friends rather than on his studies and went so far as to create his own social club. When his club did not impress or attract anyone, Bucky sought other solutions and inadvertently stumbled upon a unique method of impressing other young men with his skill as a worldly gentleman of interest to women. That episode exemplifies the reckless abandon with which he often entered into risky activities before considering all the possible consequences.

At the time, his older sister, Leslie, was moving and asked Bucky to care for her Russian wolfhound during the process. While walking the elegant dog one evening, Bucky passed the stage door of the local theater prior to a Ziegfeld Follies type of show and found that all the young actresses would stop to pet the wolfhound and chat before entering the theater. Thus, the stage door of the *Passing Show of 1912* quickly became a regular destination on Fuller's evening constitutional with the dog, and those walks soon resulted in dinner dates with the actresses. Fuller was particularly enamored of a young starlet named Marilyn Miller, and when the show moved to New York City, he withdrew most of the money his mother had deposited in the bank for his tuition, lodging, and other expenses, skipping his midyear exams to visit Marilyn Miller.[68]

During that short trip, the infatuated Fuller was determined to impress Miss Miller, and in a rash gesture, he invited her and the entire chorus line of the show to a dinner party at Churchill's, one of New York's finest restaurants.[69] Years later, he recalled that "the champagne flowed like the Hudson River" that night. However, the following morning Buckminster Fuller, the playboy, realized he had squandered more than his entire year's college allowance on a single evening. In fact, his bill was so large that his extended family was forced to cover the excess balance.[70] In a single night, he had not only ruined his college career but also blatantly displayed his true feelings toward the frugal New England attitude of his mother by using her money to consort with young actresses, a class of women whose reputation was, at best, dubious to his family.[71]

When Bucky's actions were reviewed by the Harvard administration, the combination of having spent all his money and missing his midyear examinations left them no choice but to expel him. Their official explanation was that Bucky had missed too many classes, but it was patently clear that their action was a direct response to his irresponsible attitude and lack of respect for Harvard's dignified traditions.[72]

Bucky's scandalous behavior compelled his mother to convene a council of his uncles and other family men for advice, and when Bucky expressed his thoughts and feelings to the panel of Fuller elders, he was admonished as a young man who should be seen and not heard. He was further instructed to disregard his own thoughts and experiences and follow the dictates of the more worldly, older men. The men "advised" Bucky to shape up in order to make something of himself and earn a respectable living in the "cold, hard world." Because the younger Fuller felt his elders loved him and were concerned with his welfare, he attempted to follow their recommendation.[73]

To support him in returning to a more acceptable mode of conduct and gaining an appreciation of how hard and cold the real world could be, the uncles decided to send Bucky to work far from the comforts of Boston.[74] Their selection was a job as an apprentice mechanic at a new mill being erected by the Connecticut Canada Textile Company in Sherbrooke, Quebec.[75] Although the mill was owned by one of their distant relatives, Bucky found no favoritism when he arrived in the small French-Canadian village on a dreadfully cold day in February 1914.[76]

Most people would have considered such an assignment cruel punishment, but young Bucky, ever in search of new adventures, perceived the desolate unopened mill as an exciting proposition. When he first laid eyes on it, standing in a recently cleared field amid huge piles of earth, dirty snow, and empty machinery crates, Fuller knew he faced another arduous challenge.[77] That job provided him with many valuable experiences, including his first formal contact with modern industrialization and the initial use of his engineering talent on a large scale.[78]

Fuller was also happy to be far from the societal pressures of Boston and Cambridge. At Sherbrooke, he found no unreasonable customs; rather, he discovered himself spending most of his time with a crew of Lancashire and German engineers and mechanics who had a job to do and judged a man by his work rather than his appearance. Their specific assignment was installation of the textile machinery, which had been shipped from England and France. While evaluating that machinery for the chief engineer, Bucky began to recognize and appreciate the invisible, yet significant, differences between seemingly similar pieces of equipment. He found that the cotton-weaving machines from France appeared exactly like those from England, but that the French equipment was superior in both metal quality and precision of parts. Because of such unseen differences, the British equipment was also greatly damaged in transit and British parts were continually being broken during the installation process.[79]

As a result of his obvious interest and aptitude, Bucky was assigned the task of finding ways to repair or replace damaged parts.[80] Since the manufacturing factories were several weeks' travel by ship, he quickly set about learning what local resources were available and soon found himself engaged in a self-tutored course in metallurgy, stress engineering, and the operation of small forges and machine shops.[81]

Having spent years improvising new ideas on isolated Bear Island, Bucky was extremely successful at his new assignment. Yet, exactly how he was able to communicate with the French-speaking Canadians who repaired and produced parts for him remains a mystery,[82] but in later years he often demonstrated that same ability to establish some form of intercommunication with people who did not speak English. During that period he also came to realize just how useful the universal language of mathematics could be in conveying his mechanical requirements.[83]

Regardless of how those results were achieved, they amazed the chief engineer, who suggested Bucky keep an engineering notebook of his work. As fate would have it, that notebook with its detailed sketches supported Fuller in producing even more amazing modifications to the equipment. Utilizing his mind in conjunction with the local resources, Fuller was even able to design and produce several parts which were far superior to the original ones.[84]

After a full winter of "punishment" at Sherbrooke, Fuller's family and the Harvard administration allowed him to return to college the following fall. Everyone believed he had learned his lesson and would become a model student, but that hope was quickly foiled when he was again expelled for "lack of ambition" and a tendency to spend more time pursuing girls and parties than good grades.[85] That second expulsion marked the conclusion of Buckminster Fuller's formal education. He possessed the ability to succeed, as proven by his high school grades in science and mathematics courses. However, as he would later recount to numerous genuinely fascinated audiences that often included educators, his boredom with a system that concentrated on memorizing facts rather than examining new ideas contributed greatly to his failure.[86]

Having succeeded as a millwright and an inventor in the more practical environment of Quebec, Fuller felt constrained by an educational institution which he perceived as discouraging innovation and independent thought. Accordingly, he again began pushing the limits of his instructors' patience as he had done earlier at Milton and questioned not only their understanding of a subject but the traditional beliefs and methods they employed.

When he left Harvard for the second and final time, Bucky was excited to attain his freedom and quickly moved to the bright lights of New York City, where he took a job with the meat-packing firm Armour and Company.[87] During the ensuing two years, Bucky found that his job greatly constrained his social life as he worked in twenty-eight of the company's local branches and held several different positions. Those incessant changes in location and assignment were not, however, the result of work difficulties. Instead, Fuller was frequently transferred because Armour management recognized him as executive material and sought to train him in all aspects of the business.[88]

As an informal "management trainee," Bucky was not, however, exempted from the arduous working conditions of that era. Because storage facilities were primitive at best, fresh meat was delivered early each morning, and a daily work schedule beginning at 3 A.M. was the norm. Fourteen-hour workdays were also common, and Bucky was lucky if his workday was finished by 5 P.M.

Even bound to that rigorous schedule six days per week and confronted with endless hours of demanding physical labor, loading and unloading quarters of meat, Bucky throve at Armour.[89] He was excited to witness and learn about New York City in the early morning as the metropolis was prepared for another day, and he began to appreciate the elaborate systems required to satisfy basic human needs when people lived in tightly packed, large communities.[90]

Although such living conditions are common today, they were rare in 1915, when the majority of the World's population was still agrarian. Through that intimate involvement with turn-of-the-century New York City resource supply, Fuller gained an appreciation of an environment which was to become dominant during his lifetime, and he acquired that understanding well before most people were aware of the problems inherent within large cities. The New York City influences were most palpable in his later designs for futuristic, densely populated living environments. Within those intricate community designs, Fuller focused primarily on the experience of individuals and improving their daily lives through effective resource management.

CHAPTER 2

Experiential Education

 A great deal of Fuller's own formal and informal education was predicated upon experimenting with and discovering the elegant simplicity of universal patterns he found in mathematics. Bucky's love of geometry, in particular, began with his kindergarten exposure to the construction tools of dried peas and toothpicks and continued through Milton Academy and Harvard. It did not, however, stop when he left Harvard.

As an apprentice at the Sherbrooke textile mill, he was intimately involved with geometry. Although his Canadian experience was supposed to provide a lesson in acceptable conduct and the results of rebelling against the system, Bucky found it to be a practical apprenticeship rather than a punishment. At the mill, he continued examining the shapes and patterns which so fascinated him, and, in the process, he was able to solve many of the problems he encountered.[1] In fact, he was able to employ his geometrical insights to improve upon designs which had been used satisfactorily for decades.

Fuller also continued to examine and reexamine the natural geometric patterns and shapes he first discovered on Bear Island. He felt that all of Nature was predicated upon a geometric coordinating system which, if discovered, could be of great benefit to humanity. Consequently, Bucky set out to uncover it. Over the years, that quest for Nature's coordinating system became one of his principle missions. His early ideas percolated with his exploration and, by the late forties, he felt he had discovered many of the fundamental elements of that system, which he called *Energetic/Synergetic Geometry*.[2] Later, he abbreviated the title of his new discipline to *Synergetic Geometry* (also known as *Synergetic Mathematics*), and in the late seventies, he was able to compile all his findings on the subject into two massive volumes, *Synergetics* and *Synergetics 2*.

Although detailed and complex, those books reflect Fuller's love of simplifying phenomena so that they could be easily understood by everyone. Within the *Synergetics* volumes Fuller employed two techniques which he had discovered were extremely helpful in explaining his ideas: the construction and display of actual models and the use of narrative story-like examples. He felt that one of the best methods of understanding both the physical and metaphysical phenomena which he felt dominated all Universe and the principles that control those phenomena was the construction and display of models.[3] He also found that geometry provided an excellent vehicle for experimentation with basic shapes as well as the development of models, and that geometric models and principles could be applied to other, more specific areas.[4]

Over the years, Fuller would present thousands of lectures and displays using geometric models as a basis for more detailed explanation. Invariably, he discovered that his audiences, like most people, had ignored or abused the immense possibilities geometry provides because they had been led to believe mathematics and science were arduous disciplines relegated to a small band of specially trained individuals. Bucky also found that many people's formal education distorted the context of science and mathematics to such a degree that those areas were clouded and removed from the realm of everyday human experience.[5] He discovered that most of us examine and understand phenomena which we personally experience, and that the more we encounter those phenomena the better we comprehend them. Yet, although we humans are inundated with the products and innovations of science and mathematics, our education tends to hamper most of us from appreciating the foundation of those results.[6]

Today, a majority of people are exposed to formal science and mathematics only in schools, which generally focus on abstract principles that tend to confuse students rather than enlighten them. Furthermore, the relationships between those principles and everyday individual experiences are ordinarily not explored.

Fuller's own formal education did little to train him in the exploration process which supported his work so successfully over the years. Although the content of his formal education was different from that of today's children, its context was very similar to the specialized training provided for most modern children. Later in life, remembrances of his restrictive educational experience continually reminded Bucky of just how debilitating our educational system can be for curious young minds.

He felt that, like almost all human organizations, including governments, industries, and religions, the primary function of the educational system is self-perpetuation and acquisition of fame, fortune, and power for the people who control it. Rather than contributing to the wisdom and well-being of everyone, he believed organizations attempt to indoctrinate individuals into accepting traditional dogma, restrict the perspective of members, and influence individuals to believe rather than think for themselves.[7]

After only a few years of exposure to that educational system, an individual becomes very confused. Then, true learning and independent thinking become even more difficult. Fuller often related a personal example which vividly illustrates such confusion. He would recount how he had been taught fractions using a technique similar to the one still employed in many schools today, and how that basic instruction created problems in his advanced mathematical education.[8]

One of Bucky's first teachers explained that although fractions were numbers, they represented a portion of either a single item or a group of identical items, and therefore, a fraction had to contain the same items on the top and the bottom. The teacher continued by emphasizing the fact that a fraction with dissimilar elements, such as elephants on the top and peas on the bottom, was illogical and unworkable. Because it was consistent with his personal experience, requiring elephants or any other item on both the top and bottom of a fraction seemed logical to young Bucky. Thus, he accepted the axiom, but years later that axiom confused him when he began studying another aspect of mathematics: trigonometry.[9]

In learning trigonometry, he was instructed that the sine and the cosine both represent ratios (known as fractions) between the length of a triangle's edge and the degrees in the arc of one of its angles. Although Fuller could not initially pinpoint his discomfort with that dictum, he felt uneasy about it, and he soon discovered that the concept of sine and cosine appeared to be illogical when compared to both his experience and his earlier study of mathematics.[10]

Bucky specifically recalled his previous work with fractions and the fact that they could not accommodate elephants (edge lengths) on the top and peas (degrees of angles) on the bottom. When he questioned his teachers about that seeming disparity, they agreed that ratios were in fact fractions and that sine and cosine represented unequal elements on the top and the bottom. They also directed him to overlook that apparent contradiction and accept their authoritarian contention that such ratios were reasonable.[11]

Fuller's childhood query illustrates a primary reason why, even today, many students find trigonometry difficult. Although few people then or now have been able to pinpoint their difficulties with trigonometry, Fuller felt that problems arise as a result of attempting to equate seemingly dissimilar elements within ratios. Because of his insatiable curiosity, Fuller was not satisfied by his teachers' edicts and explanations, and he continued in his personal quest to understand the logic of mathematics.[12]

Years later, Bucky found that his teachers were, in fact, correct in their dictum that sine and cosine were reasonable ratios, when he discovered an explanation which corroborated his teachers' mandate without contradicting his earlier lessons or personal experience. In that discovery, Fuller once again realized the importance of examining large systems rather than minor, special-case instances. In trigonometry, the larger system is spherical trigonometry which includes plane trigonometry.[13]

While studying spherical trigonometry, Fuller discovered that the edge

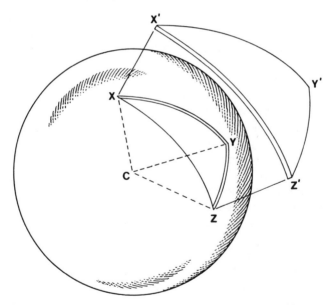

Fig. 2–1   Sine and cosine are actually relationships between the angles formed by lines from the center of the sphere (C) and the vertices of the spherical triangle (X,Y,Z) rather than between the edges of the triangle and its angles. For better perspective, the section of the sphere which is the spherical triangle is shown removed from the sphere itself (labeled X', Y', and Z').

of a triangle drawn on a sphere (XZ) is also the arc of the angle formed at the center of the sphere (C) by lines joining that center point to the ends of the triangle's edge (lines CX and CZ). With that understanding, he determined that the sine and cosine ratios actually denote relationships between the center angle and the surface angle of a triangle drawn on a sphere rather than between the edges and the angles of flat triangles, as most students are taught. Such a ratio of angles to angles was completely logical to Fuller, and it again reminded him of the importance of searching for the largest perspective possible in every aspect of his work.[14]

Fuller's lifelong examination of human educational systems led him to the conclusion that trigonometry, like the majority of formal education, does not respect students' feelings or their sense of what is natural and logical. Authoritarian teachers expect their students to accept definitions such as sines and cosines without question or thought, even though such information appears to contradict rules taught earlier, or the child's innate understanding of Nature's orderliness.[15]

Educators with the most honorable of intentions stifle rather than stimulate children's minds, just as their own minds and creative thinking processes were obstructed by the dogma of former generations. Fuller felt that even though phenomena can be reduced to an array of disconnected elements, by doing just that our educational system tends to discount the significance of the relationships between elements. Because they provide the essential operating material for human minds, those relationships are indispensable to human beings.[16]

As both a student and a teacher, Fuller later felt that modern education should concentrate on the significance of an individual's experience and the relationships inherent within those experiences.[17] By shifting the educational system toward such a focus, society could provide students with much more than an array of unrelated and seemingly useless facts. Children could then graduate into the ''cold, hard world'' with an understanding of principles and how principles are applied, and those realizations could be adapted to many different practical situations.

Provided with such an appreciation of general principles, individuals could more easily think for themselves as well as apply their formal and informal education to the tasks of daily life. People would also be able to examine isolated elements of their environment within a context of lucid whole systems that relate to one another, just as Fuller did once he had overcome the limitations of his formal education.

*

Through years of examination Bucky concluded that the teaching of abstract scientific principles in ways which contradict the direct experience of individuals was a significant problem.[18] Nowhere did he find that phenomenon more conspicuous than throughout the field of geometry, and he sought to help alleviate the problem by frequently speaking about the relationship between personal experience and geometry.[19]

One of his favorite narratives addresses that problem directly using three elements which were consistently involved in Fuller's life: triangles, children, and geometry. At some point during almost every one of his lectures, Bucky would recount the absurdity of teaching children that a triangle is a shape which always appears on something the teacher designates as a "flat plane."[20] Always the curious experientialist, Bucky dearly wanted to see or feel just one such flat plane, and he consistently questioned teachers and experts about that issue. Even though, as a boy, he realized that no one could ever show him the single component which geometry teachers predicated so much of their work upon, by questioning, he continued to challenge people to reconsider their position.[21]

As he examined the situation, Fuller discovered that years of inaccurate instruction in plane geometry led people to assume that triangles occur naturally in this condition, even though a two-dimensional "flat plane" simply does not exist. Through years of trial and error, Fuller found that he could not even conjure up a flat plane in his imagination. Every time he attempted the process, his image always contained some depth because it could not be created any other way. He found that although of some limited use as a concept, discussing "flat triangles" as if they were tangible spawned more problems than it solved.[22]

From experience, Bucky also learned that although the two-dimensional triangles studied so intently in schools do not occur naturally within Universe, triangles are a fundamental component of all structures. The natural triangles which provide the foundation for structures are, however, never two-dimensional and rarely occur in the "flat" configuration studied by children. Instead, most indigenous triangles are spherical.[23]

Spherical triangles are triangles whose three sides constitute arcs of a sphere rather than straight lines (Fig. 2–1). Because those naturally occurring spherical triangles are primarily tiny segments of much larger structures such as trees or rocks, humans rarely observe them. Even when specialists studying microscopic objects or sections of organic material have uncovered spherical triangles (as was the case with the shells of viruses), they generally have not translated their knowledge into practical information which can be taught to children and applied to the problems of humankind.

Fuller examined our educational system in great detail throughout his lifetime and found that even in the 1980s children were still being taught the same geometric concepts of shape he had been exposed to at the turn of the century. He felt that geometry and other subjects taught as abstract courses tend to indoctrinate children into a system which supports specialization.[24] That system also induces individuals to invalidate their personal experiences and believe themselves to be remote and removed from other individuals, societies, and Nature. Once he had attained a comprehensive appreciation of most subjects studied in schools, Fuller set out to share his insights with interested individuals.

In speaking about the inadequacies of our educational system, Fuller would explain that in their first formal exposure to geometric shapes, children are taught that a triangle is an area bound by a closed line of three edges and three angles.[25] Similarly, a square is defined as an area bound by a closed line of four equal edges and four equal angles. Most shapes used in elementary geometry are likewise defined as areas bordered by connected lines, and those definitions support the indoctrination of individuals into a system which views only the inner area of a shape as being that figure. Fuller elaborated upon that observation to conclude that people educated within such a system would perceive only the inner area of any boundary or limitation to be important.[26]

Humanly conceived systems (i.e., cultures, religions, societies, and so on) categorize everything outside the borders of "our group or locality" as separate from and less significant than the "us" who reside within the border. Whether the limiting system is a family, city, sports team, corporation, or anything else, Fuller felt that such divisive systems have been utilized throughout history to obstruct new organizations of individuals and maintain the status quo of power within the hands of a few. He also felt that such systems indoctrinate people into the current "you *or* me" mentality which threatens the very existence of all Earth.[27]

Since he had been ostracized from social groups as a youth, Bucky had a direct, personal experience of that separation, and his remembrance of that feeling contributed to his philosophical inclination. Having known the trauma of being relegated to the outside, Fuller sought ways to spare others from the pain of similar castigation. Even within something as fundamental as elementary geometry, he found children being exposed to propaganda which could eventually influence them to oppose other human beings for being different or outside the accepted boundaries of a particular group.

Fuller did find that the individuals perceived as different were rarely ostracized from a particular society because they championed revolutionary new ideas or thought for themselves. Rather, they tended to be viewed as outsiders because of commonly held beliefs which were predicated upon antiquated guidelines. Even society's worst villains, like the people judging them, remained within relative limits imposed and defined by leaders and ancestors.[28]

Bucky believed that the few true radicals who attempted to live well outside those limits did realize, as he did, that such restrictions were the product of earlier generations who had experienced an entirely different set of circumstances. The environment of those previous eras then dictated boundaries for future generations and were held to be important by descendants, even though circumstances had usually changed radically.[29]

For example, during the early stages of the industrial revolution, most people believed that the natural environment was unlimited and was therefore the perfect disposal site for industrial waste. That belief, although untrue for any generation, was championed primarily by industrialists, who realized that dumping waste into rivers or other convenient locations was far less costly than recycling or determining if a substance was harmful. Such views have been "sold" to the general public for decades, and only in recent years, when the results of those greed-oriented actions has become more and more obvious, have some individuals begun to truly rebel against the power of the enormous bureaucracies which created the problem.

Still, even the most radical environmentalists tend to attempt working within the limits of the societal system in which they live. They further try to influence governments rather than realizing that the systems which created the problem are predicated on conditions which no longer exist. Therefore, the systems themselves are obsolete. Despite years of making little or no progress in influencing the system or the powers who dominate it, individuals who have been strongly indoctrinated into a particular society remain convinced that the system can work for the greater good of its members. They refuse to realize that any traditional system is, and will most likely continue to be, strongly biased in favor of the gigantic industrialists, lawyers, and financial manipulators who dominate most institutions.[30]

Although those conventional systems and the actions they supported were appropriate during earlier times when basic resources were scarce, Fuller felt that they were simply of no use to modern individuals and were, in fact, more of a hindrance than an aid in solving the problems which confront our global society.[31] He was particularly distressed by our educational system because it is predicated on specialization and divisiveness, which he believed are counterproductive in modern society. Even during the 1930s, Fuller was convinced that a fully cooperative era was dawning in which one individual's well-being could not be fulfilled until every person on Earth was provided with the necessities of life.[32]

At the time, Fuller also determined that the "you *and* me" era which is now becoming a key element of modern society would become a genuine possibility in the early 1970s. He announced that approaching transformation after he had documented that the advances of technology were producing a situation in which the resources on Earth would become sufficient to sustain every human being. The date he prognosticated for that change was 1970.[33] Following the realization of that predicted shift, Fuller felt educational systems, as well as most people's traditional beliefs, would became even more outmoded than ever before.[34]

In considering societally administered formal education and geometry as taught in schools, Fuller found the system to be extremely restrictive and not aligned with Nature's procedures or his personal experience.[35] For instance, his experiential focus led Fuller to appreciate the importance of the simple fact that a triangle, or any other figure, must be represented upon some substance. Even an imaginary triangle is envisioned scratched on the ground or drawn on some material as basic as paper or as sophisticated as a computer screen. No matter how resolute an individual is in attempting to create or imagine a concept (be it a triangle, love, or anything else), the human brain will always envision a specific experience or item.[36] Neither children nor adults can relate to the conceptual examples of ideal shapes used in elementary geometry classes without translating those examples into special-case instances. Human beings require finite, tangible illustrations for effective learning, but because a plane cannot be experienced or depicted, such examples are impossible within the realm of traditional plane geometry.

After years of being frustrated by the conventional educational system, Fuller set out to discover the experiential geometry employed throughout Nature. He theorized that because Nature's operation is tangible, it has to function within the confines of a limited, yet viable, coordinating system, and that understanding Nature's coordinating system could be of enormous benefit to humanity as well as to his work.[37]

He felt that once people understood the simple efficiency of Nature's system, they could consciously apply its principles to numerous and varied aspects of life, be they design, disease control, or education. Bucky believed that when people truly understood and appreciated Nature's ways, human behavior would enter into a process of mirroring Nature's operation of perfect harmony and maximum efficiency.[38]

He felt able to vigorously pursue and uncover the essence of Nature's coordinate system because of his ability to understand a fundamental relationship between the human brain and the human mind. Fuller defined human brains as biological organs which relate only to specific finite experiences and, consequently, which always provide a particular example when information is requested.[39]

Human minds, on the other hand, he defined as unlimited metaphysical phenomena which search for relationships between specific examples and uncover the generalized principles upon which phenomena are predicated. An individual's mind does communicate and work in conjunction with his or her brain.[40] In the triangle scenario, the person's mind scans the various triangle images presented by the brain in search of relationships and may eventually uncover the common mathematics which describe the generic shape "triangle."

With an understanding of those mathematical principles, the individual's mind can further elaborate upon relationships and apply the general knowledge discovered to other experiences of triangles. However, even though an individual may understand a principle independent of limitations such as size, color, or time frame, that principle is still forever visualized with explicit characteristics. In the case of a triangle, it always appears with dimensions, texture, location, and so on.

Fuller observed that throughout history many philosophers and thinkers have attempted to modify the human thought and visualization processes and have failed. Consequently, he theorized that minds deal only with concepts and relationships, and that brains accept and understand only specific instances and items.[41] Like many curious individuals before him, he came to this realization only after years of unsuccessfully attempting to function in an idealistic, conceptual mode. Once he clearly worked out the operational relationship between mind and brain and the fact that human beings can function only within the limits of tangible components, Fuller shifted his focus to specific inventions and procedures from which he could extract generalized principles.[42] By working in that manner, he was able to operate more effectively while developing models and inventions which could be displayed and appreciated by others.

Fuller felt that every human being is born with an innate understanding of the relationship between brain and mind as well as a natural feeling for and attraction to the most effective modes of human operation.[43] As a result of those characteristics, young children are constantly learning by employing a system which has always been part of the human experience: trial-and-error experimentation. Such experimentation is predicated upon special-case instances which humans eventually generalize in order to understand their environment even more completely. Children's natural mode of operation leads them to test specific phenomena, such as fire, only once or twice before learning the universal sensation "burn." The traditional educational system, however, induces children into believing they should discontinue experimentation and accept the "wisdom" of their elders and the abstract, remote concepts taught in schools.[44]

Bucky found that, like himself, most young children have difficulty adapting to the predigested education imposed upon them, and that they question the specifics of conceptual illustrations taught as truth. He also discovered that over time, innate curiosity is usually socialized out of children as they are compelled to unquestioningly adapt to society's rigid systems, standards, and ideals.[45]

He often illustrated the problems inherent within our system by using the familiar triangle. He would recount a scenario in which an adult asked a young child to draw a triangle and received the response, "Where?" This is the query of an operational experientialist accustomed to dealing in specific conditions and elements.[46]

To facilitate the child's own experimentation, she is asked to draw her triangle on the ground, and when the figure is complete, the knowledgeable adult, who operates experientially as Fuller did, tells the child she has, in fact, drawn four triangles. Since the girl has, like most people, been trained within the bounds of our restrictive educational system, she believes her three connected lines in the earth constitute only one triangle, and she asks for an explanation. It is at this point that the child is faced with the formidable prospect of considering the situation from an expanded perspective which encompasses a much larger system of operation. That system is the surface of the Earth.[47]

By drawing a triangle on the spherical surface of the Earth, she divided that surface into three areas: the small area within the three lines,

Fig. 2–2   Drawing a seemingly insignificant triangle in the dirt.

the even smaller area on which the lines are drawn, and the enormous area outside the three lines.[48] Because our traditional definition of shapes deals only with bounded areas, the minor area on which the lines are drawn is not considered. The other two sections constitute the majority of the Earth's surface, and they divide it into two sectors.[49]

To clarify the situation, the adult suggests that the girl study the Earth's equator circle, which also divides its surface into two major sectors, the Northern and Southern Hemispheres. Those sections are of equal area, but if the imaginary circle known as the equator is shifted farther north, the southern area becomes larger while the northern area becomes smaller.

As a further aid to understanding the system of operation, the child is asked to reproduce her triangle on the surface of an Earth globe using a large rubber band stretched between three pushpins stuck into the globe. By relocating the pins, she can shift the lines and points of her triangle to move it in a manner similar to the way the equator line was shifted north or south. Using this simple technique, the child can change the orientation as well as the size and shape of the triangle.

Obviously, such a triangle is much larger in comparison to the model Earth globe than was the original triangle when compared to the actual Earth. Still, both are triangles on the surface of Earth. Employing the globe, the girl can create a triangle which covers whole continents, and she begins to see that her triangle does, in fact, divide the entire surface of the Earth into two distinct areas.

Because it is an area bound by a closed line of three edges and three angles, the territory inside the three lines clearly fits the definition of a triangle. Nothing in that definition limits the size of the angles or the curvature of the sides. Consequently, even though the area viewed as "outside" the three curved lines, which delineate the first triangle, is bounded by much larger angles, it still conforms to the definition of a triangle (i.e., an area bounded by a closed line of three edges and three angles).[50]

After considering this explanation, the child realizes that she did, in fact, create at least two distinct triangles on the Earth's surface, even though neither is flat. Both triangles are spherical, and one is tiny while the other is huge. Because of the enormous size of the larger triangle's three angles, it is unfamiliar and not readily recognized by most people. The large "outer" triangle does not conform to the image of a triangle presented in most schools, and people rarely notice it.[51] Nevertheless, that immense triangle does conform to the definition and is as much a triangle as is the area inside the three lines.

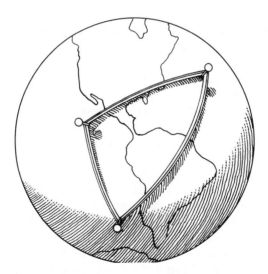

Fig. 2–3 Any triangle on the Earth's surface divides it into at least two distinct triangles, as illustrated by a rubber band on an Earth globe.

By experimenting with her rubber band and pins on the globe, the child experientially learns about the concept of spherical triangles. She may also begin to appreciate the fact that she has been educated within a system that promotes a specialized, myopic view of reality. Further examination of that system and her beliefs may even lead the child to understand how seldom she considers her entire environment.

Fuller felt that most people are deeply rooted in a similar predicament, and he perpetually sought to dislodge himself from that narrow view and mode of operation. He supported his quest with constant experimentation and evaluation, which permitted him to move beyond the limits of traditional education and the unquestioning acceptance of the "wisdom" presented by the educational system and the leaders who sustain it.[52] He was able to function successfully in that radical manner because he truly believed he was striving toward a new harmony with all Universe, and that within such a harmonious operation he would be able to function much more effectively and efficiently. He also felt that all people must eventually move beyond the restrictions of formal education and become truly aware of their environment and the restrictions they impose upon themselves and their concept of reality.[53]

Fuller believed people had to start recognizing the fact that those human beings, experiences, and phenomena outside humanly established boundaries of acceptability are as important as those within them.[54] Such a universal acceptance represents a significant step toward becoming a comprehensivist. A comprehensivist, like Fuller, considers as many aspects of a situation as possible before formulating the most effective conclusions and actions. The desired result of such thoroughness is a condition in which harmony with all Universe and support of humanity become intrinsic in every aspect of an individual's life.

Bucky often commented that the confusion of the child who drew a triangle on the Earth and learned she had actually created more than one triangle was a direct result of restrictive, "modern" education. He also understood that her response was predicated upon societal conditioning.[55]

In drawing the triangle on the Earth, the child inadvertently altered and divided the entire Planet. Having conducted this experiment several times, Bucky found that when confronted with such a realization, a child erroneously assumes some type of wrongdoing has occurred and feels guilty. He also discovered that a child confronted by that circumstance invariably apologizes and explains that she had no intention of creating a problem or causing harm. Fuller believed that such an unwarranted reaction is primarily a consequence of the child's interaction with adults.[56]

Because most people are fearful of mistakes, that apologetic reaction is typical among both children and adults who have been led to believe that mistakes are bad. Bucky concluded that our fear of mistakes is so overwhelming that most people concentrate on caution and safety rather than breaking new ground and taking risks.

Fuller believed that the humanly devised concept of mistakes is erroneous.[57] What most people consider mistakes are mistakes only in the eyes of the beholder. Fuller viewed that concept in an entirely different light. He understood that although people sometimes take actions which do not produce the sought-after effects, still lessons can be learned from any action, including ones regarded as mistakes.[58]

Because they are less "socialized," young children generally possess a healthier attitude toward criticism, and that attitude is evident in the scenario of the child drawing the triangle on the ground. When the girl understands that she has created two triangles, she learns from her experience, and, rather than chastising herself, she asks about the other two triangles of the four she has been told were created.

Fuller regarded such inquiries as invitations to support the development of another human being, and he responded whenever possible.[59] In fact, he was known to spend hours explaining ideas to a single interested individual while other matters, perceived as more important by his colleagues, were held in abeyance. When a query came from a truly interested person, Fuller knew that any concepts introduced would be eagerly examined, and he positively exploited such opportunities to their fullest.

In the triangle-in-the-dirt scenario, the relevant concepts he sought to illuminate were *concave* and *convex*. Although concave and convex always occur jointly, they are very different from one another. A lens possesses a concave and convex side, and in experimenting with those two facets their different characteristics are apparent.[60]

When light passes through a lens and exits from the concave side, it is focused and concentrated, a phenomenon illustrated by a magnifying glass on a sunny day concentrating sunlight and burning a hole in a piece of paper. The focal qualities of concave lenses are also utilized in eyeglasses, which help focus visual images more clearly within the eye. Eyeglasses mirror the normal function of the eye's natural lens and aid people with impaired eye lenses by refocusing images. The convex side of a lens has exactly the reverse effect. Light exiting through it is diffused and scattered, a phenomenon illustrated in crystal prisms, which split normally invisible sunlight into a rainbow of colors.[61]

 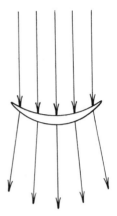

Fig. 2–4   Concave (left) and convex (right) always exist simultaneously even though concave focuses energy, such as light, while convex diffuses energy.

Once the child grasps the distinction between concave and convex and sees that her triangle was drawn on a spherical surface which has both a concave and a convex side, the second set of triangles begins to appear to her. She understands that both her large and small spherical triangles are, in fact, convex when considered from the usual position, but that when observed from within the sphere, the concave aspects of those same two shapes emerge.[62]

Within even the simple act of drawing a triangle in the dirt, an orderly pattern begins to become evident. Whenever a person tries to draw one triangle, four triangles always appear. That unique pattern and its significance is much more evident to those who, as Fuller did, consciously search out relationships and an inclusive perspective. Fuller uncovered that phenomenon and its importance early in his examination of geometry and triangles, and he was able to utilize it in the invention and development of structures throughout his life.[63]

This pattern of four occurs because drawing a triangle in the dirt is predicated upon a fundamental ''fourness'' operating throughout Universe. The conspicuous fourness which became a fundamental element of Bucky's life and work is also seen in the four sides of a tetrahedron. Once he understood just how important the fourness of the tetrahedron is in Nature's construction, he began mirroring that element in his designs and inventions.[64]

He found that because humans operate within a tangible rather than a conceptual environment, the minimum number of triangles which can be drawn is always four.[65] This occurs because no surface is perfectly flat, and any triangle drawn upon a curved surface has to possess an inner and outer pair of concave triangles as well as an inner and outer pair of convex triangles.

Over the years, Fuller devoted a great deal of time to examining the fields of mathematics, science, and human behavior as well as their inter-relationships. When he combined these three areas, Bucky found that most people rarely consider the mathematics and principles which govern all Universe because they believe themselves not to be significantly affected by seemingly abstract mathematics and principles.

Fuller felt that no creature is excluded from the timeless straightforward authority of the simple, yet highly sophisticated, eternal mathematical precepts, and he devoted a great deal of time to understanding those principles so he could operate more effectively within the confines of our environment. Through years of observation, he discovered that some people believe themselves to be exempt from natural laws because they live primarily within humanly manufactured surroundings.[66] Large cities provide them with a false confidence that humans, and not the fundamental laws of Nature, control everything, including the destiny and environment of individuals.

Spending a considerable amount of time isolated from Nature furnishes such people with the illusion that humans can dominate Nature. Fuller, on the other hand, was certain that all human actions must conform to natural laws and principles. There was no doubt in his mind that as long as human beings continue to operate within the limitations of the physical environment, we cannot escape Nature or its principles.

In fact, we humans are constantly employing Nature's generalized principles to formulate distinctions and better understand ourselves and our environment.[67] Within the context of that type of operation, Fuller perceived learning as much about Nature's principles as possible to be a logical action. He, therefore, wisely predicated his operational strategy on doing just that and was able to employ that strategy in acquiring an understanding and insight known to few human beings.[68]

CHAPTER 3

Lessons of Love, War, and Technology

Although Fuller's philosophy and strategies of life would not be fully developed for many years, even as a child he was beginning to envision Nature as a critical element of his life.

As he grew older, Bucky began to appreciate how uniquely Nature designed even the simplest of organisms, and he sought to translate that unmatched efficient design into inventions as well as a more organic philosophy of life.

That saga of a young man managing to attain a perspective which helped him to mirror Nature's designs in his inventions cannot, however, jump ahead to momentous accomplishments like the Dymaxion Car or the geodesic dome without first examining the events of his earlier life. How the rambunctious young man who had twice been expelled from Harvard matured into an insightful designer as well as a husband and father while developing a comprehensive understanding of Universe begins with the rigors of a turn-of-the-century work environment and a relationship which would endure for sixty-six years.

Fuller's true intimacy with and observation of the "cold, hard world" began when he left Harvard for the second time. In 1915, he earned a meager $15 per week working for the meat-packing company Armour in New York City while putting in six fourteen-hour days each week.[1] Still, he enjoyed his job and remained with Armour for nearly two years. During that time, the free-spending young man resolutely attempted to live within his financial means. He conserved some of his sparse funds by living with his mother, who had sold the big house in Milton and moved to an apartment in New York after her eldest daughter,

Leslie, was married.[2] When the tight quarters and his mother's dominance became too much for Bucky, he moved to another inexpensive abode, the Brooklyn YMCA.[3]

Although work at Armour occupied most of his time, Bucky displayed an incessant vitality and an ability to shift from work to partying night after night. Even while working seventy-hour weeks, he still found the time and energy to indulge his passion for dancing, especially with beautiful girls.

It was during a 1915 semiformal summer dance he attended on Long Island that Bucky was first captivated by a beautiful brunette woman with brown eyes. Questioning his friends, Bucky discovered that the beauty was nineteen-year-old Anne Hewlett, the best girl of one of his friends, Kenneth Phillips.[4]

Since many of the local young men considered Anne's sister Anglesea, nicknamed Anx, even prettier, she was the one who was constantly pursued. Always one to accept a challenge and valuing his friendship with Kenneth Phillips, Bucky danced much of that night with Anx, even though it was Anne who had caught his eye.[5] He liked Anx, and over the course of that summer, Bucky and Anx frequently double-dated with Kenneth and Anne on weekends. At times, dates consisted of simple evenings or afternoons at the Hewlett house on Martin's Lane in Lawrence, Long Island. Just being at the large Hewlett house proved to be great fun for Bucky. With an often unruly clan of five sons and five daughters, Mr. and Mrs. James Monroe Hewlett's home often abounded with friendly turmoil.[6]

A short distance down the road, Anne's grandmother spent summers living with her two bachelor sons in the ancestral homestead, Rockaway Hall. That stately mansion, which still exists and is now known as Rock Hall, is one of the most beautiful eighteenth-century homes built on Long Island, and Bucky was awestruck whenever he was privileged to visit there with the Hewlett family.

Because Anne's mother was ill from the time Bucky met her until her death a few years later, she spent most of her time quietly convalescing in an upstairs bedroom while the children and Mr. Hewlett maintained the family. The father, James Monroe Hewlett, was a noted architect and painter who eventually introduced Bucky to the finer aspects of architecture and the arts.[7]

During that first summer, Bucky's primary focus was on Anx, although he also began to find himself growing more attracted to Anne. As summer turned into fall, Bucky's feelings for Anne grew stronger, but he restrained himself because he believed that Anne was about to become engaged to Phillips. Anne found herself entertaining similar feelings for Bucky while Anx was moving on to other boys. Thus, it was Anne who invited Bucky over to dine with the family on a regular basis during the fall of 1915.[8] After several such invitations, Anne felt compelled to confront Bucky's polite but constrained attitude toward her, and she initiated the opportunity for his courtship by straightforwardly informing Bucky that she was not engaged or attached to Kenneth Phillips and did not intend to marry him. As he had had little experience with refined women or serious relationships, Bucky was both overjoyed and anxious about the prospect of courting Anne.[9]

Fig. 3–1   Anne Hewlett during the time of her first encounters with Bucky.

As Anne was the eldest child, Bucky became the first outsider to attempt penetrating the closed inner circle of the Hewlett family. Accordingly, he was subjected to a great deal of scrutiny and good-natured harassment.[10] Some of the younger children made great sport of teasing both Anne and Bucky, whom they called "The Butcher Boy" because of his job at Armour. Despite those obstacles, Bucky's tenacity as well as his warm personality, playful nature, and love of Anne soon won over the entire Hewlett clan. In fact, the family so enjoyed his company, they bestowed on him the title of "honorary Hewlett."[11]

Although the multihour lectures Fuller delivered during the last decades of his life provide clear evidence of an ability to speak eloquently when necessary, as a young man Bucky was quite shy and often found himself tongue-tied. Those traits were especially evident when he attempted to express his emotions to Anne in words. Hence, when he discovered another means of demonstrating his feelings of affection, Bucky quickly jumped upon it.

He found that Anne loved roses, and he began sending her a bouquet of red roses from Weir's Florist on Fulton Street every week. In typical Fuller style, although he was attempting to remain on a budget, the financial ramifications of that splendid gesture were not well thought out. Because it seemed like a respectable way to display his love and win over both Anne and her family, Bucky spent five of the fifteen dollars he earned each week on her roses.[12]

Eventually, Bucky worked up the courage to make a small request of Anne. He asked for a picture. Because Anne had no formal photographs, she looked through the family albums until she found an old photo of herself standing beside an elderly gentleman on the deck of a ship. Bucky gratefully accepted the picture and promptly trimmed the old gentleman off. Thus, Mark Twain's picture went into the wastebasket. Years later, Bucky would reminisce that he regretted that thoughtless gesture of his youth, which left him without the photograph of Anne and Mark Twain.[13]

During the summer of 1916, Bucky and Anne's formal engagement was announced and was followed by a series of gala parties honoring the couple.[14] The impending entry of the United States into World War I, however, significantly overshadowed the festive nature of those events.

When the United States declared war upon Germany in 1917, Fuller, like most young men, volunteered for military service with patriotic fervor. Later, he would recount that World War I did to his former college classmates what his own innate curiosity did to him. Both phenomena removed them from Harvard. Of the seven hundred men entering as freshmen in 1913, only forty-five graduated four years later. The majority of the student body members responded to the prevailing circumstances and dropped out of college to serve their country in the "Great War."[15]

Fuller's experience with boats and the sea was an enormous advantage when he endeavored to enlist in the navy. He first attempted to join the military during the fall of 1916 at an official army training camp. That encampment had been established and commanded by former President Theodore Roosevelt to train young men for the defense of their country in the event of the United States' entry into the War. Because of Bucky's extremely poor eyesight, he was rejected from that army training center.[16]

The American demand for servicemen did, however, increase as the war escalated, and Fuller tried again the following year.[17] That second effort was much more cleverly planned and demonstrated the ingenuity which would later propel him past many obstacles. Bucky first persuaded his relatives that volunteering the family's small cabin cruiser to guard the coast of Maine was a greatly needed patriotic gesture. He then offered the navy not only a seaworthy ship but a trained crew to man the vessel. Within weeks, the tiny USS *Wego,* commanded by United States Navy Chief Boatswain R. Buckminster Fuller, was patrolling the Maine shoreline.[18]

With a crew of six, including Bucky's brother Wolly and best friend Link Pierce, the tiny *Wego* was extremely crowded. In fact, the *Wego* may have been the smallest regularly commissioned vessel in the entire United States Navy.[19] Sailing out of Bar Harbor, she soon became nicknamed the "State of Maine Navy."[20]

Although Fuller and his crew were motivated by the best of intentions, their only significant contribution to the war effort resulted from the tiny ship's inadequacies rather than her superiority. When local yachtsmen noticed Bucky and crew patriotically guarding the coast and harbors, they followed suit and volunteered their vessels for similar duty. In dire need and uncertain as to where German U-boat submarines were patrolling, the navy accepted those offers and appointed each of the ships' captains to the rank of boatswain. Having been commissioned before the rush, Bucky was a chief boatswain. Thus, the twenty-two-year-old with the slowest and smallest ship became the senior officer present and commander of the Bar Harbor flotilla. Regardless of Bucky's rank or position as commander, the *Wego* could not keep pace with the expensive yachts under his command, and when the diminutive fleet sailed out on maneuvers, Bucky and crew were soon trailing far behind the other boats.[21]

Fig. 3–2   Fuller (right) in uniform as the commander of the *Wego* with his first mate, Lincoln Pierce (left).

On one such occasion, the *Wego*'s engine began to sputter while it was lagging well behind the other ships. To make matters worse, the crew was caught in a coastal fog. Because Bucky and his crew were familiar with the area, they were able to maneuver their ailing ship into Machias Bay so they could make repairs and anchor safely for the night. To ensure security, Bucky posted a rotating watch through the night. At 2 A.M., when the fog lifted, the sentinel on duty spotted a rapidly flashing light at

Fig. 3-3 The U.S.S. *Wego* with Fuller at the helm.

the end of the pier and saw it answered by similar flashes from Cross Island. He summoned the other crew members, who were below decks working on the engine, but no one on board could decipher the message being sent. Having some military knowledge and a vivid imagination, Fuller assumed that a message was being transmitted by secret code. Following that display the crew observed a large sardine trawler sailing out of Machiasport to Cross Island. Because it was anchored without lights, the *Wego* went unnoticed during that operation.[22]

Upon returning to Bar Harbor, Fuller immediately reported the surreptitious incident to naval authorities, and several large gunboats working in conjunction with naval intelligence agents converged on Cross Island the next day. Despite all the commotion, it was not until after the war concluded that Fuller and his crew learned just how accurate his vivid imagination had proven to be. When secret information was declassified, they learned that while stranded in the bay, they had inadvertently been the first to stumble upon and uncover a refueling depot for German U-boats and a transfer point for German spies entering the United States. They were also told that the sardine trawler had actually been covertly transporting German spies and equipment to that depot on the night the crew spotted it.[23]

The only drawback Fuller encountered in commanding the Bar Harbor armada was that he could no longer spend time with Anne. Because his chief boatswain's salary of $1800 per year was more than double what he earned at Armour, Bucky felt affluent and able to support a family. With such financial affluence and a strong desire to be together, he and Anne decided to marry that very summer.

In what would be the second of a long series of significant events occurring on his birthday (the first being his father's death), Bucky and Anne were married on the south lawn of Rock Hall on July 12, 1917, with full military honors.[24] As with everything during that era, the war dominated the Fullers' honeymoon. Immediately following their wedding, they caught the late train to Boston and a ship back to Bucky's command post at Bar Harbor. When they arrived at the dock, the newly wed Mr. and Mrs. Fuller were greeted at the dock by the entire contingent of captains under Bucky's command. Bucky was especially touched when those men presented the couple with a beautiful silver basket engraved with the names of each ship's captain.

Bucky and Anne rented a small cottage overlooking the harbor, but his duties kept him at sea for the majority of those first weeks. Shortly after that tour, the navy decided that the *Wego* was no longer useful and removed her from active service.[25] The ship's crew did, however, retain their positions, and Bucky was given command of a larger ship patrolling the outer reaches of Boston harbor. The *Wego* was returned to the Fuller family, and it provided many more years of service at Bear Island. In fact, it served as a general freighter and ferry until it had to be retired in 1957.

Wartime conditions led to rapid promotions, and the able Fuller soon found himself commanding the even larger patrol boat *Inca,* which was assigned to the Naval Air Station at Newport News, Virginia. The only American aircraft which contributed significantly to World War I were navy planes, but experienced pilots were a scarce commodity during those early years of aviation. Newport News was the navy's primary testing and training facility for both pilots and new aircraft. Reckless courage was the order of the day among the inexperienced young pilots being trained there as they eagerly displayed their readiness for combat duty.[26] Tragically, fatal accidents occurred every day.[27]

Planes landed and took off from the water using large pontoons, but naval aircraft design was so primitive that they often flipped upside down during landings by overzealous, unseasoned pilots. When such an incident occurred, rescue boats, including the *Inca,* would rush to the scene within one or two minutes. Upon reaching an overturned aircraft, most of the crew would dive overboard to rescue the pilot, who was hanging upside down under water, trapped by his own seat belt. Although the rescue crews were speedy in their attempts to save lives and always carried a doctor with them, pilots frequently drowned before they could be cut loose.[28]

After witnessing a few of those tragic deaths on the decks of his ship, Fuller decided to employ his mechanical talents in an attempt to solve the problem. He worked late into the night for weeks trying to create a quick release mechanism for the pilots before he was seized by an entirely different possibility.[29] Following that insight, he designed a huge winch, much like those on the backs of large towtrucks, to be installed on the rear of the rescue boats.

After convincing Commander Pat Bellinger, the commanding officer of the air station, that the idea had merit, Fuller was authorized to have a prototype of his winch constructed and installed on the *Inca*. Using it, Bucky and his crew could quickly hook onto an overturned aircraft and hoist it out of the water in less than two minutes. Although the pilots remained hanging upside down in the open cockpit while the planes were being raised, Fuller felt that they would still be afforded a much better opportunity for survival than the old methods provided.[30]

After several practice runs retrieving previously crashed planes, Fuller felt that his crew was ready for a real test and requested an opportunity to demonstrate his invention in an actual emergency. Once the plan was approved by Commander Bellinger, the crew did not have to wait long for a crash. On their first attempt, the pilot was on the *Inca*'s deck within minutes of his crash.[31]

Although the pilot was not breathing, the *Inca*'s doctor laid him on his stomach, pumped the water out of his lungs, and began giving him artificial respiration as the crew looked on. Then, after several minutes of treatment, which Fuller would later recall seemed like an eternity, that pilot opened his eyes and promptly vomited.[32] That act provided Fuller with what he later recounted as one of the happiest moments of his life, and within days other such winches were built and installed on all the rescue boats, saving hundreds of pilots' lives.[33]

In what would turn out to be an unusual outcome for Fuller, within only a few weeks he was rewarded for his achievement of inventing that lifesaving device. Accolades, acknowledgments, and honors for his innovations and ideas would later be years or decades in the making. But his early accomplishment was noted by Commander Bellinger, who quickly nominated Fuller to be trained as an officer at the United States Naval Academy at Annapolis.[34]

Prior to leaving for Annapolis, Bucky experienced a chance encounter with the cutting edge of technology which provided him with a momentary glimpse into the future. Dr. Lee De Forest, the man who had invented the triode tube in 1907, arrived at the Newport News Air Station to conduct radio experiments just as Fuller was achieving notoriety. Dr. De Forest's tube was the single most significant component used in the newly developed radio transmission of voice messages, and he was attempting to perfect ship-to-plane voice transmission for the navy. As luck would have it, the *Inca* was assigned to Dr. De Forest for his experiments, and Fuller became the quintessential host for the distinguished scientist. Not only did he provide De Forest with assistance, but the ever curious Fuller closely studied the emerging technology of radio. Then, one bright autumn day, when the *Inca*'s crew had loaded De Forest's heavy equipment on board, they set sail for a historic test with Bucky on the bridge continuously asking questions of the distinguished guest beside him.[35]

As the *Inca*'s crew looked on, one of the navy's largest Curtiss flying boats took off filled with another set of experimental radio equipment. Then, after a great deal of adjusting dials and switches, Dr. De Forest spoke into a crude microphone, and through a mass of static, a distorted voice was heard to reply. It was the first time the voice of a person on a ship was transmitted to the crew of an airplane.[36]

Fuller's stint at the Naval Academy proved to be one of the most arduous, yet fulfilling, times of his life. Because of the war, the navy had expanded its shipbuilding program immensely in just a few months. Government officials wanted to enlarge the fleet quickly and needed officers to command those ships. Accordingly, the navy initiated a three-month officer-training program to establish a corps of new officers, who soon became known as ninety-day wonders.[37]

Bucky was nominated and, to the surprise of almost everyone, accepted, even though he met few of the qualifications. Aside from his poor eyesight and diminutive height of five foot six inches, he did not fulfill the requirements of having been overseas or possessing a college education or its equivalent. Still, he satisfied the precondition of being sponsored by a senior officer, and that officer, Commander Bellinger, carried a great deal of influence in the navy. Bellinger also made certain that those deciding on which candidates would be admitted to the Naval Academy noticed

Fuller's superior record commanding patrol boats and his inventive ingenuity.[38]

Thus, Fuller was propitiously accepted for three intense months' training at Annapolis in early 1918. That time and his subsequent naval duty would prove to be one of the most fascinating and expansive periods of Fuller's life. It was during those months that his broad World view and understanding of comprehensive management and operation began to develop.

Training at the Naval Academy occupied almost all of Bucky's waking hours, and even the time he allowed himself for sleep was extremely limited. Still, he found some moments for Anne. She had moved into an inexpensive hotel room nearby, and although Bucky was confined to living in the barracks with the other cadets, he could usually slip away for a walk after evening parade.[39] Those walks were Fuller's only diversion from his naval studies. All other time was devoted to a comprehensive training schedule designed to provide officers with all the skills they needed to operate autonomously in handling any situation at sea.[40] At the time, wireless communication was still in its infancy, and ships at sea could not contact other vessels or commanders on land who were not within the fourteen-mile visual range of the horizon. Accordingly, neophyte naval officers were trained to deal with a wide variety of problems, including taking command if the ship's commanding officer was incapacitated.[41]

Such training was not unique. Since men had first sailed from ports on primitive rafts, the ship's captain had provided the final authority at sea, and even though he was suspicious of traditional authority, Bucky respected the necessity of such command for the survival of a ship at sea. From his personal experience, he understood that decisions had to be made quickly at sea or entire crews could be lost. Hence, he was willing to accept and follow the authority of the ship's captain even though such subordination contradicted his burgeoning belief in the importance of independent thinking.[42]

The tradition of the sovereign authority of a captain at sea·had, over the centuries, been developed into a highly sophisticated system by the time Fuller started his formal naval training. At the onset of World War I, most of the knowledge concerning naval warfare and World domination through the employment of naval forces had been handed down and retained by the nation which had "ruled the waves" for centuries: Great Britain.

Great Britain, however, was on the verge of losing control of the seas to Germany and needed more ships and weapons. The only country capable of providing the armaments which could defeat Germany was the World's emerging industrial giant, the United States, and in desperation, proud British government officials were willing to enter into agreements which would lure the United States into World War I in support of England. One of those agreements was to assist the United States in building its navy up to a size and strength comparable to Britain's. As part of that effort, the British supplied the United States Navy with their superior technology and training methods as well as their knowledge of ancient naval history and tradition. Accordingly, British naval wisdom greatly influenced the Naval Academy and the other sites where Fuller was trained and later assigned to duty.[43]

Fuller's indigenous fascination with the sea intensified during his naval career. When combined with his incessant curiosity, that enthusiasm supported his benefiting greatly from the naval history and technology he was taught. Those lessons would eventually be incorporated into his comprehensive philosophy and his understanding of the history of human development.

Because naval warfare was such an essential component of Fuller's philosophy and understanding of human history, he often utilized it in his stories about the development of humanity. The following is one such example from his naval studies which Fuller frequently related during his lectures. It provides some insight into the ideas that eventually guided Bucky to formulate and champion a unique philosophy predicated upon peace and cooperation rather than traditional warfare.

At the United States Naval Academy he learned that the basic naval warfare strategy employed during World War I dated back to 1184 B.C., when Greece conquered Troy. At that time, an entirely new form of battle was initiated, and it dominated most human activity for thousands of years. The fall of Troy marked the first time shipbuilding and navigation had achieved a sophistication which allowed a nation to employ large supply ships in supporting distant troops for years on end.[44]

Prior to that time, monarchs had maintained power by fortifying themselves and their subjects within huge castles such as the one at Troy. When an attacking force was seen advancing into a kingdom, peasants burned their fields and buildings, poisoned their water supplies, and retreated with all the supplies they could carry to the safety of the local castle. Those castles were well fortified and supplied for such emergencies. They could support local residents for months or years at a time, and safely inside them, the villagers and the king's troops simply waited for the invaders to leave. Local residents believed that dwindling supplies of food, water, and other necessities would force the invaders to abandon any hope of conquest and leave after a few weeks. Prior to the invasion of Troy, that had been the outcome of almost every such confrontation.[45]

Because those castles were heavily fortified with thick, high walls which would withstand assaults even when defended by a minimum number of soldiers, invaders seldom attacked. Instead, they conformed to the traditional battle strategy known as the siege. A siege consisted of simply surrounding a castle and waiting, but since they tended to be well prepared, the local residents could usually wait much longer than the invaders. Encroaching armies were further hampered by the size and speed of the pack animals or small ships which provided lines of supply from home. Thus, a siege generally resulted in the invaders' leaving or perishing from lack of provisions.[46]

At Troy, however, "modern" technology ushered in a new era of warfare in which the invaders' disadvantage was lessened. The primary innovation which Fuller felt altered that superiority and the course of human activity for future generations was the Greeks' use of ribbing technology to create much larger ships. Such large-ribbed ships maintained an ample, continuous line of supply for their troops holding Troy under siege.

The Greek victory marked the dawn of "line-of-supply" warfare, a tactic which remained dominant into modern times. Since the fall of Troy, Fuller observed that monarchs and military leaders had shifted their focus from establishing larger and stronger protective fortifications to defending supplies en route to and from their kingdoms. This was particularly true of supplies for armies at war. With that change in strategy, power became contingent primarily upon a nation's ability to protect lines of supply.[47]

Since most land passage involved traveling through often unknown hazardous territory, including rugged mountains, deserts, and forests, a nation's initial inclination was to use the much more efficient and economical sea routes whenever possible. Sea routes were far less expensive because goods did not have to be transferred from one type of vehicle to another or to be hauled over rugged terrain.[48] As the use of ribbed ships provided an enormous increase in the amount of goods or weapons transported, nautical transportation became the key to both military and economic dominance. Accordingly, the nation which controlled the most marine traffic became the dominant global power, a factor which would eventually become a key component in Fuller's comprehensive understanding of global politics and resource allocation.[49]

Through centuries of military conquest, that position of naval and global supremacy shifted from Greece to Venice and then to Portugal. When Spain defeated Portugal at sea, the Spanish dominated. Then, Great Britain conquered the Spanish navy during the 1500s. It was during World War I, while Great Britain dominated the sea-lanes, that the line-of-supply military strategy reached its pinnacle. That was also the period in which young Buckminster Fuller enlisted in an American navy which was being trained by the accomplished military strategists of the British navy.[50]

At the onset of World War I, navies were the dominant military force on Earth. Most significant battles were fought at sea, but such battles were generally unmanaged because no rapid communication existed between ships and the governments which owned and commanded them. Messages farther than line of sight had to be transported by a courier on a ship. Thus, the commander of a ship or a fleet of ships at sea wielded enormous power, and his word was law. The captain's undisputed autonomous authority caused monarchs and heads of state to select ship captains and admirals with utmost care. Men chosen for those prestigious positions had to be unquestionably trustworthy as well as cunning and intelligent.

Ship captains, like all naval officers who might have to assume command of a ship in an emergency, were specially trained to retain a comprehensive outlook at all times and to perform their duties with no direct support from superiors. Their sweeping viewpoint had to encompass both the internal and the external operations of the ship or fleet under their command. An officer was expected to be familiar with all internal aspects of his ship's operation, including such diverse elements as the morale of his sailors and the maintenance of boiler rooms. Externally, he had to understand and consider the global position of his nation and the worldwide implications and consequences of his independent actions at sea.

Those demanding responsibilities resulted in significant differences between the training of naval officers and army officers. Army officer training generally taught men to consider only local battles and conditions when making decisions because they almost always had some direct contact with their superior officers. The higher level officers were expected to understand and carry out the overall battle strategy which had been approved by government leaders, but line officers were to follow orders rather than think for themselves.

As a United States naval officer during World War I, Fuller was among the last men provided with the comprehensive training previously furnished to all officers sent to sea. He received the sweeping tutelage which had become a tradition throughout powerful navies but which was phased out in the years following World War I as radio communication with other vessels and commanders became practical. Once radio communication was established between ships at sea and government leaders on land, the ultimate autonomous authority of a ship's captain was terminated. Instead of having to make decisions in every situation, he was expected to request and follow orders from superior officers, just as army officers had been doing for centuries.[51]

Serving prior to the institution of such communication, Fuller experienced the last moments of an era far different from modern naval operation. He was assigned to every type of naval vessel. He also spent time in shipbuilding yards, learning the construction and business aspects of sailing, and as a naval attaché with the state department, learning diplomacy. The navy also provided him with a brief course in law so that he could formulate just legal decisions at sea. Because the war was raging, each of those assignments was short and intense.[52]

Fuller's multiplex naval education was diametrically opposed to the popular notion of education prevalent throughout land-based institutions and other military organizations. At that time, large universities were creating specialized programs in which the brightest people were selected from undergraduate programs and educated as specialists in individualized fields.

Bucky closely examined that aspect of education and found that the people who dominated power structures, whether land-oriented or nautical, had always attempted to maintain as comprehensive a perspective as possible while keeping the general public divided and specialized. He felt that that strategy of "divide and conquer" had been successful for thousands of years and had been perpetuated through the support of universities.[53]

Fuller witnessed institutions of higher learning singling out the students with tremendous potential. Those were the people who could most easily perceive and utilize the comprehensive knowledge retained within the power structures, and the power structures did as much as they could to keep those students from appreciating and using that knowledge. Bucky believed that such naturally gifted students were coerced with incentives of money and fame into specialized fields, where they would not be interested in or have access to comprehensive knowledge and perspectives. The handful of people in power could then combine and employ the new discoveries made by university-trained specialists to realize even more domination and wealth.[54]

As a naval line officer, Fuller was trained comprehensively so he could assume command if his supervising officer was killed. The naval order-of-command succession was very precisely delineated, and a junior officer was educated so that he could continue advancing into higher and higher positions of authority during battle. That succession was so automatic that had he been the highest ranking officer alive in the fleet, Fuller would have been expected to assume command of it, and he was trained to do just that in an instant.[55]

The method of promotion in the navy was also different from that of the army, where upgrading was based on an automatic rotation. In the navy, on the other hand, promotion of officers was predicated on selection. That policy allowed senior officers to choose and rapidly advance young men who they felt would obey orders and support the power structure when commanding or serving on an autonomous ship at sea.

Fuller's naval training was a primary element responsible for the comprehensive perspective which would eventually support his fame and fortune. That training helped him understand the importance of life at sea and how it differed from life on land. Bucky found that because of the rigors experienced at sea, nautical technology was always on the leading edge of innovation. He therefore studied maritime cultures and their innovations in hope of applying what he learned to similar land-based applications, which were generally much less technologically advanced.[56]

In studying warfare, he discovered a key difference between maritime and land-based human behavior. As with other activities, nautical battles adopted the latest technology well before land warfare. For instance, even though the fall of Troy in 1184 B.C. marked the end of the effective protection of fortifications, that antiquated scheme of attempting to wait out a siege was still being used in land battles as late as World War I.

Despite the fact that new, more effective lines of supply were constantly being established as history progressed, ground troops continued surrounding an enemy's city or military position, trying to cut off supplies, and waiting. During World War I, the land-based military siege strategy became somewhat more advanced and mobile as troops on both sides were transported to a remote site where they "dug in" to hold a position. That tactic became known as trench warfare, and the result was soldiers living for weeks or months in the trenches they excavated on either side of an imaginary line designated as "the front."

Still, Fuller found that trench warfare was a form of siege, and the majority of the troops' time and energy was devoted to maintaining their trench homes, shouting at one another across the front, and waiting for something to happen. Eventually, the leader of one of the battling factions would become so frustrated by the stalemate's devouring his side's resources that he would order his officers to launch an attack. Attacks amounted to nothing more than sending troops charging across the front in an attempt to kill enemy soldiers in their trenches. If successful, the attacking forces commandeered the enemy troops' trenches and sent them retreating to a new site, where they excavated another line of trenches for their safety. If they were fortunate, the retreating troops simply moved back to the trenches they had occupied prior to their last successful attack.[57]

The result of those tactics was that the front moved slowly from one location to another while the war consumed enormous numbers of men and stockpiles of resources.[58] Trench warfare also created acres and acres of land congested with long rows of trenches and regiments of troops who simply moved from one trench to another every so often in an attempt to acquire territory and kill the enemy.

In his examination of maritime cultures, Fuller found that naval battles, on the other hand, were far more sophisticated and fascinating because they continually employed the leading edge of technology. Following the defeat of Troy, rulers who controlled prosperous fortified city-states near water began developing maritime military strategies to protect their expanding trade. They invested vast amounts of resources in building larger, more stalwart ships for transporting merchandise as well as protecting their kingdoms and ships at sea.

As sea battle strategies shifted from defending immobile fortifications to the establishment and defense of supply routes, the global economics which would become such a significant factor in Fuller's work changed dramatically. Since the majority of the important trade routes were nautical, the most obvious changes could be seen in the area of marine warfare and transportation, which Bucky examined diligently.

From the time sailors ventured out on crude rafts, sending ships to sea has been a perilous undertaking. Even as recently as the nineteenth century, the duration of an average sea voyage was several years, but as a result of harsh conditions at sea, many ships never returned. Following the inception of ribbed ships, which were much stronger than anything previously known, armament became a key component of many sailing vessels, and as heavily armed ships began appearing along sea routes, theft and enemy attack began to take a toll on sailors and their cargoes.[59]

Fuller found that as early as during the Greek and Roman dominance of maritime lanes, building a single, large, oceangoing ship was a monumental task requiring massive resources and capital, and that tradition continues even today.[60] Still, the people who initiated such construction usually built and maintained an entire commercial and military fleet. To build such ships, a powerful individual needed almost unlimited authority within a community, be it city, kingdom, or nation. Hence, a person instituting boatbuilding and sailing in a community tended to be the ruler of that area, or at least a very wealthy noble, and tended to be able to control the thousands of workers who were required to assemble at one location for months in order to construct a ship.[61]

Because of all the labor and resources required just to construct a single ship, it was only natural that once those huge vessels were finally launched, sailing dates, destinations, and cargoes were maintained in strict secrecy. Vessel owners and crews were supported in maintaining that secrecy by a fundamental characteristic of traveling upon a spherical surface. Since the Earth is spherical, a ship more than fourteen miles away remains below the horizon, hidden beneath the Earth's curvature.[62]

Fuller learned that when a ship did successfully return to its home port after a long voyage, it invariably provided enormous, instantaneous wealth to its owners. This occurred because ships brought resources from different locations together in a synergistic process which increased the value of those resources.[63]

For example, vessels would transport a commodity, such as tea, which was considered ordinary in its indigenous location of the Orient, to a place where it was not natively available, such as Europe. Because of its rarity, the resource immediately became exceptionally valuable when relocated to the new site. Through his examination of that type of nautical product relocation, Fuller began to appreciate the true significance of information about resources, and the seeds of his strategy of supporting more people with fewer resources were planted.

Bucky observed that throughout history, foreign resources had often been combined with local substances and knowledge to produce even more valuable products. For instance, even the Greeks and Romans created crude alloys by combining metals imported from various lands. Alloys have consistently contributed a great deal to even the most rudimentary technology. Not only did the melding of elements create wealth for shipowners and merchants, but more significantly, newly fabricated products advanced humanity's development in ways which Fuller would later attempt to mirror.

The years of waiting for merchant ships to return with profits and rare commodities did not go unnoticed. Port cities kept watchtowers in operation at all times, and even ordinary citizens tended to keep an eye on the horizon in search of incoming ships. The population's incessant interest in ships, the sea, and trade, as well as in how they would benefit from a ship's arrival, is evident in the language of today when people speak of waiting for their "ship to come in."

In his examination of maritime history and how it influenced technology, war, and our current perception of wealth, Fuller found that from the onset of naval trading centuries ago, the instant a ship left port it became a target. Few individuals or nations possessed the resources required to build large ships, but most nations and wealthy individuals could construct smaller vessels. Owners would then dispatch those small, fast ships to sea in search of larger trading ships to plunder. Sending ships on brief missions to observe a neighboring kingdom's ports for a few weeks was far more economical and effective for small, impoverished nations than attempting a risky trading journey requiring several years.[64]

The small, armed ships would intercept larger ships returning home after lengthy voyages while they were still out of sight of, and communication with, their home port. Being smaller, those vessels could often overtake and capture the larger ships, which carried few armaments in an effort to devote as much space as possible to valuable cargo. Such maritime theft epitomized the high-seas piracy which flourished throughout human history and endured in various forms until World War II.[65]

Realizing that those in power could not enforce laws at sea, many renegades became pirates, attacking and plundering merchant ships at will. Rulers of the land did attempt to expand their authority onto the sea; however, without thousands of fighting ships to patrol the oceans, they could extend their power only as far as they could shoot a projectile from a coastal fortification.[66]

Even as late as World War II, that distance of fire amounted to no more than three miles, and three miles became the agreed-upon jurisdiction of a country's sea ownership. As nearly three quarters of Earth is covered by water, a vast amount of territory was relinquished to the pirates, who freely roamed the seas and spent their lives at sea. In fact, they were so independent of any regional or national affiliation that Fuller considered them the first global citizens.[67] He later compared those pirates to others who chose to live outside the conventions of society as dictated by power structures. And, he suggested that such individuals are, in fact, essential to forging the new realities so valuable to human development.

Still, the problem of pirates became continually worse, and the most powerful monarchs had to establish navies to protect their trading ships and coastlines. In response, some of the more adventurous pirates banded together to defeat the ships dispatched by inland rulers. Pirates then continued battles by sailing into the unprotected ports, defeating the land-based troops, and proclaiming themselves as the new rulers.[68]

Through such actions, pirate leaders were transformed into monarchs. In Fuller's perception of history, those former radicals then entered into the process of becoming the traditional authority, while leaving innovation to a new generation of freethinking pirates.[69]

As a result of that scenario of persistent battles and shifts in dominance, the individuals with the most powerful naval forces were legitimized as rulers while those with less potent navies flourished upon the "open seas" outside the jurisdiction of the law.[70] Accordingly, the outsiders became known as *outlaws*. They also became the people who sought to gain more power, and consequently they developed the newest technology in an effort to defeat stronger opponents.[71]

Although the balance of naval power shifted frequently between warring factions, the most powerful individuals were easy to identify because they ruled the nations with the strongest ships and navies. Even though he realized that such competition and greed did result in the development of innovations which benefited humanity, Fuller also argued that all warring factions remained competitive pirates at heart.[72] He felt that humankind had recently entered an era in which such competition and greed would have to cease or humanity would destroy itself.

In order to facilitate such a shift, Bucky supported and actually became a new breed of outlaw. Rather than opposing traditional laws and established bureaucracies, his class of modern outlaws seek methods of circumventing the establishment and demonstrating that every person can make a difference. Because of his belief in such action, Fuller championed the persistent spirit of individual initiative which flourished among the outlaws who had lived independent lives on the high seas centuries earlier.

*

At the onset of World War I, Great Britain held the position of ultimate power as the dominant naval force on Earth. Because it is an island nation, the country itself served as an unsinkable flagship for the British fleet. The large islands provided numerous secure ports where ships could be protected in relative secrecy from perils such as weather and enemy ships. With its strategic location overlooking much of the western coast of Europe, Great Britain was also able to oversee a majority of the World's ports and establish an unrivaled empire predicated on naval might.[73]

Having escaped Britain's domination less than two centuries earlier, the United States attempted to keep out of World War I and to focus its attention on domestic issues. Rather than seek global conquest and power, as most other large countries were doing, Americans concentrated their efforts on a native frontier: industrial development. Leaders in the United States were quite content to build bigger factories and accumulate more profits. They had little desire to enter a military effort that might drain resources and profits and could result in defeat.

German leaders, on the other hand, were extremely interested in wresting global domination from the British, and they knew that such a takeover had to be accomplished within the traditional area of dominance and battle: the sea and marine commerce. The Germans realized that they could not rival the naval domination of the ocean's surface which Great Britain had acquired over centuries or the massive British navy, and they sought a unique new strategy for controlling sea lanes. Rather than challenge British vessels solely on the surface, the Germans developed submarines and began attacking vessels from the surreptitious depths of the sea.

That ingenious strategy was successful, and it advanced an entirely new form of warfare. Using submarines, the Germans were able to sink many more ships than the British, and they began closing nautical supply lines to England. Because it is an island nation and aviation was not a practical means of transportation at that time, the blockade of naval trade routes was far more critical to Great Britain than to most other countries. The Germans also challenged Allied troops in land battles throughout France and neighboring areas; however, those inland offensives were also directed toward nautical domination. The primary German objective was to advance through France in order to ferry troops across the English Channel and conquer the British Isles, which remained strategic to control of the seas.

The British realized they were trapped in a precarious position which could be alleviated only by massive support from the largest industrial power on Earth, the United States. They needed more ships and weaponry to offset the Germans' submarine forces, and they made unprecedented agreements to draw the United States into World War I on their side.

At the naval academy, Fuller personally experienced and greatly benefited from one of those agreements. As compensation for industrial assistance, the British government had agreed to share their superior naval technology and training with the United States. They had also agreed to help the United States enlarge its navy to a strength and sophistication comparable to those of the powerful British navy.[74]

As a naval officer during that era of British support, Fuller was exposed to an unparalleled comprehensive training which had been developed over hundreds of years and had previously been reserved for British naval officers.[75] That training provided him with a firsthand experience and understanding of successful global strategies and operating procedures. It was that training and exposure to global management strategies which also established the foundation for the comprehensive operating strategy which served Fuller so well in tackling many diverse issues.

It was during his World War I naval training and experience that Fuller first began to observe and appreciate a rapidly increasing shift from visible to invisible technology. Years later, he would continually assert in both lectures and writing that humanity had entered into an era in which 99.999 percent of the activity which influences even the most mundane aspects of daily life could not be perceived by the human senses.[76] He would share some of his insight into that invisible reality when he recounted his observations of the sophisticated naval technology which came into existence during World War I.

Even naval battles of past centuries had presented a preview of modern warfare in which most battles are simply a matter of comparing technology. That phenomenon had been undeniable for centuries, and it was so conspicuous during World War I that Bucky found many experts claiming that almost anyone observing a skirmish at sea could determine which ship would win after only a few barrages of artillery fire. Some naval experts even asserted that when the World's powerful countries dispatched their navies against one another, within the first few minutes of combat, an observer could ascertain who would control the Earth for the next twenty-five years.[77]

At the time, such statements were fundamentally correct. The nation dispatching the best engineered and designed ships and weaponry invariably won the right to dominate sea-lanes, and the superior warship could be determined with the firing of only a few shots. A ship firing farther with greater destructive power and accuracy would nearly always hit an enemy ship long before that vessel was within the effective range of its guns, and that superior ship would win most maritime battles easily.[78]

Naval engagements simply resolved the question of which nation's military engineering skills were most evolved at a given moment, and that country then dominated maritime traffic until another nation developed better weaponry. To achieve or sustain such domination, heads of state persistently demanded that the finest minds under their jurisdiction concentrate on creating weapons which could effect greater destruction, over longer distances, in shorter periods of time, with greater accuracy and less effort.[79]

With so much attention focused on the production of better weaponry for domination of the sea, it was only natural that the underlying technology of ships and their armaments was cloaked in utmost secrecy. Although such clandestine naval innovation had initially been developed for and employed in destructive weaponry, over time it was also adapted to life-supporting endeavors.

In fact, a great deal of the technology which now affects nearly every human activity was first developed for maritime military applications. Items such as refrigeration, air-conditioning, radio communication, and alloy metals all had their initial practical application in naval military operations, and Fuller was fortunate to have been exposed to some of those "wonders of technology" during his naval career.[80]

Prior to the employment of such innovations, the effectiveness of any ship or weapon could be judged by its size and weight. However, with the shift to the invisible technology, two ships of the same weight and design might appear identical, but their intangible modifications rendered them very different. Those vital variations were not discovered by an enemy until a battle occurred, and by then it was too late for the captain with inferior firepower to do anything but attempt to evade the destruction or capture of his ship.[81]

One example of such innovation, which Fuller studied and often spoke of in great detail, occurred when nations first began to arm their ships. At that time, cannons and cannonballs were fabricated primarily from heavy lead, and the immense weight of that weaponry limited the armaments a ship could practically carry to sea. That weight component was particularly significant in cargo ships, where heavier weapons reduced the capacity for transporting freight and, accordingly, lessened profits. To remedy the problem, the rulers of nations with powerful navies assigned their best scholars, scientists, and engineers to the task of creating lighter, more effective weaponry. By mandating that action, those in power instituted a period of great progress in the development of alloy metals which persists today, although that technological progress is now focused primarily on space rather than the sea.

When Fuller closely examined the historical methods used to develop and implement better naval weaponry, he found that the first phase of that process was melting down and combining metals in the hope of producing lighter alloys for weaponry.[82] New metals were then tested for use in cannons and cannonballs. When outfitted on a warship, the new weapons did not alter the vessel's appearance. They did, however, lessen the ship's weight, allowing it to sail more rapidly or to maintain the same weight and carry more weapons. In either case, the newly armed ship was provided with an advantage over similar vessels.

Lighter metals also permitted the manufacture of larger cannons, which were combined with the lighter cannonballs to extend a ship's effective firing range. That technology enabled a warship to hit an enemy vessel before the enemy ship was within its effective firing range. Powerful nations then found themselves not only engaged in naval battles, but, more importantly, struggling in a competition to increase their knowledge and to employ resources more efficiently.

That productive competition, rather than the conspicuous conflict of war, would prove to be the most important influence of World War I on Bucky. During the war, he began to understand that within such rivalry the seeds of future prosperity could be found, and later he was often found engrossed in serious explorations of just what benefit could be obtained from seemingly destructive situations and innovations.

CHAPTER 4

Years of Deterioration

 Upon graduation from the United States Naval Academy, newly commissioned Ensign R. Buckminster Fuller was assigned to the transport *Finland* and given a week's leave. He very much wanted both to spend more time with his bride and to vacation on his beloved Bear Island during that period. Since they had been married less than one year and had spent little time together, Bucky felt a true honeymoon was in order. He did not, however, take his wife's preferences into account. Thus, they spent the week on rugged Bear Island, where the refined Anne felt quite out of place.[1]

Certainly, Bucky's self-assured, cocky aplomb helped him overcome obstacles and lack of support for his work and ideas over the years; however, that same characteristic had a negative side which invariably cropped up over the years. His self-confidence often resulted in his letting the feelings of others—especially women—go unnoticed. Although he was not intentionally unloving or inconsiderate, Bucky often believed that his view of reality was best and that everyone would naturally want what he wanted.

That attitude was especially true when he was a young man, but even in his later years, Bucky, like so many talented eccentrics, was frequently so consumed with his ideas that he did not seriously consider the thoughts or feelings of others. Even though on an altruistic level he felt that everyone's opinions were equally valid and that each person had something to offer, the practical side of Fuller often prevailed and brushed aside the feelings of some individuals in his attempt to be as supportive of as many people as possible.

An arrogant, chauvinistic attitude was not uncommon among American men at the turn of the century, and Bucky was no exception. The problem was most noticeable in the low regard many men of that era held for women's feelings, and again, Fuller conformed to the conventional pattern. Although he worked on eliminating attitudes of prejudice within himself throughout his life, Bucky never truly broke free from the traditional chauvinistic posture engendered by his turn-of-the-century upbringing. Even in his later years, Fuller's universally insightful lectures were tarnished by references to "girls" or "gals" when he spoke of women.

Despite that characteristic, Anne remained faithful to him, fulfilling the qualities and attitudes of her own upbringing. She had been raised as a lady in the genteel refinement of Long Island and followed her husband to their Bear Island "honeymoon" even though she found the idea oppressive. Unlike Bucky, Anne did not enjoy the rugged outdoors, bathing in icy water, using outhouses, and cooking on an old wood stove. Yet, despite her aversion to the wilds, she faithfully followed her husband and accepted his bidding without question, as she would continue doing for the next sixty-five years.[2]

Upon returning to New York, Bucky learned that the *Finland* had been torpedoed and was being repaired in drydock. Then fortune shone upon him, as that chance incident resulted in his being reassigned as a communications officer on the staff of Rear Admiral Albert Gleaves in Hoboken, New Jersey. Admiral Gleaves commanded all naval cruiser and transport operations, and after only a few weeks, the Admiral took a liking to Bucky, appointing him "personal aide for secret information." In that position, Bucky intimately participated in the most extensive and successful transport effort until that time.[3]

Admiral Gleaves and his staff were responsible for the safe passage of two million troops, and the supplies required for their support, from the United States to Europe despite a legion of German submarines and ships patrolling the Atlantic Ocean. Even though that mission was extremely hazardous, the convoys were so well planned that not a single soldier traveling to Europe was lost to enemy action at sea. A few ships were sunk on return trips because those voyages were not escorted by destroyers and cruisers but, on the whole, that operation was considered an enormous success.[4]

Sailing schedules were planned in great detail and secrecy. As the individual responsible for maintaining secret information, Fuller was expected to know the location of every American ship at sea and to send a report to the chief of naval operations each day. He was also the person who communicated with those ships on a daily basis, using a crude radio system which had just become practical. Because of the major consequences of his work, Fuller quickly began to observe a regimen of always transmitting precise messages.[5] He realized that a misunderstood message could result in fatal actions, death, and destruction. A completely unintelligible message, on the other hand, would result in a request for resending rather than disaster. Accordingly, Bucky resolved that it was much better if his communications were completely incomprehensible rather than misunderstood.[6]

Because of that decision, he began disciplining himself to send extremely precise communication which included as much information as possible using a minimum of words, and from that time on, Fuller's language was very exacting. He perpetually worked on employing only the most accurate words and paying particular attention to not being misunderstood.[7]

As a result of the vast amount of planning involved in that operation, Bucky began to consider seriously the details of large-scale patterns and global strategies for the first time.[8] That assignment also spawned an idea which would transform his perception of reality and support his initiative to contribute as much as possible to the well-being of all humanity.[9]

Fuller's wartime idea was to concentrate the same intensity of planning and thought on supporting life as was being focused on the destructive power of war. Years later he externalized the idea and named his creation the "World Game." World Game is an educational exploratory experience in which participants consider problems from a comprehensive perspective.[10]

Although Fuller chose to use the word *game* in its title, World Game is actually a serious undertaking. The title first came to mind during his naval tour of duty when he encountered officers who discussed the management of global resources for destruction and domination, calling their planning *war games*. Bucky felt that if the same energy and commitment of resources were devoted to constructive purposes, great advances in human development would ensue. Thus, he coined the term *World Game* as an alternative to military war games.[11]

During his naval career, Bucky also became aware of the navy's precise system of filing communications chronologically and the benefits that system afforded in preserving information for future use. He found that, like most people, he was usually able to remember the approximate time frame of an event, and from such recollections, specific documents filed chronologically could usually be located without much effort.[12]

Like many young children, Bucky had been a collector. However, his collection was somewhat unusual. At the age of five, he had begun accumulating articles from newspapers, magazines, letters, and other miscellaneous sources which focused on his primary interests, the wonders of "modern" technology. At first, the collection was simply a scrapbook of fascinating inventions; however, as he grew older, Bucky began to include his reactions as well as his observations of others' reactions to the technology that so interested him.[13]

In the navy, Bucky began to understand just how valuable such a record could be if it was maintained in a fashion which was easily accessible. The method used for filing naval communications provided a perfect model for that system, and Fuller adapted it to his work. Thus, he began preserving all his personal written material plus a great deal of information gleaned from other sources.[14]

Over the years, that collection would be expanded to include copies of his lectures and pieces done about him in all media, including audio and video recordings. He labeled the archive his "Chronofile," and today it remains an enormous resource for humanity. Fuller's Chronofile helped him determine and understand large-scale patterns operating throughout Universe and to recall his feelings about the rapid changes occurring around him.[15] The Chronofile could also provide a personal and encompassing view of human development to future generations who might someday discover and explore it. In it they would be able to examine the experiences of one individual living through and being deeply involved in the technological and cultural innovations of most of the twentieth century.

In the fall of 1918, Ensign Fuller was reassigned to sea duty with the transport fleet he had been monitoring for Admiral Gleaves, and on November 11 of that same year, the Armistice was signed and World War I ended. His naval career was, however, far from complete. He loved both the navy and the new technology which he experienced in the navy.[16] That exposure continued when he was transferred to the post of communications officer on the *George Washington*, a ship being reconditioned and refurbished to carry President and Mrs. Wilson to the Paris Peace Conference. As communications officer, Fuller worked closely with a group of technicians installing a state-of-the-art, long-distance wireless telephone.[17]

The bulky equipment he had worked with only months earlier on the *Inca* had a maximum range of seventy miles, but this new equipment was said to facilitate the transmission of transoceanic voice messages. Although few people believed that such a development was possible within only months, Fuller remained both curious and open-minded. Naturally, when the *George Washington* sailed to Paris, the inquisitive communications officer was assisting with the new technology as well as observing every nuance of its operation. He was even present on the momentous day President Wilson made the first true long-distance telephone call, speaking from the ship anchored in the French harbor at Brest to the government wireless installation at Arlington, Virginia.[18]

When the *George Washington* was decommissioned in early 1919, Admiral Gleaves asked Fuller, who had been promoted to the rank of lieutenant junior grade, to rejoin his staff aboard the flagship USS *Seattle*. The Admiral had just been given command of the entire Asiatic Squadron and wanted Bucky to accompany him on an extensive tour of the South Pacific.[19]

That offer loomed as an exciting possibility for the adventurous Fuller, but a more important factor took precedence. On December 12, 1918, shortly after he had returned from an Atlantic crossing, Anne had given birth to their first child, a daughter they named Alexandra.[20]

Alexandra was frail from birth, and her health was complicated by the fact that serious epidemics were claiming thousands of lives in a matter of days or weeks. Although Bucky loved the sea, when the choice arose between staying with his daughter and wife or following the adventure of the navy, Bucky chose his family.[21] Anne may not have relished the possibility of a vagabond life following Bucky from port to port, but she probably would have done so for him. What persuaded Bucky to resign his commission and stay with his family was the realization that the fragile Alexandra could not possibly survive so much traveling, and that if he sailed off with the fleet, he would most likely never see her alive again.

Bucky's intuition was correct. Even as he was processing out of the navy, Alexandra contracted and nearly died from the influenza epidemic which swept the country.[22]

Although that was a difficult period for the small family, Bucky was able to obtain a new position with his former employer, Armour and Company. Because of his experience, the company made him assistant export manager at the handsome salary of $50 a week. Yet, even with his excellent salary, the family faced financial difficulties because of Alexandra's ongoing illnesses. Following her bout with influenza, she came down with spinal meningitis and then polio, and even though cures for those diseases were not then available, Alexandra demonstrated the Fuller tenacity by surviving them all.[23]

She did, however, require a great deal of constant care, which Bucky and Anne personally provided along with the assistance of a series of home nurses. It was during those long sessions sitting with their sickly daughter that Bucky began to notice an unusual phenomenon which would greatly influence the rest of his life. He and Anne frequently passed the time watching over Alexandra by discussing topics which were far too complicated for a young child to understand, and during one such conversation, Bucky was startled to hear Alexandra suddenly utter words which had just entered his thoughts.[24]

That phenomenon continued and escalated in intensity to the point where Alexandra sometimes offered sophisticated answers to questions Bucky had scarcely formulated in his mind. After weeks of observing and considering her actions, Bucky and Anne began to believe that Alexandra's physical restrictions were, in fact, contributing to the amazing events they were witnessing.[25]

Bucky theorized that because of her severe physical limitations, Alexandra had begun to develop in ways which were different from most children. He observed that her primary mode of expression and learning was metaphysical rather than physical, and he felt that one of the most conspicuous indications of her growth in that area was her rapid development of telepathic communication.

Fuller would continue his examination of that phenomenon for the balance of his life and came to believe that everyone is capable of telepathic communication.[26] He hypothesized that Alexandra had mastered telepathy quickly because she was young and had not yet been conditioned to disregard or disbelieve metaphysical possibilities. He also came to feel that every human being is born with the innate capacity to operate within the realms of the metaphysical, but that most people rapidly unlearn those abilities as they grow older and become more socialized.[27]

Following his experiences with Alexandra, Fuller was convinced that telepathic communication is not a mysterious power possessed by a few gifted individuals. He also reasoned that if telepathic communication did occur, it had to be governed by the generalized principles which are operational throughout all Universe, and that by discovering and studying those principles, he could better understand telepathy.[28] Since so few people embraced or accepted telepathy or the metaphysical, Fuller speculated that those generalized principles were largely unknown or misinterpreted. After all, once an individual grasped the principles responsible for a phenomenon, he or she would naturally accept and utilize that phenomenon.

Fuller initiated his quest to learn about the principles governing the metaphysical following his telepathic encounters with Alexandra. He first examined telepathy and its relationship to known wave-pattern phenomena. At the time, human knowledge of invisible wave patterns was primarily limited to radio waves which were at least one mile long. Shortwave radio was not discovered until the 1930s, and the even shorter waves, such as those used in television and satellite transmission, were not discovered until very recently.[29]

Fuller's explorations in the 1920s and 1930s led him to theorize that such short waves existed, and he further concluded that even shorter ultra-ultra-high-frequency waves were also a possibility.[30] Those very short waves became the foundation for his ideas concerning telepathy and the generalized principles which govern it.[31]

In the years following his initial consideration of telepathy, Fuller addressed thousands of audiences and often recounted that during every talk, he experienced meaningful nonverbal communication between his eyes and the eyes of audience members. Although he did not claim to understand all the implications and details surrounding that communication, he did feel that it was related to telepathy and could be explained by the generalized principles governing waves.[32]

In the last years of his life Fuller was not surprised to learn that scientists had determined that very-short-wave energy could, in fact, be transmitted from one location to another. Years earlier, he had speculated that the human brain functions in a similar manner, transmitting and receiving short wave messages through people's eyes. According to Fuller's hypothesis, those invisible messages are received through one person's eyes and transferred to his or her brain, where they are deciphered. Once inside the brain, wave pattern signals are translated into tangible images which the brain can understand. Because human brains deal exclusively in sensory information (as delineated by the Fuller distinction between brain and mind) and those wave-pattern transmissions cannot be consciously experienced by the human sensory system, most people assume that the data received through unseen short waves emanate from some magical source, usually referred to as the paranormal.[33]

By examining that phenomenon within the context of humanity's most advanced scientific data, Fuller came to believe that telepathy is actually the result of information transmission via invisible short waves. Accordingly, he felt it could be explained by the generalized principles which govern all wave patterns.[34] He hypothesized that individuals who are more sensitive to and accepting of such transmissions convert them into what humans regard as conscious awareness. Because such people perceive wave energy transmissions in almost the same manner they perceive information transmitted over more conventional channels, such as speech and sight, they react to both forms in a similar manner.[35]

Since Alexandra was not trained to distinguish between conventionally accepted means of transmission and metaphysical telepathic transmissions, she naturally accepted both as authentic. Consequently, Fuller believed that she responded as quickly to his thoughts as she did to his words.[36]

Although Alexandra's illness placed a constant strain on the Fullers, both the Fuller and Hewlett families helped the struggling couple by contributing funds toward nurses and medicine as well as assistance whenever possible. With that support, Alexandra's condition gradually began to improve. Unfortunately, the Fullers found that their daughter's improvement tended to be a cyclical phenomenon. One week she might be able to move about the house under her own power, but only a few days later she might be bedridden again. During the periods of relative healthiness, Bucky and Anne would allow themselves some diversion and particularly enjoyed going out with old friends.

Anne's father would take them out on what they called the "Bohemian circuit," socializing with some of New York's "struggling artists." On one such foray, Mr. Hewlett took them to a tiny restaurant which had just opened south of Washington Square. In the café, they encountered only two other people: the owner and proprietor, a woman named Romany Marie, and a pale young man introduced as Eugene O'Neill. The entire evening was devoted to conversation with those two unique individuals. Although O'Neill soon became well known as a major American playwright, it was Romany Marie who would significantly influence Bucky, becoming his close friend and confidante during the most difficult years of his life.[37]

Other diversions from the hardships of Alexandra's illness included the regular evenings spent at the Hewlett family home. The family was always putting on shows to celebrate birthdays or other occasions, and Bucky loved being at the center of the action. He had always enjoyed parties and would dress up in ridiculous costumes and recite comic verses he had written especially for an event.[38] That gift for writing witty verse which enlivened a gathering remained throughout his life. For decades, his presence generally lifted the mood of parties, and those who were privileged to share Bucky's lighter side remembered it dearly.

Despite the family's best efforts to maintain Alexandra's health, New York winters took their toll on the baby's well-being. Apartments and houses of that era tended to be drafty and cold. To alleviate Alexandra's having to suffer through those conditions, Bucky and Anne decided to send her to a warmer climate during the winter of 1921–1922. As export manager at Armour, Bucky had many foreign connections, and he was able to rent an inexpensive house in Bermuda for Anne and Alexandra.[39]

While his wife and daughter were wintering in Bermuda, Bucky sought upward mobility in a new job. An old friend of his had been appointed president of the Kelly-Springfield Truck Company and offered Bucky the prestigious position of national accounts sales manager. As the salary was considerably more than he was making at Armour, Fuller jumped at the opportunity.

Then, in the first of what would be a lifelong series of unpleasant confrontations with big business, working for other people, and pursuing endeavors for financial gain, that apparently logical decision turned sour within a few months. What neither Fuller nor his friend knew when they began working for Kelly-Springfield was that its parent company, J. P. Morgan and Company, had, months earlier, made a decision to close down the company.[40]

Even though Fuller increased sales dramatically during his three-month tenure as sales manager and the company was on an upswing, Morgan's earlier decision was uncompromising. Thus, in the summer of 1922, Bucky found himself unemployed, and he blamed his problem on the inconsiderate masters of industry, whom he would later refer to as "the great pirates."[41] Only with the perspective of many more decades and numerous jobs would Fuller eventually realize that his temperament was simply not suited to working for someone else, and that he was far more successful when he was self-employed and in total control of his time and endeavors.

*

Still, as a young man with a family to support, Bucky was extremely depressed, and he began finding solace in something which had been a part of his wilder youth: alcohol. In his earlier years, he had frequently partied and got drunk with friends, but when Alexandra was born, he made a conscious decision to stop that behavior and become a better parental role model. The combination of his financial problems and his daughter's failing health, however, caused Fuller to turn back to drinking as an escape from reality.[42]

When Alexandra returned from Bermuda stronger and healthier than ever, Bucky felt encouraged. He was also pleased that support from a network of family and friends was beginning to materialize. His former naval comrades rescued Bucky from his financial dilemma by offering to have him temporarily restored to active naval duty in the New York area, and he quickly accepted an interim assignment commanding Eagle Boat Number 15 during naval-reserve summer training.[43]

Number 15 was part of a squadron of Eagle Boats (i.e., a ship regarded as a hybrid between a destroyer escort and a PT boat) commanded by a very rich naval reserve officer, twenty-six-year-old Vincent Astor.[44] Astor was a tall, thin young man who had inherited approximately $70 million on his twenty-first birthday and who appeared to be the antithesis of the short, impoverished Fuller. However, in actuality, the two young men found that they had a great deal in common and quickly became close friends. Astor then chose Number 15 as his flagship for the summer.[45]

Both Fuller and Astor were avid sailors who took their military duties seriously but were inclined to devote an equal amount of energy to partying and having a good time. Accordingly, the two men truly enjoyed touring New York harbor with their summer flotilla. They even went so far as to present the New York police commissioner with an honorary appointment in their squadron. In return, the police commissioner designated them honorary captains in the New York City Police Department, a distinction which included an impressive gold badge.[46]

That post actually did provide Bucky with two benefits. First, he was able to function as a police officer during the New York World Series, when he was permitted to put on a uniform and patrol the stands. Bucky's "beat" was the nonsmoking section of the stadium, where he made sure people did not smoke during the game. The other benefit was the ability to engage his passion for fast cars without incurring tickets or fines. Whenever a New York police officer would stop him for speeding, he would simply show his police badge, and the unwitting policeman had no choice but to salute and apologize for the inconvenience.[47]

When the naval reserve flotilla was disbanded in September, Fuller and Astor remained close friends, with Astor taking Bucky to society parties and Bucky returning the favor by showing Astor the Bohemian life of Greenwich Village. It was at one such event that Astor offered Bucky the use of his private seaplane for the month of October.[48]

Astor and Fuller had used the plane to fly to various parties during the summer, and Bucky was fascinated with the machine. Because Astor had to travel to Europe for the month of October, his private plane and pilot were free, and thus, the impoverished Fuller became master of one of only two such custom-built flying boats.[49]

Since Alexandra's health continued to improve, Bucky and Anne were able to leave her with family and nurses for days while they flew to weddings and other gatherings up and down the eastern seaboard. Naturally, a couple using Vincent Astor's private plane was the delight of the social set, and the Fullers found themselves written up extensively in society columns. The most joyous of those October plane trips for Bucky was their journey to Bear Island. Staying for over a week, theirs was the first plane ever to spend more than a few hours on Penobscot Bay. Every day during that stay, Bucky flew to a different local site, and wherever the plane touched down, crowds gathered.[50]

Then, October ended, and Bucky was abruptly returned to a far less glamorous reality. It had been a joyous month, but he was deeply in debt and without a job or any prospects.[51]

Once again Fuller became depressed and despondent, and a few days prior to the Yale–Harvard football game he decided to take his mind off his troubles by attending that game with friends. On a beautiful November day, Anne and Alexandra walked him down to the station near their Long Island home to catch the train for Boston and the game. As canes were in fashion and his football knee injury had left him with a slightly wobbly leg, Bucky walked with a cane.[52]

Throughout his life, Bucky never forgot that day and the walk recounting football stories to his nearly four-year-old daughter. He described to her the scene of singing school songs, cheering for the team, and the heroic athletes on the field. He also described a picture of thousands of fans cheering from the stands and waving school pennants attached to small canes which were sold at the gates.[53]

Before he got on the train, little Alexandra looked up and asked, "Daddy, will you bring me a cane?" Bucky promised he would bring back the souvenir as he set off for an enjoyable day of football and friends.[54]

Harvard won that day, and Bucky spent most of his time lost in drink, camaraderie, and parties, forgetting his troubles as well as his family on Long Island. When he arrived in Pennsylvania Station in New York the following afternoon, Bucky telephoned Anne, who could barely speak. She told him that Alexandra had suffered a relapse and was in a coma. Stunned, Bucky caught the next train to Long Island. Arriving home, he found Alexandra still unconscious and a doctor doing all he could to save her life.[55]

Bucky could only sit near her bed looking on helplessly as the doctor and nurses continued their work well into the night. Eventually, the situation calmed down, but Alexandra's condition did not improve. Then, in the early hours before dawn, she opened her eyes and smiled up at Bucky. As he bent close to his daughter, Bucky heard her tiny voice ask, "Daddy, did you bring me my cane?"[56]

Fuller could only turn away in shame and agony. In the furor of drinking and celebrating, he had forgotten his daughter's simple request. Following her question, Alexandra closed her eyes for the last time and died in her father's arms a few hours later. Bucky never forgave himself for that incident, which, even in the last years of his life, would bring tears of remorse to his eyes.[57]

A portion of the guilt he felt did resolve itself in a 1967 incident. After decades of pragmatically studying spiritual and metaphysical phenomena, Fuller felt he was able to appreciate and interpret the interconnectedness of that event and the death of his daughter over a half century earlier.

In 1967, he was spending time on the island of Bali, where he felt quite at home. His Balinese hosts also felt that he had some deeper connection to them and their homeland, and they invited him to join in visiting a prominent artist friend who lived in a remote house far up in the hills. The only means of reaching the artist was to hike up a volcanic mountain, which was constantly being pounded by rainstorms. At the age of seventy-two, Fuller found that the journey was not easy, and his native friends helped him whenever possible.[58]

Returning down the slippery slopes following a tremendous rainstorm was even more difficult, and one of his Balinese friends cut down a bamboo strip and fashioned a staff for Bucky. When they returned home,

Bucky's guide asked if he could have that staff, and Bucky gave it to him. The following morning the guide approached Fuller and said, "All of us, we Balinese, are saying that you are not a stranger. You were here long, long ago and you have just come back to us. So I have set aside a room in my house and put your staff in it. Nobody will ever go into that room again because your cane is there."[59]

That traditional honor of permanently maintaining a room for him moved Bucky very deeply. He was even more touched by the fact that his guide and friend chose to keep a symbolic cane in that room. That gesture provided Fuller with an enormous sense of relief which he later described in detail.[60]

Fuller recounted, "I said to myself, this is a very mysterious thing. All those years and all those things that have happened since little Alexandra died. Now somehow I feel intuitively that this is a kind of message. That she has her cane at last."[61]

Alexandra's death in the late fall of 1922 marked the onset of what Bucky would later describe as "a winter of horror." Only months later, Anne's mother succumbed to illness and died, devastating the large family, especially Anne's father. As the eldest daughter, Anne felt obligated to assist her father, and the Fullers were soon living in the Hewlett family homestead while Anne managed the household and cared for her young brothers and sisters.[62]

Meanwhile, Bucky was becoming ever more distant and depressed. That attitude did not make him much of a companion, much less a prospective employee, and he remained unemployed.[63]

Fuller continued to dwell on his responsibility in Alexandra's death. During months of solitary contemplation, he even convinced himself that if he had only provided her with a home which was not so damp and drafty, she would have survived. Learning of other family members and friends who were dying from the epidemics regularly sweeping the country only reinforced Fuller's misguided notion that drafty houses were primarily responsible for the problem.[64] Accordingly, when he found an opportunity to do something about redesigning housing, Bucky jumped at it. Anne's father, James Monroe Hewlett, had been working on a new form of building construction over the years, and he had eventually invented a technology which produced structures of enormous strength and insulation capacity at a fraction of the cost of conventional construction techniques.[65]

Mr. Hewlett had spent a great deal of time attempting to interest others in his idea, with no success. Bucky was not so difficult to convince. Being extremely naive about business and having no experience with starting a new enterprise, he perceived Mr. Hewlett's idea as a real opportunity. By introducing it to the general public, he could dramatically improve housing while also earning a living. With that honorable intention as his primary goal, Bucky told Mr. Hewlett that he would be responsible for implementation of the new technology.[66] Thus, Fuller entered into the first of what would be many new ventures focused on providing less expensive, more efficient shelter for the benefit of humanity.

The foundation of Mr. Hewlett's idea was a new type of brick formed out of composite material and connected with concrete pillars rather than mortar. Each brick was sixteen inches long, eight inches wide, and four inches high. The bricks also contained two four-inch vertical holes, evenly spaced in from the ends. The bricks were to be laid in alternate layers like ordinary bricks but without any mortar to hold them in place. Instead, the four-inch holes, which alternated between left and right as the bricks were stacked, were filled with concrete after each layer of bricks was laid. Thus, a strong shaft of concrete held the completed walls in place.[67] Those shafts were also aided in supporting the structure by a layer of mortar or plaster, which was often used as wall covering and bonded to the bricks.[68]

When Bucky took on the operation, it was merely an idea. He had no production strategy, and, in fact, Mr. Hewlett had not even decided what material would be most suited for fabricating the bricks. Still, Bucky jumped in with his typical enthusiasm and was soon tackling every obstacle, including selecting the best material and method of production for the bricks.[69]

After weeks of testing in the barn behind Rock Hall, which he appropriated for his factory, Fuller decided to fabricate the bricks out of shredded wood bonded in high-pressure molds with a bit of magnesium-oxy-chloride cement. With their two holes and shredded-wood composition, the finished products looked more like oversized dominoes of shredded wheat than bricks. That appearance was no coincidence. In examining possible methods of formation, Fuller had earlier invited Scott Perky, the son of the man who invented shredded wheat, over to dinner so that he could learn about the technology of shredded-wheat manufacturing.[70]

Despite Bucky's enthusiasm and a superior method of construction, the project followed the course of most new businesses and progressed slowly during the first months. Bucky continued to retain the support of his father-in-law throughout that difficult time. Mr. Hewlett recognized

Bucky's talent and ability and encouraged him to think for himself and try new ideas.[71] Without that support, Fuller might well have surrendered to the pressures of society and settled for the mediocrity of simply earning a living. With Mr. Hewlett's trust and backing, Bucky met all challenges and invented the machinery to produce the bricks. Thus, in conjunction with Mr. Hewlett, Fuller was granted his first patents. Those patents covered both the machinery and technology for creating what they named the Stockade Building System.[72]

They chose to christen their company Stockade because when walls of their buildings were completed, they were supported by the concrete pillars which had been poured in the holes row by row as the bricks were laid. The resulting series of concrete poles reminded the two men of the poles which had been the basis of earlier stockade forts, and they adopted that name.[73]

After months of preparation, Stockade Building Systems was ready to open its doors for business. To raise money, stock in a parent company was sold to people interested in organizing subsidiary local Stockade Building companies for construction of buildings in specific areas. Many of the investors in the parent company, Stockade Building Systems, were relatives and friends of both Bucky and Mr. Hewlett.[74] The investors saw an opportunity to make money and wanted to be a part of Stockade's imminent prosperity. They did not, however, realize that Stockade had as its president a man who would one day admit that he was a terrible businessperson: R. Buckminster Fuller. At the time, Bucky was quite naive and had no idea that he possessed such a radical aversion to the business environment or would fare so poorly in it. He truly believed that if he produced a good product at a reasonable price, he could contribute to society while earning enough money to support his family.[75]

Five factories were opened in the East from New Jersey to Illinois. As president, chief engineer, and inventor, Bucky was expected to support those outlets in every aspect of their operation, including sales, marketing, and management, as well as engineering. He was, therefore, away from Anne and home a great deal. Unlike modern traveling businesspeople, who are able to fly home for weekends, Fuller was at the mercy of the transportation of that era. Traveling by car and train, he would sometimes be on the road for weeks or months at a time.[76]

That freedom encouraged his continuing penchant for drinking and partying. During those escapades, he found some escape from his sorrow over Alexandra's death as well as the companionship of his drinking friends. Drinking also tended to help him forget the seemingly endless problems that he discovered were inherent in supporting a new business founded on an unfamiliar technology.

When he was home, the pressure of the business and his drinking only served to augment Bucky's sometimes short temper and often created enormous tension between him and Anne. Bucky had always been apt to fly off the handle in a spurt of emotion, but his generally jovial nature tended to obscure that trait. However, the pressures of entrepreneurship and the death of Alexandra helped move that characteristic into greater evidence.

In one representative incident, he displayed the level of indignation which often boiled just beneath his calm exterior during those years. He and Anne were spending an evening at the Hewlett house, and she was innocently flirting with a young man who was also a guest for the evening. Bucky sat silently observing the scene for quite some time, and although he said nothing, the expression on his face made it clear that he was jealous and becoming angrier by the minute. Then, without any warning, he leaped up, stormed out of the house, and began walking the twenty miles back to their New York apartment. He stubbornly continued his hike over country roads throughout the night and reached the Queensborough Bridge at dawn. Even though he had very little cash with him, no one, including Anne, heard from him for two days.[77]

Then, without a mention of the incident, Bucky cheerfully walked back into their apartment, and nothing more was said of the incident. Always the poised lady she was raised to be, Anne calmly went about her business, and no one ever found out where Bucky had slept or how he had lived for those days.[78]

After months of shuffling from one factory or construction site to another, Bucky finally decided to move to Chicago to oversee the start-up of a Stockade plant in Joliet, Illinois. Leaving Anne to live near her family on Long Island, he rented an apartment in Chicago, which he seldom used since he was constantly at work or out drinking and carousing. As prohibition was the law of the land, Bucky quickly found his way into the Chicago speakeasy scene. One of his favorite hangouts was a speakeasy he passed during his daily drive between Chicago and the manufacturing plant in Joliet, where he met another regular customer who was about to gain notoriety, Al Capone.[79]

Always searching for a party and good time, Fuller also spent a great deal of his time in brothels. In the later years of his life, he would recount his prowess in such establishments over the years while exposing his naiveté by saying, "I'm sure I went to over a thousand of them. They seemed to be the only place where people really talked straight to me; those girls. . . . They were terribly interesting as people. I enjoyed talking to them."[80]

In 1926, Bucky's life reverted to a more normal regimen when Anne joined him in Chicago. Although Stockade Building Systems was still struggling, as president of the company, an officer of one of the franchise companies, and general manager of another, Bucky was earning a handsome salary. Still, the work was arduous at best.

Bucky was constantly confronted and harassed by the individuals and organizations dominating the construction industry. The benefits provided by Stockade buildings were unmistakable, but the construction establishment was, and still is, concerned primarily with finances and maintaining the status quo.[81]

Stockade bricks weighed only two pounds, could be tossed up to a second-floor scaffolding, were virtually unbreakable, and required no mortar. Because of those advances, solid buildings could be erected in a fraction of the time required to construct conventional buildings. The bricks were also fireproof and waterproof.[82]

The problems Fuller encountered in selling his new technology to others were epitomized by the reception that those fully fireproof bricks received from the establishment. Fuller permitted engineers, local government officials, and anyone else interested to test sample walls built of Stockade bricks. No one could get the bricks to burn, but their appearance continued to cause suspicion and challenges. Because they were fabricated from waste wood shavings, the bricks had a wood-like appearance, and people believed they would burn easily even though that was the antithesis of the truth. Once the wood shavings were mixed with concrete and molded under pressure, the final product was more akin to a lightweight rock than to a piece of wood, and over time, the walls tended to become petrified, making them even more solid and fireproof.[83]

Still, the wood tone haunted Fuller, until he finally acquiesced to popular opinion. Rather than suggest that officials examine the results of tests or test the bricks themselves, Fuller simply added a bit of gray dye to the brick formula. That simple change produced bricks which, although identical in construction to the earlier ones, looked like cinder blocks rather than shredded wheat and were viewed as fireproof.[84]

The fireproof question was only one of many Fuller confronted during the four years he championed the Stockade System. Most of his problems emanated from trade unions, architects, and contractors. Those three factions composed the majority of the building-industry establishment, and, as such, they opposed any innovation. In fact, Fuller found that an innovation such as his, which would increase construction efficiency, was actively suppressed by those in power, who feared change would threaten jobs and profits.[85]

Accordingly, almost every time Stockade Systems landed a contract, local officials demanded that the company construct a sample wall for testing before they would issue a building permit. As chief engineer, Fuller was generally the person who traveled to the locality and attempted to convince officials that although his new technology was radically different, it exceeded the requirements of their building codes.

Unconvinced and fearing the unknown, most local officials forced the issue and demanded a test wall. Thus, Bucky was coerced into hiring a crew, shipping in supplies, constructing a wall, and allowing the concrete to set for a few days. Only then could he schedule a time when all the necessary individuals could witness a test of the wall's strength, water and fire resistance, and general performance. In every instance, the tests proved that Stockade System's walls were vastly superior to anything being used at the time. However, the time and money required to run a test in every community and sometimes for each building in the same community were an insurmountable handicap.[86]

Interestingly enough, nearly forty years later, another of Fuller's inventions had to surmount similar skepticism. In the 1960s, following his invention of the geodesic dome and its subsequent popularity, Fuller observed that most of the independent individuals attempting to construct domes were confronted by local officials leery of such construction. Time and time again, potential dome builders were impeded by building codes and officials steeped in tradition. Even forty years after his Stockade experience, the construction industry establishment continued to be, by and large, unwilling to consider any technology which might upset their domination or threaten their profits.

Despite such obstacles, Fuller persisted in his quest to revolutionize the construction industry. Between 1922 and 1927 Stockade erected 240 buildings but did not generate any profit, even though most businesses were thriving in a period of great economic prosperity. Because Mr. Hewlett held a majority of the company's stock, he maintained controlling interest in the company, and Bucky was free to continue his attempts at radically improving housing and construction.[87]

However, in 1927 his own financial difficulties forced Mr. Hewlett to sell his stock in the company. Within weeks Stockade Building Systems became a subsidiary of Celotex Corporation, whose primary motivation was akin to that of other conventional companies: making a profit. Celotex management took one look at Stockade's financial records and called for a complete overhaul of the company.[88] The first casualty of that transition was Stockade's controversial president. Hence, Fuller again found himself unemployed and with no prospects for supporting himself and his wife. As the Fullers' lifestyle had matched his larger salary, they had no savings to fall back upon.[89]

Bucky's idea was, however, validated by Celotex. The corporation adapted his formula and methodology, applying it to another insulating product, ceiling tiles. Using Fuller's ideas, they created and profitably marketed the acoustic ceiling tiles covered with small holes seen throughout the World for decades.[90]

Bucky had devoted five years to persuading others that his invention was beneficial; however, by 1927, the rigorous opposition to his ideas seemed so incessant that it nearly overwhelmed him. With no job or money, he again fell into depression and despair.

In addition, Anne was once again pregnant. On August 28 of that same year, she gave birth to another girl whom she chose to name Allegra after the daughter of Lord Byron, a woman whose book of letters to her father Anne had been reading.[91]

Once again Bucky felt the responsibilities of parenthood upon him, and he vowed to be a more supportive father. He also began assessing the first thirty-two years of his life and observed that he had drastically failed in every attempt to work within the business community and earn money.[92] Thus, Bucky decided he needed to shift direction. He had to support his wife and new child. The primary question he posed to himself was not how to do that, but whether his small family would derive more value from him alive or from the cash they would receive from his life insurance policy should he suddenly die.

CHAPTER 5

Emergence from the Ashes

 In a scene reminiscent of an overly melodramatic movie, during 1927, Bucky found himself unemployed with a new daughter to support as winter was approaching. With no steady income the Fuller family was living beyond its means and falling further and further into debt. Searching for solace and escape, Bucky continued drinking and carousing. He also tended to wander aimlessly through the Chicago streets pondering his situation. It was during one such walk that he ventured down to the shore of Lake Michigan on a particularly cold autumn evening and seriously contemplated swimming out until he was exhausted and ending his life.[1]

With the cold Chicago wind pounding at him, Fuller thought about his wife and new daughter as well as all the family and friends he had once again disappointed. Several of those people had invested in his Stockade Systems venture, had lost money, and were vilifying his reputation. They equated the fact that he was a poor businessman with his overall character, and having heard others constantly slander him, Bucky was beginning to believe their accusations.[2]

Although neither his nor Anne's families were excessively wealthy, Bucky felt that if he removed the burden of his presence from everyone's lives, the two families would band together to support his wife and child. He also knew that his insurance policy would provide them with sufficient funds for several years. In weighing only that narrow range of personal issues, suicide appeared to be a viable alternative for Fuller that depressing day. After several hours of internal debate over the question "Buck-

minster Fuller—life or death,'' Bucky was suddenly seized by an experience which would transform his life and his overall perception forever.[3]

As he would later recall, he experienced what he thought was the
most significant of many seemingly mystical events which occurred
throughout his life. During almost all of the lectures which followed that
experience, he would recount that he had unexpectedly found himself
suspended several feet above the ground enclosed in a ''sparkling sphere
of light.'' During the incident, time appeared to stop while he listened to a
voice speaking directly to him.[4] Although the experience was difficult for
the pragmatic Fuller to accept as reality, he remembered not resisting and
allowing himself to savor it fully. Because of that, he was able to remember the exact words he heard. The voice declared, ''From now on
you need never await temporal attestation to your thought. You think the
truth.''[5]

It continued, ''You do not have the right to eliminate yourself. You
do not belong to you. You belong to Universe. Your significance will
remain forever obscure to you, but you may assume that you are fulfilling
your role if you apply yourself to converting your experiences to the
highest advantage of others.''[6]

Although we will never know if that ''voice'' was simply Fuller's
inner thoughts or some higher guidance, it did result in a vastly transformed individual. Even the harsh conditions of that lakeshore could not
impede Fuller's metamorphosis, and he began considering the insight he
had received and charting a new course of action. That operating strategy
would not only change his life, but it would dramatically influence all
humanity.

Following that experience, Fuller began to examine his life and
believe that he had become entangled in his latest amalgam of problems
by listening to and accepting as truth what others told him.[7] He realized
that his family and friends loved him and wanted the best for him; however, he was also beginning to feel that their ideas were severely prejudiced by the restricted teachings of traditional leaders who sought to
maintain the status quo.[8] Because he had attempted to follow those traditional guidelines and failed miserably, Bucky wanted to search for new
ideas and methodologies which were flexible enough to focus on the
reality of the moment and could be of use to himself and future generations. Accordingly, he resolved to accept nothing other than his own
experience.[9]

The only exception to that resolution were the few individuals whom he experienced as operating in a manner similar to his new strategy. Fuller listened to those people because he felt that they were not attempting to persuade him that their ideas were superior or that they were seeking personal gain.[10] In other words, he consciously chose to return to a position of always questioning—a process which is natural to every child and remains available to all human beings throughout our lives.

Believing that smaller, special-case issues were embodied within larger issues, Fuller usually tackled the most ominous questions or problems first.[11] Hence, the first thing he thought about following his new resolve was God. As he relied heavily upon pragmatic, experiential logic, God (which he would later refer to as the Greater Intellect or Greater Integrity operating throughout Universe) had truly become an enigma to Bucky. Standing on the shore of Lake Michigan, Fuller asked himself if there was anything in his personal experience which forced him to admit that there was a Greater Intellect than human beings operative in Universe.[12]

Because so many of his ancestors had been Unitarian ministers, Fuller had been provided with rigorous Christian training, even though the majority of that teaching had been within his mother's Episcopal church.[13] He remembered being taught by adult relatives and their friends whom he had trusted and whose ideas he had accepted on faith. However, standing by the lake, he seriously considered the possibility that such traditional ideas might be inaccurate.[14]

Examining the depths of his own experience and insight, Fuller was overcome by the massive amount of evidence that he felt supported the existence of a Universal Intelligence.[15] His primary confirmation lay in the perfect design and orderliness he had found within every aspect of Nature. Everything he had studied seemed to be governed by an exquisite set of interaccommodative principles which he felt could only be the work of a Greater Intellect.[16]

Looking back upon the first thirty-two years of his life and all that he had learned, Fuller was also amazed by the perfection and integrity of events which had earlier appeared to be disastrous mistakes. Those mistakes had served to guide him, and he felt that they could have appeared at the exact moment required only as a result of divine intervention. Bucky's perspective was beginning to expand radically, and that heightened perspective could not exclude a Greater Intellect or the perfection of the actions of that Greater Intellect as manifest through all creatures. As a human being, Fuller was particularly interested in the role of people in such an organized pattern. Under that intense scrutiny, Bucky acknowledged and accepted the existence of God.

From his earlier years' experiences, it is evident that he had always accepted the existence of a perfect plan operating throughout Universe. Bucky was constantly discovering the interrelationships between phenomena and elements, but on that autumn day he mindfully examined all his experience and accepted the fact that there was a grand plan to everything and that he must, therefore, be a perfect element of that plan. By reexamining his experiences within that context, he was able to perceive the uniqueness of his first thirty-two years and their possible significance. He remembered experiences such as his Sherbrooke mill training, working for Admiral Gleave, learning about telepathy from Alexandra, and the frustration of trying to erect quality Stockade buildings, and he could only conclude that he was a storehouse of experiences which could prove valuable to future generations.[17]

Although Fuller did not realize it at the time, he was beginning to comprehend the dawn of a new era which the general public would not begin to acknowledge until the 1970s. He was grasping the fact that even though most people believe wealth is found in material objects, the nonphysical commodities of knowledge and wisdom are the essence of true wealth.[18] Only recently have most people begun to understand and accept that maxim as humanity advances into an "information age" predicated upon knowledge. Once he realized that he was a unique repository of valuable information, Fuller began viewing himself as an irreplaceable link in a chain of human knowledge dating back to the dawn of civilization. He also concluded that he did not have the right to end his life and break that chain.[19]

Walking home, Fuller decided to share his insights and knowledge with others so that they would not have to suffer the pain and mistakes he had experienced. He also remembered the harsh reality of his predicament. Cold and tired, he realized that he still had to do something about supporting his wife and child. He could not simply abandon them and walk the countryside sharing his wisdom like some itinerant preacher.

Reality set in even more quickly as Bucky entered their rather lavish apartment overlooking Lake Michigan. Among other expenses, the rent on their home was due, and he had no idea how he would meet the payment.[20] He did, however, possess a new view of the perfection of Universe, and that perspective included a redefined fiscal philosophy which would serve to support him and his family financially for the next fifty-six years.

Fuller felt that the Greater Intellect operating within a perfect Universe would support the tasks it wanted to have accomplished. Logically, Bucky deduced that if he were participating in such a task, he would be supported and his needs would be met. Although his faith in that perfection was often tested when projects running low on support were almost abandoned, Fuller found that if a project contributed to the welfare of humanity, the needed support would eventually appear. Usually, a test of his faith caused that support to emerge at the last possible moment prior to disaster, and from an unconsidered source.[21]

The impending disaster of Autumn 1927 was the Fullers' imminent eviction as well as the question of procuring the necessities of life. Upon returning home from his Lake Michigan experience, Bucky immediately told Anne about his insight and plan to devote himself to serving humanity with no attention to earning a living. Although initially skeptical, Anne soon began to appreciate the wisdom and insight of Bucky's idea as well as just how staunchly he felt moved to change his life. The melding of that understanding and her provincial New England upbringing, which deemed that a woman should follow her husband, also helped Anne in deciding to support Bucky's outlandish scheme.[22]

Bucky decided that more than anything he needed time. Time to clear his mind of everything he had been told and to start out fresh; time to assess his capabilities and to determine how best to utilize them; and time to study the great independent thinkers of the past and to postulate a plan for independent thinking and contributing to humanity.[23] Although he had selected an extremely ambitious assignment, Bucky's excitement and overwhelming enthusiasm would eventually carry him through the seemingly impossible.

Still, during that autumn, the conflict between high ideals and the practicalities of daily existence haunted the young family. Deciding to forgo the convention of earning a living was one thing, but putting that conviction into practice was quite another. Although Bucky once again felt he was a failure, the four-year Stockade experience provided him with new insights and reinforced his earlier attitude toward money.[24]

When he compared his work with Stockade to his naval experience, Fuller immediately noticed that in the navy his sole objective had been to successfully operate his ship, but in business he had been confronted by two conflicting motives. The first motive was successfully erecting quality buildings, and the second was making money.

That discrepancy combined with his later vision to spark Fuller's developing the unique economic philosophy he espoused throughout the balance of his life. His interpretation is exemplified in his often quoted statement that "An individual can make money or make sense; the two are mutually exclusive."[25] That comment has often been mistakenly interpreted as demonstrating a socialist bent, but nothing could be further from the truth. Bucky felt that every human being was entitled to being supported and that financial gains should not be a factor in selecting areas of work. However, in a more pragmatic context, he believed, and proved through his actions, that a person would make money if he or she concentrated on actions which made sense and supported the greater good of all humanity.[26]

Fuller had chosen to operate Stockade so that his product and operation made sense to him; the result had been financial losses every year. Stockade's product was useful and of value, but Fuller's attention had not been intently focused on the marketing or sales aspects of the operation. Rather, as a technically oriented, pragmatic individual, his primary concern had been continuing to refine and develop the quality of his product.

He had also flaunted a lack of respect and a lack of affinity for money which would become one of his trademarks. He perceived money as the means of exchange rather than as an end in itself. Accordingly, he never attempted to accumulate money or profits. When money flowed into a Fuller enterprise, it was immediately used to meet expenses, and any "profits" which remained also flowed back out, as he simply applied them toward developing a more efficient product or his next project.[27]

One of the most significant of those projects was Fuller's invention and popularization of the geodesic dome years later. That creation would probably not have occurred without Fuller's Stockade experiences. During the four Stockade years, Bucky discovered exactly how difficult it is to implement new technology, especially in the fields of construction and housing.

In 1927, however, the practicalities of daily life nagged at the young family, and their first order of business was finding a much less expensive apartment. Bucky combed Chicago until he located a tiny, cheap apartment hotel in a slum section of the city, where he rented a single room with one closet. Their tiny home did have a small alcove with a stove and sink, where Anne cooked what meager rations they could afford.[28]

As Bucky was withdrawing from society to formulate his grand strategy, maintaining the basics of life fell on Anne's shoulders.[29] Although the Fullers sometimes survived on one skimpy meal per day, she managed to feed the family on the sparse cash they usually received at the last minute from unconsidered sources. For example, two deaths in their families resulted in small inheritances, and a few old friends stopped by to pay forgotten debts.[30]

Meanwhile, Bucky entered a period of intense contemplation. Rather than spending his time drinking and carousing or working to earn a living, he went to the library and studied the works of other independent thinkers such as Mohandas Gandhi and Leonardo da Vinci.[31] He also took a vow of silence, pledging to not say anything until he truly had something significant to communicate. He did speak to Anne and Allegra, but for nearly two years he was silent to the rest of the world. It was during that period of silent meditation and study that Bucky was able to formulate the self-disciplines which would serve him so well during the balance of his life and that would support his contribution to society.[32]

One of Bucky's early resolutions was to make his contribution to humanity primarily through a unique concept he had observed in Nature: precession. *Precession* is a word Fuller first found used in physics and engineering. He later recognized the phenomenon of precession to be a

Fig. 5–1   Bucky, Anne, and Allegra, 1928.

generalized principle applicable to any project or area. Precession, as defined by Fuller, is the effect of bodies in motion on other bodies in motion. It was originally used to describe the action of spinning gyroscopes, where a force applied to the spinning axis always results in a motion whose direction is skewed from the original force.[33]

In applying the generalized principle of precession to his life and work, Fuller employed and relied upon what most people regard as the "side effects" of an action rather than upon what is considered the direct action.[34] For example, while working to improve plumbing, he developed a prototype toilet bowl and later an entire bathroom. Rather than initiate a public crusade to reform bathroom fixtures and the sanitation industry, Fuller simply produced his inventions and allowed the public and the manufacturers to discover his unique ideas. Such indirect action may appear ineffective within the context of the short-term results that motivate most of society, but Fuller found it to be far more efficient in producing maximally effective results with a minimum of effort.

Bucky's first step in operating precessionally was to find something which he felt needed to be done and which was not being attended to by anyone else. He would then undertake that task with the faith that he, his new project, and his family would be supported if his enterprise was, in fact, needed.[35]

Taking lessons from his Stockade experiences, Fuller resolved never again to attempt marketing, selling, or impressing other people with his ideas and inventions. Instead, he simply identified a future problem, developed an invention to solve that problem, and waited for the issue to reach the critical moment of massive public awareness.[36]

Following his 1927 transformation, Fuller seldom concerned himself with direct support, such as payment for inventing a particular artifact. Rather, he focused his attention on a project with the knowledge that seemingly indirect assistance would be provided by a regenerative Universe which always supports required actions.

That mode of operation did not provide him with the luxury of a boss or an instructor to correct his work or a regular salary. The Greater Intellect did, however, guide him by furnishing financial and moral support for those projects which, in retrospect, could be seen to have contributed to human success and development.[37] The sustenance Fuller received was usually so indirect that he could almost never categorically state that a specific compensation was for the accomplishment of a certain task. He also found that support for his projects generally appeared only at the last instant and from unexpected sources.[38]

To sustain the operation of his seemingly tenuous precessional system, Fuller maintained a policy of responding to as many invitations as possible. By doing that, he kept a diversity of options open and made himself available to even more support.

Fuller also resolved that minimizing the emotional aspects of decisions and actions would greatly enhance his efficiency. Regardless of the fear or joy of a particular moment, he always sought to act on behalf of all humanity with as little regard as possible for his personal feelings.[39]

Because he seldom followed the mainstream of society, Fuller's style of action generated criticism from others as well as insecurities within himself. As a result of constant scrutiny (from others as well as from himself), later in life Fuller would remark that he habitually felt like a child who had meandered into a forest and arrived at the critical moment of realizing he was lost. That sense of perpetually being on the edge of the unknown caused him to continuously reexamine the elements of his changing environment with great sensitivity and care.[40]

Some people did criticize the resulting deliberate action as inflexible and overly controlling. This was particularly conspicuous when others involved in a project sought to move more rapidly or in a direction of which Bucky did not approve. In instances such as when others involved with Stockade wanted to look for new methods of expanding profits rather than developing the product, Fuller's will usually prevailed. However, time and time again, such confrontations resulted in criticism.

Critics usually did not, however, appreciate the fact that during the early years of his life, Fuller had thrown caution to the winds and had been dominated by others on numerous occasions with devastating results. Being expelled from Harvard twice and having his first child die in his arms after realizing that he had not kept his promise are only two of the more vivid instances from which Fuller learned the importance of considering his actions thoroughly and seriously.

During his youthful exploration, Bucky endured many major learning experiences—mistakes—which taught him valuable lessons but also impeded the contribution he was able to make to his fellow human beings. Once Fuller determined his life mission, he felt obligated to do everything possible to maximize his effectiveness, and to deliberate with caution prior to undertaking important actions. The most conspicuous such deliberation was the nearly two years of virtual silence he spent during 1927 and 1928, but Bucky also carefully considered what could occur prior to starting work on the Dymaxion Car and the prototype Dymaxion House, among others of his projects. In both those future endeavors and other instances, Fuller was, in fact, so cautious and dominating that he, and he alone, retained the power to veto the actions of any other involved parties.[41]

Another significant self-discipline Fuller employed throughout the remainder of his life was the direct result of his self-imposed 1927 hiatus. During that period, he learned to maximize his efficiency by taking great care in what he said and to whom he spoke. In the years that followed, Bucky would continue closely watching what he said and to whom he spoke, and he soon found that people really listen to someone only when the listener solicits information.[42] The invitation may be subtle, such as a casual question or reading a book, but Fuller was certain that when an individual takes the initiative to request or invite information, he or she is seriously interested and will provide the best audience.[43]

Whenever he was invited to speak, Bucky felt that the question of interest was satisfied in advance and that his audience sincerely wanted to hear what he had to say. Because that audience had solicited him to provide information, he knew that individuals attending were sincerely interested and would make every effort to grasp his thoughts and words.[44]

Even though he spoke to audiences ranging from a single individual to thousands of people, hundreds of times each year, communication via the written and spoken word was not Fuller's true talent or primary mode of contributing his message to others. He was a thinker, not a communicator, and he acknowledged that fact at the beginning of many of his talks when he told his audience that he was not utilizing notes, nor had he thought about what he would say prior to speaking.[45]

Thus, he christened his talks "thinking-out-loud" sessions. Such a mode of speaking appears to contradict his discipline of considering his words carefully; however, Fuller's thinking-out-loud sessions were not intended to be taken as direct communication from one person. Rather, he felt he was acting as a channel for a flow of information as well as sharing exactly what was occurring within his mind at a given moment. Because of that practice, during many of those sessions, he would change and update insights as he himself broadened his perspective.[46] Through years of practicing that method of verbal communication, Bucky came to be a master at such oration. During his 1927 self-evaluation, however, he was far from that mastery, and he sought an operating strategy more compatible with his innate talents and his observation of Nature's patterns.

In examining possible options at that time, Fuller discovered that Nature was constantly transforming the environment in ways which supported the improved behavior of those creatures living within their environments, including human beings. He then resolved to mirror that strategy of operation by designing and developing inventions which alter environments and make them more beneficial. He referred to those inventions as "artifacts."[47]

Initially, Fuller focused on discovering and understanding the structural, mechanical, chemical, and technological principles Nature employs in transforming environments. Once he grasped those principles, he designed and built artifacts which supported our eternally regenerative Universe, just as Nature does. Since he was a human being living on Earth, his particular focus was supporting eternally regenerative human success for every individual on Earth.

One of the most critical elements of Bucky's philosophy and operating strategy arising from his period of silence was his understanding of the metaphysical aspects of human existence, which were to become the focus of much of his time and energy. The groundwork for that understanding had been laid during his earlier years, when he had learned to appreciate and respect the wonders of Nature, and it had been further enhanced by his experience of young Alexandra's life and untimely death. Following his 1927 withdrawal, Bucky devoted a major portion of his life to studying and gaining an understanding of the metaphysical phenomena which he felt embody the essence of all human existence.

Some of the first metaphysical issues he examined arose from his unparalleled experiences with Alexandra. Following her death, he had attempted to purge himself of those painful memories, but by contemplating his experiences in a more scholarly and less personal context, he discovered a virtual wellspring of inspiring ideas. For instance, Bucky determined that when two individuals successfully communicate with one another, something people refer to as "understanding" occurs, but understanding cannot be found within the physical body of either person. Understanding exemplifies the numerous synergistic phenomena which transpire between humans but cannot be materially isolated. He felt that such phenomena, like everything which transpires between human beings, are metaphysical, yet they have the potential for influencing our physical environment.[48]

Because he was convinced that all Universe is controlled by a set of generalized principles which govern both physical and metaphysical (i.e., nonphysical) phenomena, Fuller decided to examine the physical in his study of the metaphysical. He selected physics to help him in that study because physicists had developed a precise definition of the physical. He then employed that definition to understand the metaphysical.[49]

Bucky found that physics defines the *physical* as "anything composed of energy." When the energy is associative (i.e., powerfully connected with other energy), it is designated as matter, and when it is disassociative (i.e., moving with less connection to other energy), it is called *radiation*. All energy perpetually shifts between these two forms, but in either case, measuring it with humanly conceived instruments is frequently possible.[50]

Fuller was convinced that metaphysical phenomena were also energy-based phenomena, but that they were so subtle they could not be measured by any devices of his era or be consciously experienced by human senses. He also felt that the consequences of metaphysical events could be measured.[51]

For example, a metaphysical phenomenon, such as information transmitted between beings, could not be quantitatively gauged, even though researchers might experimentally determine which people are more sensitive to the information they receive. Each participant in a metaphysical event (such as an exchange of information) does, however, experience and appreciate the consequences, quality, and significance of the event (such as the transmitted information), and those experiences can be quantified. Because of those findings, Bucky concluded that although metaphysical phenomena remained beyond the range of the senses of almost all the people he met, those phenomena were still a product of the same energy transmission patterns explained by scientists.[52]

After thoroughly studying the metaphysical, he began to appreciate it as the most significant component of what humans designate as "life." He also found that in order to operate most effectively he needed a comprehensive definition of life which focused on metaphysical phenomena and still satisfied his experiential mode of operation.[53] The creation of such an experiential-metaphysical definition appeared to be a difficult task because metaphysical phenomena are, by definition, not directly experienced by the human senses.

In developing his definition of life, Fuller first considered the characteristic of awareness and, in particular, the awareness one being has of another being. He then developed the following hypothesis: When one living being is consciously aware of another living being, communication occurs, and that contact can be regarded as one experiment which, if successful, experientially demonstrates the existence of "life." Fuller further ascertained that although communication is evidence of life, communication does not transpire between physical entities.[54]

In the Fuller hypothesis, the physical entities—be they humans, cats, or any living beings—are simply vehicles for messages, just as a telephone is a vehicle for the transmission of messages. The beings (i.e., life forms) which initiate contact are metaphysical phenomena Fuller designated as "pattern integrities." He further speculated that although those pattern integrities cannot be directly detected by our physical sensory equipment at its current level of development, they are, in fact, the essence of life.[55] Because pattern integrities are not material, Fuller found the concept difficult to describe using our limited vocabulary and worldly impression of reality. Yet, it is important to translate that metaphysical phenomenon into the clearest possible physical representation in order to understand Fuller's impression of the essence of life.

Over the years, Fuller devoted a great deal of time to creating such a representation. He spent long hours finding ways to mirror metaphysical reality in easily recognizable physical phenomena and demonstrations, and he came to appreciate the significance of the phenomena he called "pattern integrities"—which include but are not limited to human beings—in even the most mundane tasks.

While formulating his theory of pattern integrities, Fuller also realized that metaphysical phenomena are always conceptual, a fact which establishes them as independent of physical limitations such as size and time. Being devoid of the tangible components that most humans experience and view as reality, he speculated that pattern integrities are extremely difficult to understand, much less accept.[56] Human beings are accustomed to dealing with what they believe to be the solidness of material reality, and without such familiar forms, people tend to become easily confused. Since both physical and metaphysical phenomena, including pattern integrities, conform to the same set of strict rules (i.e., generalized principles) which Fuller sought to uncover throughout his life, he theorized that the metaphysical could be modeled and described by tangible events and objects which can be detected by the human sensory system.[57] Using such representations, Fuller was able to more easily understand the important metaphysical concepts which continuously influence all aspects of human existence.

He felt that all pattern integrities exist independently of local phenomena or events (i.e., human bodies, planets, flowers growing), and that humans can discover their existence only by experiencing the local phenomena or event which results from them.[58] For instance, human physical bodies are the local phenomena created by pattern integrities which are the essences of individual human lives.

Fuller often recounted the following graphic example to explain more completely the concept of pattern integrity. He first asked his audience to imagine splicing together three sections of identical diameter rope, each composed of a different material. The first one-foot section would be nylon, and it would be spliced to a one-foot section of manila rope. A third one-foot segment of cotton rope would then be attached to the end of the manila segment, creating a single three-foot rope composed of three materials.[59]

If the unspliced end of the nylon section is grasped and manipulated into a horizontal circle followed by a second perpendicular circle, a slip-knot is formed. Because the rope is interfering with itself, that configuration can also be considered an interference pattern within it. When the ends of the composite rope are pulled, the knot constricts, modeling the manner in which energy interferes with itself and creates the physical environment that humans perceive as reality.[60]

To better understand the self-interference which creates physical mass, Fuller studied the phenomenon of celestial bodies. In his examination, he found there is a perpetual attraction between the spherical islands of mass we designate as planets and stars. When two such masses exert a pull upon one another, as the ends of the rope were pulled, an energy interference (modeled by the knot) occurs between them. As the pulling attraction increases, the interference is tightened until the energy becomes so bound up and compressed that it can be perceived as what we call mass.[61] In this way, planets, stars, and all matter have been, and continue to be, created. Similarly, as the energy-interference knots loosen, mass is converted back into its more disassociative form, radiant energy.

Fuller often remarked that this ostensibly simple phenomenon is the basis of Einstein's theories. Einstein discovered and mathematically described the situation Fuller later modeled with his rope demonstration. In his work, Einstein calculated formulas which explained that energy and mass are essentially the same. He further theorized that those two fundamental components shift from one state to another in human perception just as water transforms into ice or steam.

When energy changes appearance to become mass, it interferes with itself, slowing down in speed and tying itself into knots that humans designate as solids. Energy constricted into mass can also unknot itself and become what we call radiant energy, such as sunlight or heat.

When the slipknot of the spliced rope is relaxed, it can slide along and move from the nylon section to the manila section. If the process is continued, the knot can be slipped from the manila segment to the cotton segment. Finally, the knot can be maneuvered off the end of the cotton section, and it disappears.[62]

After sharing that illustration hundreds of times, Fuller determined that when discussing the slipknot most people talk about the rope as if it were the knot. The rope is not, however, the knot. The rope is a limited local phenomenon which provides a vehicle for the invisible pattern ("knot") to become visible to human beings.[63] Fuller used the rope in that demonstration to explore the intricacies of the pattern known as a knot. He also explained that the slipknot's initial appearance was not on the rope, but as a metaphysical thought within his mind. His mind transmitted the idea *slipknot* to his brain, which directed his hands to create it.[64] Through that outwardly simple, yet highly complex, procedure, Fuller found himself responsible for the manifestation of a knot pattern formed upon the rope. Yet when he spoke about the demonstration, he emphasized that the initial step in the process of manifesting the knot (and any creation) was the metaphysical act in which a human mind discovered a pattern and imagined its physical counterpart.[65]

That simple rope exploration also demonstrated that the knot was not nylon, even though it was upon the nylon section that it first made its presence known to people other than the individual who first thought about it. The manila and cotton rope segments were also involved in the physical manifestation of that pattern knot, but neither of those materials was the actual knot.[66]

In Fuller's theory the knot was, and remains, a pattern integrity.[67] People are able to physically detect and observe that pattern integrity because it interferes with and shows itself upon an item (rope) which humans can perceive via their limited sensory system. Fuller felt that there are, however, many more phenomena that cannot be experienced by human physical senses than ones which can be detected with the human sensory system. He believed that only a small percentage of the pattern integrities operating throughout Universe make themselves readily known to human beings through a limited number of humanly apprehensible objects and phenomena.[68]

*

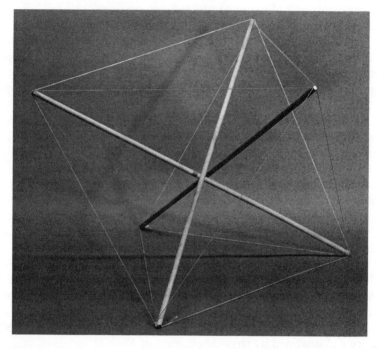

Fig. 5–2   Model of a simple tensegrity structure. Notice that none of the struts touch, yet they are held in place by the taut wires.

Bucky began investigating patterns and the concept of pattern integrity during the 1920s. In the 1940s, he initiated a more intense exploration of how such patterns could be modeled into complex structures, and that exploration resulted in the invention of tensegrity structures, the basis of geodesic domes. He coined the term *tensegrity* by combining the words *tension* and *integrity*.[69] Tensegrity structures exhibit an ideal balance of tension in which each element is perfectly aligned in the exact location where it is most ''comfortable'' while providing maximum strength to the overall framework. These structures also model the natural energy patterns that Fuller believed were present throughout Universe and that provide a three-dimensional representation of the interactive interference patterns he found in all natural construction from trees to the cells of human beings.

Although Bucky believed he was, in fact, unearthing a fundamental natural pattern as early as 1927, he was not able to confirm the accuracy of his ideas to his satisfaction until the late 1940s. At that time, he successfully constructed large structures to demonstrate principles which he had previously explored only through model building and mathematical calculations. During that era, practical tensegrity structures and geodesic domes proved structurally sound and, in fact, stronger per pound of material used than any other type of structure, providing him with tangible proof that his ideas were valid.[70] His work was also corroborated years later when virologists discovered that the hard shells of viruses are actually microscopic geodesic spheres.[71]

Each tensegrity structure, be it one of Fuller's models or a natural formation, relies on closure for its strength. The characteristic of closure causes every interference pattern (i.e., section) of the structure to support itself and the structure as a whole by interfering with and consequently buttressing its neighboring patterns.[72]

Such closure and interdependence are difficult to explain and understand because tensional construction and engineering are rarely employed by human beings. Even today, architects and engineers seldom utilize, much less understand, the properties of tension which Fuller championed throughout his life. Consequently, relatively few people have direct experience of man-made tensegrity, and without firsthand familiarity, individuals find the concept of tensegrity hard to grasp, even though it is a fundamental component of Nature.

Most people have, however, witnessed a form of tensional interdependence in the fragile structure of a house of cards. Such structures exhibit unmistakable closure and interference. In order to remain standing, each card is tenuously dependent on its neighbors, and only when all the cards are in balance does the house stand without external support. Although such an equilibrium is difficult to create and maintain by the use of cards, tensegrity structures are expressly designed to continuously distribute loads and pressures and to maintain that perfect balance.

Fuller began describing that phenomenon as *synergy,* a word he first found employed in the field of chemistry. Synergy is defined as the combination of elements to create results that are not evident when one examines the elements singly. Tensegrity structures are perfect examples of synergy. Although Fuller did not coin the term, by using it consistently throughout his lectures, he did bring it into common usage.[73]

The synergistic cooperation evident throughout tensegrity structures results in a structural integrity in which each section is harmoniously integrated with its neighboring sections. Within tensegrity models, interference creates a mutual synergistic support system which is the foundation for structures much stronger and far more efficient than would be imagined from observing the pieces individually. Integrated interference also produces a structure in which a pattern can be followed completely around the framework from any point and always returns to that same point.[74]

Seemingly simple interference patterns not only generate incredibly strong and efficient structures, they also produce a uniquely flexible stability which provides geodesic domes with their durability.[75] That same flexible stability is often maximized in nature's designs, but society has yet to apply it widely in construction or other endeavors as Fuller did in his work.

Another graphic model of pattern integrity Fuller often discussed can be observed by dropping a stone into water. That simple act produces a circular wave pattern which emanates from the point of contact and is the same for any object or liquid. Prior to serious studies of that phenomenon, most people believed that waves were part of the water or other liquid in which they were detected. However, after more comprehensive examination of wave phenomena, the everyday appearance of similar waves in different liquids at distant locations led people to realize that waves were not the substances in which they were found.[76] Scientists eventually established that waves are a universal phenomenon which demonstrate similar characteristics when passing through any substance.

For example, the falling stone also creates waves in the air before it strikes the liquid. At first, individuals find this phenomenon difficult to accept because they cannot see air or patterns in air. However, if the falling object is large enough and the observer is extremely still, he or she might detect a slight breeze as it passes. That breeze is the surge of strong airwaves which the object creates as it falls through the air. Modern researchers have added smoke to air in order to study gaseous wave patterns in wind tunnels, and in doing so, they have determined the universality of all physical wave patterns. Just as the knot was a pattern integrity which provided proof of its existence to Fuller and his audiences through a local vehicle (i.e., the rope), so a wave is a pattern integrity which is evident to humans only via local substances such as air or water.

Once Fuller grasped the universality and magnitude of such patterns, he formed his belief that human beings themselves are one of the more complex pattern integrities. In fact, he often stated that human beings represent the most complex pattern integrities known to humankind.[77] Bucky felt that a person is much more than the local physical vehicle in which each of us is manifested for personal expression on Earth. Like the knot or the wave, he viewed individuals as metaphysical phenomena providing evidence of their existence to other metaphysical beings (i.e., people and animals) by operating through local vehicles perceived as physical bodies.[78]

The concept of human beings as pattern integrities is not easily understood or explained, and Fuller sought graphic illustrations to support its understanding. The following was one of his most often used such illustrations. Most of us have had the experience of watching a movie or a videotape played in reverse. In such instances, a person appears to be flying backward out of the water and up to a diving board or looks ungainly as he or she rides a bicycle in reverse. Viewers understand that those scenes, which appear to reverse time, do not portray actual conditions. The scenes do, however, provide an imaginative model which Fuller utilized to examine the everyday processes essential to life.

He envisioned a scenario in which his breakfast flowed in such a reverse manner. First, the food which he had eaten and was becoming part of his physical body flowed backward through the digestive process, came out of his mouth, and returned to his plate. The dishes containing Fuller's meal were then carried back to the kitchen, where the food jumped into pans and was uncooked. The food continued its reverse journey and reentered packages. Eggs were restored to their shells and placed back in cartons, and milk miraculously flowed into its carton. The packaged food then returned to the supermarket and from there proceeded to processing plants. It moved in bulk from processing plants to farms, where vegetables entered into a process of reverse growth, returning to the ground. During that reverse growth, plants were reduced to the basic elements from which they had been formed, such as rainwater, solar radiation, and minerals.[79]

By imagining that backward flow, Fuller realized that just a few weeks earlier, a large part of what he viewed as "Bucky" had actually been basic recirculating elements and energies. The physical bodies that most of us believe constitute our existence might easily have been a cloud floating over the Himalaya Mountains or radiation traveling from the sun only weeks earlier.[80] For Bucky, that astounding revelation supported his appreciation of just how fallacious it was to identify himself or anyone else with physical appearances. It also created some contradictions in his personal life, as he, like all mortals, found himself bound to a physical body which he sometimes considered less than ideal.

Through years of observation and study, Fuller did come to understand and appreciate the processes which create and maintain human physical bodies. Through those processes, numerous elements are continually brought together. Water, air, proteins, and other elements move closer and closer toward one another until they reach a critical proximity and combine with other elements in a configuration recognizable as a human being.

Although Fuller felt that the essence (i.e., spirit, soul, pattern integrity) of human beings is eternal, he also realized that the process of elements' combining to produce a specific human body does not continue indefinitely. Individual human beings integrate physical elements only as long as the particular pattern integrity known as a person is operating with the parameters that humans acknowledge as living.[81] A person is the most compact and intricate association of elements known to humanity. Other compact associations of elements result in less complex configurations, such as trees, cats, and automobiles.

Fuller felt that in all those closely packed associations elements are tied into what could be regarded as knots which we perceive to be the components of physical reality.[82] Within the hierarchy of physical components, the elements which come together are themselves composed of knots of less complex elements, and those knots are, in turn, composed of knots of even less complex elements.[83]

That hierarchy of elements continues breaking down into less and less intricate structures until a minimal basic component is realized. That fundamental component is the essential energy known to be the quintessential substance of the entire cosmos. As explained earlier, Fuller hypothesized that energy ties itself into knots, and that when those knots are tightened by the pull of increased tension (such as gravity when bodies move closer toward one another), they form the matter fundamental to physical existence and biological life.

Fuller believed that all element knots, from the most elaborate to the very primitive, assemble themselves around pattern integrities. The fact that the knots are patterns which integrate elements led him to coin the phrase ''pattern integrity'' to describe the phenomenon essential to all manifestations of life.[84] Pattern integrities are nonphysical (like the idea of a knot) and, therefore, cannot be perceived by unaided human senses. They do, however, gather elements into their fields of operation at times, creating physical manifestations so that they can be perceived by human beings. Fuller thus felt that the most complex of those manifestations recognized by humans are human beings themselves.[85]

Although Fuller did not believe that civilizations identical to ours exist throughout the cosmos, he did feel that the cosmos is teeming with many different forms of life. He was also confident that humanity is on the threshold of becoming aware enough to recognize some of those life forms.[86]

One of the easiest means of examining the logic behind that thinking is through his concepts of pattern integrity and element knots. In his element-knot scenario, configurations of elements which are physically near enough to one another to be considered element knots are composed of smaller, less complex element knots which flow through them. Those smaller element knots are, in turn, composed of even smaller element knots. For instance, animals are made up of organs, which are made up of cells, which are made up of molecules, which are made up of atoms. Each of those gradations is a more refined element knot which flows through the larger knots.[87]

Fuller's reverse breakfast narrative provides a more detailed example of such a flux. Bucky felt that he, like all human beings, was, and would continue eternally to be, a unique, everlasting pattern integrity.[88] In order to be experienced by other unique, eternal pattern integrities, he felt that people allowed less complex element knots, such as air, water, and food, to flow through their individual pattern, and that those elements combined to produce the physical bodies which are visible to other pattern-integrity humans operating similarly on Earth.

The elements which form a physical body at any moment are not, however, stationary. They are perpetually moving through a body, just as the rope was sliding through the slipknot. A slight shift in perspective helps one to better understand that motion.

Fuller originally regarded the knot as moving along the rope, but he later determined that the rope could be seen as moving through the pattern integrity knot. Like a snake, the rope slides through the pattern knot. Similarly, elements slide through human bodies. Bucky hypothesized that because we human beings are a very complex species in the hierarchy of known life, we require more chains of intertwined elements than less sophisticated entities, such as trees.[89]

Each individual element strand is intertwined (i.e., knotted) with other element strands to form a virtual fabric of elements recognized as a

Fig. 5–3   As in all pattern integrities, less complicated filaments combine to form more complicated strands, which eventually produce yarn or rope that can be knotted into fabric.

person. Human bodies are composed of material strands containing elements such as oxygen and water knotted to create more intricate chains of
elements, which are, in turn, intertwined with other such strands to produce even more complex element chains. That hierarchy continues until
the physical configuration of a human being is complete.[90]

In an analogy applicable to all physical reality, a person's body can
be likened to a sweater. Humans can see and accept an article of clothing
called a sweater. However, without special tools and detailed examination, they do not perceive the microscopic fibers which are the fundamental material of the sweater. Furthermore, we humans cannot see the atoms
and molecules which are knotted together to form those microscopic
fibers. The most delicate fibers are spun (knotted) into longer strands in a
process which eventually results in thread. That thread is then combined
with other threads to form yarn, and the yarn is knotted or woven into the
sweater. Yet, in the true spirit of synergy, neither the yarn nor any
component of it could be recognized as the item known as *sweater*.

Sweater itself is actually an idea, a metaphysical phenomenon according to Fuller's theory of life. Hence, it initially appeared within the
metaphysical minds of human beings. After having expanded their
awareness enough to discover a new concept, human beings consciously
manifest that idea as a physical artifact. Only then can the material elements which result in the creation of a sweater be brought together to
create the actual item. Fuller theorized that that entire process, like all
human discovery, occurred under the guidance of an expanding human
understanding, another metaphysical phenomenon.

In considering the analogy of human bodies as living sweaters composed of element threads, it is important to note that the component element fibers and the levels of constituent elements which compose them are in perpetual motion. They advance through the living patterns and are immediately replaced by other elements. Customarily, new elements replace the same type of elements, but they are, in fact, different elements. Hydrogen replaces hydrogen and oxygen replaces oxygen, but the new elements are substitutes.

Fuller observed that that same movement and change also occurs on a much larger scale as evidenced within our physical bodies, where he observed that human beings constantly shed old skin and waste products.[91] We are also converting resources into the energy which allows us to move, talk, and perform all the other activities considered a part of living.

Fuller felt that the elements which constitute physical reality continue moving throughout Universe as they are replaced in a specific pattern knot (be it a human, a rosebush or any physical object) by other moving elements.[92] In other words, what individuals perceive as trees, roses, or even human beings are simply strands of elements and energies tied into forms which conform to a particular pattern integrity, and those elements are perpetually moving toward becoming components of other pattern integrities. As with all phenomena, the elements flow in the direction of least resistance. When an element abandons a particular pattern, the closest pattern integrity which offers the easiest access automatically becomes its next destination.

Fuller's pattern integrity hypothesis covers all aspects of the life process. He explained that because they are eternal, the pattern integrities (knots) through which the elements pass remain intact throughout the process. The physical manifestation of a pattern integrity may accommodate more elements and may appear to grow, but the basic pattern never changes. It simply integrates more strands of elements as it develops.[93]

Once that physical representation of the unseen metaphysical pattern integrity reaches maturity, the process of accommodating additional element strands reverses itself as the biological dog or tree advances toward death. When physical death occurs, element strands continue to flow from the biological form, but they are no longer replaced by other elements.

Because of a pattern integrity's ability to integrate ever-shifting elements without assuming the configuration of those elements, an oak tree continues to look like an oak tree from sapling to full growth and a poodle always looks like a dog. Even though both of those living beings may employ the exact same molecules of moving elements during different periods of their lives, they do not assume the form of those elements. Instead, the less complex elements conform to the structure of the more complex pattern manifestations. Water may be recycled through both the dog and the tree as well as through thousands of other living beings, but it is the water which assumes the form of the pattern through which it passes. Water never transforms a living pattern integrity or its representative form into another pattern integrity or form by its presence. Within the hierarchy, the less complex element always conforms to the pattern integrity of the more complex element.

Fuller's initial consideration of the knot-and-rope analogy was from the perspective of the knot's sliding along the rope rather than the rope's sliding through the knot. Because people tend to resist change and movement, when faced with such a variable scenario most individuals assume that the position of one of the two components is stable.

By assuming that one element is unchanging, people are able to visualize and understand the relationship of the second component to the immobile component. Because both elements are moving in relation to one another, considering either element to be in motion is, in fact, valid. By examining both possibilities in the open-minded manner which characterized most of his life, Fuller attained a great deal of insight overlooked by most people. He also came to understand that the underlying concept of all Universe being in perpetual motion was the basis of Einstein's theory of relativity.[94] Einstein hypothesized that everything is relative to everything else, and that all Universe is perpetually transforming.

When Fuller's knot-and-rope demonstration is considered in light of Einstein's relativity, the physical aspect of humans and other life-forms can again be appreciated as complex knots sliding along multiple strands of elements which cluster in a specific configuration for a moment and then move on. Fuller theorized that, following each such moment, the pattern integrity ''person'' also moves on along the various strands of elements. In his scenario of life, everything is in motion, and change is the norm.[95]

In accordance with his pattern integrity ideas, Bucky was certain that, like all human beings, he too was a perfectly designed, eternal pattern integrity. He then logically theorized that he always was and would continue forever to be that pattern integrity.[96] Like everyone else, Fuller integrated resource elements into equipment (i.e., physical bodies) in order to operate within the biological limitations of Earth's planetary environment. Having mastered that essential concept, Bucky began to theorize logical conclusions about other possible patterns of life in Universe. As he felt that all humans are metaphysical pattern integrities accommodating the specific environmental ingredients found on Earth, he reasonably deduced that other pattern integrities could be operating similarly in other areas of Universe.[97]

Just as human beings select and utilize element strands from the Earth's biosphere to integrate into their physical bodies, other pattern integrities would most likely do the same for the particular environment in which they chose to manifest. In other words, beings which could "live" on other planets and in other dimensions would be physically similar to humans only if their biospheric conditions were analogous to Earth's, and that is an extremely remote possibility.[98]

Fuller reasoned that, in all likelihood, limited human sensory equipment simply does not allow people to detect the majority of those other pattern integrity organisms. Within the last decades, Fuller witnessed science lend credence to his theory by confirming that at our current stage of evolution, most unaided humans see and experience less than one tenth of one percent of the total electromagnetic spectrum.[99]

When combined with people's often rigid beliefs concerning reality and human superiority, he felt it was obvious that such limitations must impede humanity from discovering life anywhere other than on Earth. Bucky believed that the cosmos may well be teeming with pattern integrities fabricating unique organisms, but restrictive human equipment (both natural and man-made) isolates us from those life-forms.[100]

He was convinced that human beings were established on Earth to do more than simply survive and are actually part of a grand universal strategy. In Fuller's understanding of that strategy, human beings are not the only species performing the mission which people were established on Earth to fulfill. Bucky found that Nature always creates one or more backups for important tasks.[101]

For instance, plant seeds are naturally scattered across vast expanses of land so that if harsh conditions at one locale damage some of them, others will continue the species. Pursuing that logic, Bucky reasoned that if the force he labeled the Greater Intellect operating throughout Universe determined the human mission to be important at the minor outpost Earth, it must be significant enough to be needed and occurring at other locations.[102]

In Fuller's theory, entities performing essential functions similar to those of human beings at other sites in the cosmos would undoubtedly be unlike the wildest expectations of most people. For instance, they might have adapted biologically to extreme temperatures that humans cannot tolerate because our bodies are mainly water. In such an extremely hot or cold environment, a species of pattern integrity beings might successfully manifest a physical vehicle by replacing the fluid humans employ as blood with mercury or other exotic substances. Fuller felt a more radical, but no less likely, possibility is that living beings at other locations manifest forms which are nothing like human physical bodies, or that they exhibit no body at all. The number of viable design combinations possible in the manifestation of living entities is infinite, yet each of those designs could successfully perform the same functions humans fulfill on Earth while remaining compatible with what we perceive as hostile environmental conditions.[103]

Fuller hypothesized that such beings would most likely have access to the same generalized principles humanity is now discovering to be operational throughout Universe. In his theory, extraterrestrial entities would appear very different from humans, if they could be seen at all. They would probably gather information and operate in ways even science fiction writers could not imagine. Consequently, evidence of their existence would be exceedingly difficult for humans to detect.[104]

Fuller argued that even if humans could discover signs of the existence of such beings, that evidence would probably appear so dissimilar from anything we would consider important that, in humankind's present state of development, individuals would probably not even take note of it. Hence, because of our relatively crude exploratory equipment, it is unlikely that humans would be cognizant of other life-forms, even if evidence of those life-forms were right next to us or under our noses.[105]

Bucky theorized that we can begin to understand the primitive nature of human beings in this area when we realize that most people cannot even detect other human pattern integrities once they stop integrating physical elements and die. Although much of humanity believes that such a shift constitutes the end of a person's existence, Fuller theorized and observed that, in his personal experience, people who truly comprehend the concept of pattern integrity also accept the fact that such a change in form is not the termination of anything other than the physical body, and it most certainly is not the end of a life.[106]

He also felt that the challenge now confronting individuals is conscious selection of an evolutionary path which expands their awareness into the realms of the unseen metaphysical. Fuller observed that the most accepted means of extending consciousness into the realms of what is commonly considered the metaphysical is through scientific experience-expanding equipment such as the spectroscope, an instrument which detects unseen electromagnetic forces. Such equipment is sanctioned as credible by society because it usually produces a physical representation of an unexperienceable phenomenon, and humans easily relate to such images.[107]

Because the concept of pattern integrity often seems complex and difficult to understand, Fuller later developed the following analogy. Using this story, he was able to help people relate the idea of pattern integrity to their everyday experiences. In his hypothetical scenario, Fuller had a friend named Ed who was continually speaking about another friend named Joe and insisting that Fuller and Joe should meet. Ed persists in talking about Joe because he feels Joe and Bucky have a great deal in common. Throughout one of his conversations with Bucky, Ed makes numerous references to Joe. Then he resolutely picks up the phone and calls Joe. After a short chat, he hands the telephone to Fuller, and the two men are introduced over the telephone. During that initial conversation, Joe and Bucky discover they not only have similar professions but are also working on similar projects. Within weeks the two men are calling each other regularly for professional support and advice, and a genuine friendship ensues. Still, they never meet in person.[108]

Over the years, their telephone relationship grows. Ed and most of Bucky's other friends die, and eventually Joe becomes his only living friend. Nonetheless, they never meet face to face, and their relationship is limited to the telephone.[109]

With the death of all his other friends, Bucky's friendship with Joe becomes extremely important to him. He even has a special red telephone with a separate line installed exclusively for Joe's calls. Other calls come in and are made from other telephones, but the red phone is solely for Joe. Within weeks of its installation, Bucky finds himself beginning to think of the red telephone as Joe. That impression is not the result of a conscious decision. Fuller simply notices that whenever he thinks about Joe, the red telephone also creeps into his thoughts. Over time that association continues to grow in meaning until Bucky becomes subconsciously convinced that Joe is, in fact, the red telephone.[110]

In sharing the story of "Joe and the red telephone," Fuller would explain that his newly discovered perception of Joe was equivalent to most people's impression of other human beings.[111] Individuals see a human body of a certain height, weight, and color and believe that that physical body is Jim or Sally, when, in fact, that body is nothing more than a sophisticated telephone. Certainly, human physical bodies are more complex than telephones. However, in Fuller's hypothesis, they are only devices which allow people to function physically within Earth's atmosphere and to interact with the physical bodies manifested by other pattern integrities.

Bucky viewed human physical bodies as sophisticated vehicles which possess the capacity to grow and at times to repair themselves, but still only as mechanisms. He felt that most humans have become confused about this vital concept by centuries of misinformation and a lack of basic knowledge. Because of that indoctrination, individuals tend to believe that they actually are their vehicles (bodies) when, in fact, people are no more their physical vehicles than Joe was the red telephone.[112]

Fuller also found that people constantly create mental associations between physical bodies and particular names. They then believe the essence of the individual answering to a specific name is the body they observe. Such an opinion is analogous to removing the red phone and believing that Joe no longer exists.[113]

In the Fuller story, if the phone is disconnected and removed, Joe still exists, but the familiar system of communicating with him is no longer operational. Once that shift occurs, someone interested in communicating with Joe must seek alternate techniques. Frequently, such options are considered impossible or fanatic by the majority of humanity during a particular period of time, and individuals who study or pursue those alternatives are labeled abnormal or eccentric. However, Fuller observed that once an unusual new method of communication is proven effective and gains acceptance, those individuals who were originally considered deviant are frequently heralded as visionaries, geniuses, and heroes.

Bucky witnessed that phenomenon during his youth when wireless communication was still being developed for practical applications and was not known to the general public. He clearly remembered his parents and their peers asserting that wireless communication was a wild fantasy, and that people who considered it worthwhile or practical for future use were impractical dreamers at best.[114]

After years of observing other types of comparable innovations being similarly categorized, Fuller felt that telepathic communication and the individuals studying it were being chastised in a similar fashion. Fuller believed that in the near future telepathy would be proven effective and would be validated by society, just as wireless communication was validated and accepted less than a century ago.[115]

Through decades of experience and conscious observation, Fuller became convinced that human beings are not the bodies people believe to be their friends, neighbors, and associates. He felt that humans are complex pattern integrities, and that pattern integrities are conceptual rather than material.[116] In examining his own actions, Bucky also saw that he had frequently been personally preoccupied within that belief system and unable to separate his own metaphysical essence from his physical being.[117]

From such observations and experiments upon himself, Fuller further theorized that metaphysical human minds deal only with conceptual phenomena, while human brains continually gather and record the sensory data from which minds discover metaphysical relationships. He felt that because brains deal only with finite sensory stimulation, people's brains support their clinging to the erroneous notion that humans are the physical bodies which individuals occupy for a span of time.[118]

He felt that human pattern integrities are extremely high-frequency wave patterns which, like all energy in Universe, conform to the generalized principles governing wave patterns. The wave patterns which are the essence of human beings are most easily detected by other human pattern integrities when people have likewise integrated enough elements to establish a biological vehicle (i.e., a body) for themselves.[119]

Years later, he discovered that other high-frequency wave patterns, such as radio waves, also operate throughout the cosmos, and that humankind is learning to beneficially employ some of those patterns which are of lower frequency, as evidenced in the proliferation of radio and television. Largely because of the advances he had witnessed since the invention of wireless communication at the turn of the century, Bucky came to believe that in the near future scientists would discover how to employ wave patterns in radio-like devices that transport objects and people from one location to another at the speed of light.[120]

Although that possibility seems incredible, Fuller found it extremely reasonable within the context of recent history. When he was young, radio and television had also been relegated to the realms of science fiction. Yet, within his lifetime, those remarkable innovations came to fruition and were developed into highly sophisticated, practical inventions.[121]

In addition, he felt that when individuals realize that human beings are not physical vehicles but are, in fact, metaphysical beings, they may also understand that humans are already able to transport themselves almost anywhere on Earth at the speed of light. Fuller believed that using computers and our expanding global communication networks, people regularly transport the knowledge and wisdom components of their essence at high speed.[122]

In his quest to appreciate and understand the essence of life, Fuller also found that scientists had discovered that each individual human body is composed of a distribution of elements identical to the distribution of the chemical elements which compose all the physical universe. He used that information to deduce that every person is actually a miniature universe in and of herself or himself. In more humanistic terms, every individual is as unlimited as Universe itself and has within himself or herself all the possibilities available throughout Universe.[123]

Fuller felt that people could view their unlimited selves in one of two ways. Either we are each a total universe in and of ourselves, or we each represent one possibility of the playing of a "game" called Universe.[124] Although he felt both of those alternatives were valid, by considering the latter perspective, Bucky was seized by the following insight.

He felt that the omniscient Greater Intellect operating throughout Universe creates individual humans as tests of various options of behavior and design in determining what each unique type of human being will yield. That perspective of life and the human experiment allowed Bucky to move away from judging individuals as good or bad. Instead, he was of the opinion that each person represents a unique test case of a possibility, and that those human characteristics which are most effective are supported and utilized as models for future living experiments.[125]

In Fuller's view, the current "you *or* me," "good or evil" separatist perspective so many of us believe in was created by humans centuries ago and is not in accord with today's environment. Years of viewing the human condition from a nonjudgmental perspective led Fuller to believe that every individual human being is important, and that no individual or species is superior to another. He also forecast that humanity is only now entering into an era in which such equality is becoming acceptable and necessary.[126]

Fuller's theory that human beings are miniature universe experiments directly contradicted Darwin's explanation of the essence of human life. Darwin was certain that biological cells were the building blocks of life, but Fuller felt that Darwin had overlooked the more subtle, yet significant, metaphysical aspects of life. Fuller viewed Darwin's theory as limiting because it overlooked the essence of life, which Darwin simply did not have enough knowledge to postulate accurately. Fuller also felt that continued acceptance of Darwin's theory could only lead humanity toward greater misunderstanding and problems.[127]

Bucky was certain that humankind must now extend its vast scientific and technological knowledge, which focuses upon the nonphysical laws of Nature, into other fields so that the general populace will begin to comprehend the truth of the human experience. He also believed that it is time to discontinue misrepresenting human beings as simply flesh-and-blood organisms who happen to possess the unique ability to think and understand. In Fuller's view, human beings are an unlimited, integral component of the total integrity of Universe. He felt that we each possess a potential well beyond anything imagined by most people, and that although we are not indispensable, we do perform a needed service at this location in Universe.[128]

CHAPTER 6

Inventions Abound

 Fuller's insight into human behavior guided him to follow a strategy of developing artifacts (i.e., inventions) which altered human environments. The intended result of his artifacts was elimination of the negative human behaviors which he believed were responsible for most problems.[1] In other words, he sought to alter the environment in positive ways rather than attempting to reform people's behaviors. The production of artifacts also created inventions upon which public interest could focus, and that interest provided Bucky with an audience eager for more information about the issues on which his inventions focused. For instance, his Dymaxion House generated an audience interested in the problem of shelter.

To support his artifact-development strategy, Bucky attempted to find an advantage which he as an individual held over large corporations and organizations. His 1927 predicament of being penniless while observing problems which needed attention provided him with a great impetus for discovering that advantage. Fuller was quite surprised to find that his advantage as an individual lay in the very behavior other people had always warned him to avoid: thinking.[2] In a flash of insight, he realized that a single human being is far more powerful than any bureaucracy, government, or industry because bureaucracies and organizations cannot think.[3]

The fact that only individual human beings can think, and therefore function autonomously, immediately became a mainstay of Fuller's operation. He, like all individuals, did not need anyone else's approval to think or act. If a person observes a situation which needs immediate attention, he or she can respond instantly.

Fuller also realized that individual human beings benefit greatly from our ability to make quick decisions, whereas organizations require meetings, reports, votes, and so on prior to making a decision. In most instances, one person, such as a corporate CEO, ultimately takes responsibility for a decision, but that decision may be weeks or months in coming. And should the decision result in problems, the responsible individual usually hides behind a corporation rather than accepting the consequences. Fuller would later frequently recount that when a ship is in danger of crashing, only an individual can make the instantaneous decisions necessary to save it, and as Spaceship Earth comes ever closer to the brink of disaster on many fronts, individuals and not organizations will provide the required solutions and quick decisions.[4]

Following his 1927 period of introspection, Bucky realized that he was an ideal individual to test the principles he uncovered and demonstrate just what a lone, average, healthy human being with ordinary faculties could accomplish when he did not conform to the common belief that earning a living was important. He had been raised during a period of enormous innovation and had witnessed a dramatic increase in information. Technology had spawned rapid advances which were filtering down to benefit the lives of ordinary people, and nowhere were the benefits of technical advance more evident than in the eastern United States where Bucky had lived most of his life.

It became clear to him that someone had to step forth and test new possible ways of living, and the insight he had acquired through extensive trial-and-error experience since birth convinced him that he was the ideal candidate for the task.[5] As he silently examined the history of the human species, Fuller also came to theorize that humanity was at a critical juncture of evolution in which our access to knowledge was expanding so rapidly that the metaphysical, pattern-integrity aspect of each of us was becoming increasingly aware of the needs of our fellow inhabitants of Spaceship Earth. He believed that the time had come for people to begin functioning in a more consciously cooperative manner, and an average individual with no financial resources who was willing to be the test subject in a lifelong experiment of contribution, faith, and survival was required.[6]

At that time, Fuller also initiated his strategy of focusing on inventions to test new lifestyle possibilities. Since he had accumulated a great deal of experience in the field of housing and felt compelled to alleviate the housing problems which he blamed for his first daughter's death, Bucky first focused his inventive mind on the issue of shelter. Specifically, he worked at perfecting the futuristic "dwelling machine," which he christened the 4D (i.e., Fourth Dimension) House, shown in Fig. 6–2.[7]

Always one to consider the largest possible scale of problems, Fuller did not simply attempt to build a better house. Instead, he studied the designs and pattern integrities he found everywhere in Nature and began applying them to a practical human requirement.[8] He also examined the broad issue of controlling human environments and how people's well-being could be advanced through the implementation of better environmental controls.[9] That focus produced several unique early inventions and laid the foundation for his future work with larger structures designed to house thousands of people as well as his most well-known structure, the geodesic dome.

Fuller realized that to be most effective, he had to ultimately convert his revolutionary ideas into tangible objects which could be examined by others, but he had no resources to build models. Thus, he started by translating his thoughts into drawings. Because they graphically reflected the depth and uniqueness of his ideas, Bucky's drawings were also perceived as radical when he began showing them to interested individuals during the latter part of 1928.[10] In fact, his work was so revolutionary that, even after more than half a century, the housing industry has not implemented most of the technological advances Fuller considered critical as early as the late 1920s.

For instance, Bucky envisioned a person being able to open doors and windows or activate mechanisms with a simple movement of the hand. He believed that such a device would not be a novelty, but, rather, means of alleviating unnecessary work within the home and providing more freedom and happiness for its occupants. Following the installation of such a device, Bucky envisioned a person carrying a small child or an armful of groceries no longer having to struggle with door handles, and, thus, life would become just that much more enjoyable and efficient. Bucky was so enamored of the potential value of such an invention that he wrote to his brother Wolcott, who worked as an engineer for General Electric, asking him about the possibility of creating a mechanism which activated motors when a beam of light was broken.[11]

Wolcott had become a pragmatic engineer working within the confines of a huge corporation, and he wrote back in an exasperated tone that no such device was even being considered by modern engineers. He went on to chastise Bucky for his constant fantasizing and wrote, "Bucky, I love you dearly. But can't you make it easier on your relatives and friends by not including preposterous ideas."[12]

Within less than a year, Wolcott began to appreciate his brother's insight when Bucky's idea was introduced at G. E. He sent the following telegram to Bucky: "YOU CAN NOW OPEN YOUR DOOR BY WAVING YOUR HAND AFTER ALL STOP WE HAVE DEVELOPED PHOTOELECTRIC CELL AND RELAY STOP SEVENTY TWO DOLLARS FOR THE SET."[13]

Fuller was confronted with similar reactions when he met with engineers from the fledgling Aluminum Corporation of America (Alcoa) to show them drawings of the 4D buildings he proposed constructing out of the lightweight, efficient aluminum alloys he believed would be developed in the very near future. One of those skeptical engineers simply laughed at the ideas and told Bucky quite coldly that he was too young to realize that aluminum was not used in buildings but was used only in items like percolators, pots, ashtrays, and souvenirs.[14]

When Fuller asked if the engineers were working on manufacturing stronger aluminum which could be used in the construction of buildings, they replied that only two types of aluminum existed: soft and softer. Those engineers were also surprised by the accuracy of Fuller's vision when, five years later, the first hard aluminum alloys became available and movement toward using lightweight aluminum in construction began.[15]

Bucky worked on in silent allegiance to his ideas and the well-being of all humanity throughout 1928. Despite frequent ridicule from family, friends, and "experts" who felt that they knew how he should live his life, he was usually willing to show his unique drawings to anyone who expressed an interest. Those drawings portrayed not only radical housing concepts, but a revolutionary method of viewing the entire environment, especially Earth.[16] Fuller felt that in order to operate effectively in the largest context, he should consider the entire Planet and search for large patterns operating throughout Universe when examining any issue. He also believed that in order to work with the entire Earth, he had to be able to view as much of it as possible on a single sheet of paper with no visible distortion.

Because of that belief, Bucky embarked upon another seemingly impossible task, producing an undistorted World map. The first artifact he created in that quest, which would continue for nearly twenty years, was his 1928 drawing of the "Air–Ocean World Town Plan."[17]

Because Bucky sought to show as much of the Earth's landmasses as possible in a single drawing of a spherical Earth globe with no visible distortion, he spent hundreds of hours determining the exact position on a

Fig. 6–1 Fuller's One Town, Air–Ocean World illustration.

globe from which the most land can be viewed. Like most of his work, that formidable task also became a learning experience as Bucky used it to acquire a true "feel" for the Earth as a solitary spherical Planet rather than a divided amalgamation of nations and regions.[18]

Eventually, he determined that a position directly above the French Riviera provided the single view from which the most land was visible. Since he was fascinated with interrelationships and the development of modern business, Bucky quickly noted that a great many of the wealthiest tycoons chose to vacation at the same location that he had determined to be at the center of the most visible landmass. In other words, the men who sought to control the World through financial domination had somehow selected the center of the most visible landmass as one of their primary recreation sites.[19] Although this may appear to be an unrelated coincidence to most of us, to Fuller it was another indication of the complete interconnectedness of all Universe and every human action.

In addition to displaying a great deal of Earth's land, Fuller also rendered his illustration with his image of future travel. In 1927 air travel was in its infancy. Lindbergh had just crossed the Atlantic, but the few commercial airline flights were regarded as exotic attractions for wealthy adventurers rather than practical travel options. For instance, regular flights across the English Channel and between Florida and Cuba existed, but most travelers used slower, less expensive ships for those journeys.

Although even the visionary Fuller did not forecast the massive, instantaneous expansion of aviation which would occur as a direct result of World War II, he did envision a network of global air travel which he embodied in the Air–Ocean World Town Plan. Fuller's air travel ideas were quite practical for flight at that time, utilizing many medium-range flights to carry passengers around the World rather than the long-range flights which connect continents today. On his illustrated map, planes were depicted completing segmented trips between distant locations such as the United States and Europe by utilizing stopovers. For example, during the United States–Europe journey, passengers would spend the night at a halfway point somewhere in remote Greenland.[20] Realizing that the locations of such stopover sites would be dictated by distance rather than facilities, Fuller also devised and illustrated a series of easily erected intermediary "ports." At those remote but necessary locations, passengers could rest while planes were repaired and refueled.[21]

Because of his extensive experience in the construction industry, Bucky knew that the type of building he required for remote locations such as the Amazon in Brazil, the Sahara, and remote reaches of Greenland, did not exist. The structures he envisioned had to be quickly erectable, easy to maintain, and nearly autonomous. Since the projected stopover sites were generally uninhabited, access to the supplies required to sustain life would not be easily available. Food and fuel would have to be flown in or produced locally with equipment flown into the site. Accordingly, Fuller was one of the first people to design buildings employing the revolutionary technology of solar and wind power.[22]

A more conspicuous problem was establishing an adequate shelter without bringing in hundreds of workers and the supplies to support them

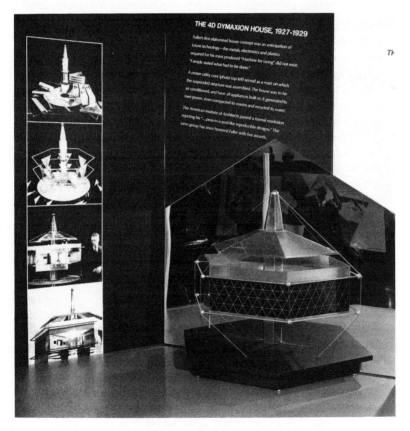

Fig. 6–2   4D House model displayed during a 1973 exhibition. Photographs of the original model being assembled and displayed are shown on the left.

for months of construction in harsh climates. Bucky perceived that "problem" as an opportunity to again study the single issue which most fascinated him and which he felt provided the best opportunities to advance revolutionary new ideas.[23] Hence, environmental control—shelter and housing—became a significant component of his global air-travel network.

By including environmental control as a part of that vision, Fuller was also able to integrate the autonomous housing ideas he had been considering for years into a "practical" application. He strongly believed that adequate housing was an innate human right, and that modern technology should be applied to providing sound housing for every inhabitant of Earth. In looking to create an autonomous shelter for his remote landing sites, Fuller simply applied his single-family housing ideas to larger buildings. Bucky's basic concept was the 4D House, which quickly became

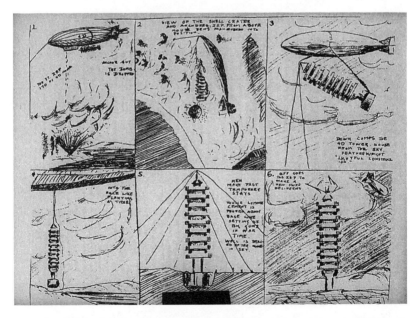

Fig. 6–3   Fuller's plan for delivering and installing a ten-deck, preconstructed building in remote locations. Panel 1: The zeppelin carrying the building drops a bomb. Panels 2–4: The building is lowered into the hole created by the bomb. Panels 5–6: Workers temporarily secure the building with wires and fill in the crater around the foundation for permanent installation.

known as the "House on a Pole" because the living area was to be suspended from cables attached to a central mast. For housing at remote sites, he expanded that idea from a single dwelling to a ten-deck tower apartment which could be easily constructed on its side in a factory and delivered by air to any location.[24]

As Fuller felt that a designer was responsible for all aspects of a project, he spent many hours creating a practical, yet ingenious, technique for erecting those ten-story buildings. First, the structure would be built horizontally in a large factory, in a manner similar to the way modern aircraft are produced. The completed building would include an attached foundation structure which was to be lowered into the ground to anchor the tower. Once the enormous unit and all its plumbing and wiring was finished, it would be attached to the underside of that era's most practical means of transporting huge items by air, a dirigible (rigid blimp).[25] The dirigible would then fly to a predetermined site, where its crew would drop a bomb, instantly excavating a crater large enough to accommodate the building's foundation.[26]

The base of the building would then be lowered, so that the structure was vertical, and the foundation would be eased into the bomb crater. Next, workers would secure the tower with temporary support cables from its apex to the ground. Pouring concrete around the base of the already operational structure was to be the final step in the simple procedure, which Fuller estimated would require less than a single day. Always relying on Nature for his best ideas, Fuller frequently compared that process to the planting of a tree.[27]

To demonstrate the practicality of global air travel, he drew several of those "planted" structures on his illustrated World map, along with small aircraft traveling between the remote sites and large cities. Even though his concept was rendered obsolete by long-distance air travel, at the time it was so realistic, yet radical and futuristic, that Fuller's illustrated map and ideas concerning such a possibility were reprinted in several magazines during the next few years as he became more publicly known.

Because he understood that improving global transportation and communication were advancing humanity toward a single World society, Fuller began referring to his illustrated map as the "One Town World" perspective. He continued refining and expanding that global perception, and in 1951 he coined a new descriptive phrase for it, "Spaceship Earth."[28]

To clarify his perspective of the relationship between human beings

to their environment, the caption on Fuller's 1928 illustration-map read as
follows:

> 26% of earth's surface is dry land. 85% of all earth's dry land shown is
> above equator. The whole of the human family could stand on Bermuda.
> All crowded into England they would have 750 sq. feet each. "United
> we stand divided we fall" is correct mentally and spiritually but falla-
> cious physically or materially. Two billion new homes will be required in
> 80 years.[29]

In working to support individual families, Bucky sought to alleviate
the massive housing problem he perceived on the horizon through another
of his 1928 inventions, the single-family 4D House. The result of his
diligent work on that project was a number of intricate drawings incorporating sophisticated solutions to the issue and demonstrating the futuristic
nature of his thinking. For instance, Bucky envisioned and designed a
method of suspending the entire House from a central mast which contained the plumbing and the equipment for generating power as well as for
distributing air, light, heat, and so on. The "rooms" of the house were
actually sections independently hung from the mast so that they could be
interchanged or removed and replaced when more advanced "units" were
produced.[30]

The beds were to be pneumatic (i.e., air beds) so they could be
inflated to a particular firmness. Rather than the usual types of rooms,
Fuller envisioned a "catch-up-with-life" or utility room containing a
laundry unit where clothes could be completely cleaned and dried within
three minutes. That room would also accommodate all the amenities of a
modern kitchen, including a dishwashing system which washed and dried
dirty dishes and returned them to shelves.

The area which most interested him, however, was the "go-ahead-
with-life" room, in which he planned to provide the residents with everything they needed for ongoing self-education. The area included a radio,
the soon-to-become-practical television, maps, globes, revolving book
shelves, drawing boards, typewriters, and other materials for learning.[31]
Although such a room may not appear very innovative in light of modern
computer and communication technology, it was well ahead of its time
and would have provided even the best scholars with more resources than
most had available in their homes or offices.

Many people felt that the most remarkable aspect of the 4D House
was Fuller's estimate of its cost. At the time, an average American automobile cost twenty-two cents per pound. Always searching for the most

efficient means of production, Bucky resolved that his House was as much a machine as an automobile and should be mass-produced in a similar manner. Utilizing automobile standards, Fuller calculated the cost of a fully furnished, two-bedroom 4D House with appliances at twenty-five cents per pound.[32] With a total weight of six thousand pounds, the complete price was to be $1,500, which translates to approximately $11,000 in modern dollars. For that price a family would be provided with nearly two thousand square feet of living space. Certainly, $11,000 would be a bargain price for any home today and even in 1928, when an average unfurnished home sold for $8,000, a completely equipped $1,500 house was perceived as beyond belief.

The 4D house was also designed to be autonomous and transportable. Once it was purchased, a family would never have to sell one house and buy another. No matter where they chose to reside, they could simply call a 4D House moving firm and have their home air-delivered to even the most remote site within a matter of days.

Conservation was always a key to Fuller's philosophy, and no resources were wasted within the 4D house. Water was to be filtered and recycled, and rain was captured from the roof for future use. Solid waste was to be packaged and sent to be recycled at central processing plants. Power was to be derived from solar and wind energy gathered and stored in the mast.[33]

Although he had purposefully resolved never to publicize his efforts or his mission, word of Bucky's eccentric ideas and monastic lifestyle soon spread, and he was frequently visited by interested observers. Since Bucky had not yet decided to speak, Anne was forced to speak with visitors, and she soon became Bucky's spokesperson. He did, however, permit most of them to view his drawings. Late in 1928, those drawings along with explanations became the subject of a major illustrated article in the *Chicago Evening Post,* and some of Fuller's 4D work was also published in the magazine *Architecture* that same year.[34]

The publicity from those extensive articles and other smaller ones attracted other inquisitive individuals, including a few who were intrigued enough to offer Fuller a small amount of financial assistance. With that patronage, he formed the 4D Company using a post-office-box mailing address and appointing himself as president. The first major action taken by 4D was to begin securing patents on the 4D House so that it would be protected for future production.[35]

In addition to his drawings, Fuller had also completed an enormous amount of writing during his years of silence. His work included more than two thousand pages of a book which focused on the future of humanity and the function of human beings. It also dealt with shelter: the importance of developing mass-produced housing and utilizing the newest technology in housing.

In 1928, Bucky began editing that writing into a shorter format, and after weeks of work, the two thousand pages were reduced to a fifty-page booklet entitled *4D Time Lock*. Fuller then presented the booklet to a group of eighteen members of the American Institute of Architects (AIA) whom he and his father-in-law had selected because of their reputation as open-minded, freethinking individuals. Those men praised Fuller's work and skill as well as the 4D structures included in the booklet.[36]

The AIA's "leaders" were not, however, so quick to accept Fuller. Still, Bucky sought to make his ideas as available to the public as possible, and because his father-in-law was a prominent architect, Bucky asked him to offer the newly acquired 4D patents to the American Institute of Architects as a gift. Although some people construed Fuller's motivation as self-aggrandizing, nothing could have been further from the truth. At the time, he believed that if he gave his revolutionary new ideas to a large organization which was extremely interested and involved in construction, those ideas would be disseminated to the public much more rapidly.[37]

Fuller's gift was summarily rejected by the AIA's board of directors, and he soon learned that his supposition that larger organizations are more willing and able to distribute new ideas was the antithesis of the truth. After examining his concepts, AIA's board resolved that Fuller threatened their very profession and issued the following statement: "The American Institute of Architects is opposed to any kind of house designs that are manufactured like-as-peas-in-a-pod."[38]

Because of their traditionally entrenched viewpoint, the board members felt that mass production would eliminate designers and architects rather than generate more opportunities for their members and all humanity, as Fuller suggested. Forty years later, in 1968, that same organization awarded the celebrated designer and architect Buckminster Fuller (who had received no formal architectural training) their first architectural design award for his Montreal Expo '67 Geodesic Dome.[39] Ironically, that design was also engineered to be inexpensively mass-produced.

Not one to be dissuaded by the criticism of those who did not understand or respect his unique ideas, Fuller used his own money to self-publish and distribute *4D Time Lock*. He then sent copies to dozens of prominent individuals who he felt were influential in determining the course of human development. Among those recipients were his friend Vincent Astor; Bertrand Russell; Harvard's president, Lawrence Lowell; and Henry Ford.[40]

From that distinguished aggregation, Fuller received what would be the first of thousands of reactions to his writing, all of which were similar. Over the years, readers have invariably been enthusiastic about the content of his writing. However, because of his intricate style, several of those reading *4D Time Lock* responded that they were certain Fuller's works contained unique ideas, but they could not detect them concealed within his complex language. In the booklet, Fuller displayed many of his drawings along with his belief that the housing industry had to become more mass-production oriented. He also continued his harsh criticism of big business and banking interests, which he felt were responsible for most of humanity's economic problems.

In addition to writing and producing drawings, Fuller initiated the development of a more tangible representation of his ideas. With the help of a group of architectural students and the finances of the newly formed 4D Company, Bucky began work on a scale model of the 4D House. That model was completed in 1929, just as Fuller decided that he had something worthwhile to say and resumed speaking publicly.[41] However, from that time on, he carefully watched his words to be certain that every word was as effective, truthful, and positive as possible. Despite that discipline and his constant attention to language, most people still had difficulty understanding the nuances of Fuller's highly complex ideas when he shared them.

Since a great deal had been written about him and his eccentric behavior during the previous year, Bucky began receiving speaking invitations. Among his engagements were speaking at the City Club of Chicago and the Chicago Home Owners Exhibition, where his 4D House model was being displayed. Such exhibits of the model resulted in more opportunities, including an offer from the huge Marshall Field department store.

Marshall Field was preparing to open a display of "modern" furniture imported from Paris, and its advertising and public relations department was searching for an advanced technology gimmick to legitimize the new line of furniture.[42] Fuller's futuristic House appeared perfectly fitted to their needs, and the public relations manager, Waldo Warren, asked Bucky to display his 4D House model next to the imported furniture. In an effort to draw even more potential customers into the store, Mr. Warren also invited Bucky to speak on the virtues of his design several times each day, and the naturally talkative Bucky, who now felt he had something important to say, quickly accepted.[43]

While formulating a promotional campaign for their exhibition, Mr. Warren and his associates were somewhat dissatisfied with the 4D designation for Fuller's "house of the future," as it was to be publicized. The publicity-conscious advertising people felt that the house should have a more dynamic, catchy title, and they hired a "wordsmith" to work with Bucky in creating a new name for the dwelling. Although the process of professionally creating the name for a product or a company is common today, it was rare in the late 1920s, and even though he liked the 4D title he had selected, Fuller was willing to try something new. Hence, he entered into the process with excitement and enthusiasm.[44]

Bucky first described the philosophy behind the House to the wordsmith, who wrote down the key sentences of his description. From those sentences, he extracted the most expressive phrases, from which he gleaned meaningful words, and from those words he found expressive syllables. Using those word fragments, the wordsmith fabricated a series of four-syllable "words."

Fuller then eliminated words he did not like from the list, leaving *Dymaxion,* which combined syllables from the words *dynamic, maximum,* and the scientific term *ion.*[45] Marshall Field's executives were so enthusiastic about the term *Dymaxion* and the success of the exhibition that when it closed, they copyrighted *Dymaxion* in Fuller's name. Bucky was also delighted with his new word, and he adapted it as his trademark. Thus, most of Fuller's future inventions became "Dymaxion artifacts."[46]

The 1929 Marshall Field display of Fuller's Dymaxion House and his burgeoning celebrity status provided Bucky with a new confidence in himself and his mission. That attitude resulted in his decision to move his small family back to New York following their annual summer trip to Bear Island.

Although his years of silent contemplation had provided Bucky with a sense of purpose, he had yet to change many of the reckless habits of his past. Feeling a powerful need to respond to an unremitting series of invitations, he made certain that Anne and Allegra were securely settled in a pleasant little house not far from her family on Long Island and set forth on an almost vagabond existence of travel, which he would maintain to some degree for the balance of his life.

With his model Dymaxion House disassembled in a huge suitcase,

Bucky would travel wherever he was invited, and when not on the road he could usually be found hanging out with other artists and intellectuals in the Greenwich Village section of New York.[47] His lecture schedule was filled with large and small venues, including Princeton University School of Architecture, Carnegie Institute of Technology, Yale University Architectural School, and the Chicago Arts Club, but his most in-depth talks were at tiny Romany Marie's tavern and restaurant in Greenwich Village, where he was a regular customer.[48]

It was there that Fuller met another aspiring young artist who would become his lifelong friend, the Japanese sculptor Isamu Noguchi. Noguchi's fame, like Bucky's, was just starting to surface during the fall of 1929 when the two men met. And, like most of America, they were suddenly caught in the throes of the Great Depression. Neither Fuller nor any of his fellow "starving artist" friends lost great amounts of money when the stock market crashed, but jobs quickly became scarce and funding for nonessential items such as lectures and works of art became almost nonexistent.[49]

Always cunning, the somewhat homeless Bucky took to sleeping wherever he could. The floors of friends' apartments, artists' studios, and even lecture halls following an address became his bedroom until he had to move on.[50] Fortunately, during his two years of seclusion, Bucky had developed a more natural rhythm of sleep which did not depend upon darkness or the time of night or day. He had examined the sleep patterns of animals, especially the dogs and cats living around the Fullers' Chicago apartment, and found that animals did not wait until night to sleep. Dogs and cats simply slept when they were tired and appeared to require far fewer hours than human beings, who were usually conditioned to sleep only at night.[51]

Always willing to experiment with a new possibility, Bucky began sleeping in a similar pattern. Whenever he was tired, he would simply lie down or close his eyes for a short nap, and he soon found that once his body had adapted to that unique system, the sleep he required was reduced to only a few hours every twenty-four hours. Accordingly, he was able to work and study many more hours and was rarely tired.[52] That ability was particularly handy during those last months of 1929 when Fuller, Noguchi, and many of their friends had to regularly scramble for food and shelter.

The Depression also affected his friend Romany Marie, and she was forced to move her establishment from Washington Square to Eighth Street just west of the old Whitney Museum. Romany Marie loved Bucky's sense of design and radical ideas, and even though she could not afford to pay him anything, she asked him to design her new tavern and restaurant. She did promise that as payment she would give him a full meal every day as long as she was alive and in the restaurant business.[53]

During the Depression era, Romany Marie's offer was quite lucrative to the impoverished Fuller. He was also open to any prospect which permitted him to demonstrate practical applications of his ideas and use his mechanical aptitude. Consequently, Bucky began work on the restaurant. He selected a strikingly modern motif in which bright colors were illuminated by powerful bulbs placed inside huge aluminum cones. When he could not locate reasonably priced furniture to match the decor, the adroit Fuller simply designed and built the furnishings himself. The result of his efforts was a thoroughly modernistic restaurant which became an even more popular watering hole for the New York artistic crowd.[54]

With his open invitation for a daily meal and lack of finances, Bucky was a regular at Romany Marie's. Still, as a gentleman concerned about the welfare of his friend, he did not eat there every day. Rather, he followed a methodical schedule of stopping by the restaurant for a meal only every other day. Bucky did, however, seem to appear at the café nearly every day to partake of the conversation, and he soon became a regular fixture at Romany Marie's. In fact, Fuller's incessant dissertations on diverse topics became so well known that many patrons came just to listen to them.[55]

Although Bucky found living from day to day somewhat difficult, he was motivated enough by his mission to forsake many creature comforts. Then, in a typical fit of Fulleresque inspiration, he discovered an "ideal" living arrangement for himself. As he was constantly seeking out the latest industrial technology, Bucky became very intrigued by a new type of warehouse under construction in Manhattan: the Lehigh Building, which was located over the Lehigh Railroad freight yards on West 26th Street. It occupied two complete city blocks and rose twenty stories high. In order to facilitate the rapid flow of goods, elevators were designed to lift fully loaded trucks to any floor. With that unique innovation, tenants could have cargoes delivered directly to their space in the warehouse with no inefficient loading and unloading at crowded loading docks.[56]

That remarkable facility was completed just as the Depression began, and it was largely empty when Bucky toured it. Upon reaching the warehouse roof, he was astounded by the panoramic view, which stretched from New York harbor on the south to the Hudson River in the west, and he decided that the roof of the building would be an exciting place to live.

Examining the flat two-square-block roof, Fuller found only two structures other than the huge tanks which supplied water to the building. One
of those structures housed the massive elevator machinery, and beside it
was a large, empty storeroom, which Bucky immediately decided would
become his future home. Within days, he had convinced the rental agent
to cut a window into the storeroom and rent it to him for a paltry thirty
dollars per month.[57]

Although Bucky's new home was ideally isolated for the raucous
parties he and his artistic friends held almost nightly, it did present some
conspicuous drawbacks. First, the elevator machinery was located next to
Fuller's storeroom-apartment, and it generated tremendous noise. More
important, when that noise stopped for the day at five o'clock, the only
access to his lofty abode was walking twenty-one flights of stairs.[58]

Still, many of Bucky's friends were willing to endure that trek in
order to attend his outrageous parties. With a concrete "yard" the size of
two football fields and no complaining neighbors, those parties were
usually loud and raucous. For example, one night Fuller and friends
invited a troupe of African drummers who had been performing at the
Natural History Museum to join in the rooftop festivities, and they played
well into the night with no complaints from anyone.[59]

By the summer of 1930, Bucky had learned to live on his wits and a
tiny amount of money. Most of what he earned went to support Anne and
Allegra, who continued living on Long Island and were visited by Bucky
on weekends whenever possible. He believed that he had freed himself
from the constraints of working at a job to earn a living until what he
regarded as a nearly perfect employment opportunity presented itself.

Officials of the American Standard Sanitary Manufacturing Company, a large producer of plumbing fixtures, had read articles about Fuller's Dymaxion House and noticed that the bathrooms were to be mass-
produced as a single unit. The idea of selling an entire bathroom rather
than individual fixtures appealed to the management at Standard Sanitary,
and they offered Fuller a research position designing the prototype
bathroom he envisioned. Although that position was a regular job, Fuller
viewed it as an opportunity to research and work on an issue which he
believed was of profound significance, plumbing. Thus, he accepted the
offer.[60]

Being a comprehensivist, the curious Fuller could not simply design
a bathroom. Rather, when he began working at American Standard, he
embarked on a sweeping examination of human sanitation which provided

him with an appreciation of the enormous influence that plumbing has on humanity.[61] While studying plumbing and its relationship to other aspects of the environment, Fuller was able to progress beyond mere pipes and fixtures to consider the complete process of human cleansing and waste disposal. As early as the early 1930s, he calculated that people were extremely wasteful in squandering several gallons of pure water with each flush which disposed of only a small amount of human waste.[62]

Although a great many people today appreciate the magnitude of this problem, at that time clean water was viewed as an inexhaustible resource, and the issue of flush toilets was of little interest. Fuller determined that a great many life-supporting resources could be conserved and many benefits provided for all humanity by merely changing people's wasteful flushing practices. He also recognized that the plumbing industry itself posed a multitude of problems he could tackle and transform into opportunities for contributing to humanity.

While working for American Standard and examining plumbing comprehensively, Fuller also discovered the primitive nature of sanitation technology. He was astounded to find that a single man had designed all the toilet bowls manufactured in the United States. He also established that no one working within the industry was aware of the origins of the toilet bowl's design and construction and that those unchanging design and construction techniques were almost never questioned.[63]

Always curious about technology and manufacturing, Fuller visited the man who was designing the toilet bowls and discovered that that man had acquired his craft from ceramic workers in England and had transferred their antiquated methods across the Atlantic intact. Bucky also found that major corporations like Standard Sanitary and Kohler regularly dispatched officials to the man's simple, one-room workshop in Toledo, Ohio, in search of production designs and information.[64]

He was amazed to discover sophisticated modern corporations rendering unquestioning allegiance to such archaic manufacturing techniques, designs, and materials. No one, from the consumer to the manufacturer, seemed to question the fact that during a period in which aircraft engineers were developing more efficient materials and designs on a regular basis, bathroom fixtures were, and generally still are, being fabricated from heavy, yet fragile ceramics and outdated designs.[65]

Upon further examination of the field of plumbing, Fuller found that the tolerances employed in the manufacture of toilet bowls were extremely large. If manufacturers could maintain a range within one-quarter inch in the dimension of an opening, they felt they were doing a competent job.[66]

Such measurements were adequate because, unlike metal fabrication, ceramic work cannot be precisely controlled, and heavy ceramic material is very difficult to work with. Although plumbing manufacturing tolerances were acceptable to most people, Fuller compared them to other industrial standards of that era, and he was not satisfied. It was a simple matter for Bucky to calculate that if an automobile or airplane were built to similar specifications, parts would be constantly rattling around, machinery would regularly fail, and very few cars or planes would be fabricated soundly enough to be operational.[67]

Fuller also resolved that exploiting enormous amounts of precious resources for the inefficient production of heavy ceramic bathroom fixtures was illogical and unnecessary. To pragmatically demonstrate a more efficient solution to that problem, he invented a lightweight stainless-steel toilet bowl similar to the ones now used in jet airplanes. Fuller's design could also be inexpensively manufactured in two halves using modern metal stamping technology. When applied to plumbing, such readily available technology resulted in a high-quality precision product which required far fewer resources than ceramic toilet bowls.[68]

Bucky then expanded upon his toilet bowl idea and designed a self-contained, mass-producible bathroom which employed the most advanced technology available. That revolutionary unit was a technological masterpiece; however, Standard Sanitary did not have the foresight to establish and demonstrate the benefits of such a product to the public prior to its completion.[69] Consequently, when the plumber's union learned that Fuller's bathroom could be installed with four simple connections in a matter of minutes and would replace all the individual pipes and connections which constituted the majority of their work, they castigated the project in their trade papers. As plumbers were Standard Sanitary's principal customers, the union's reaction brought an immediate response. Within days, Standard Sanitary's management abruptly terminated Fuller's project, hid his prototype unit in a subcellar warehouse, and hoped that the union's hostility would quickly fade.[70]

Although Bucky had devoted six months of intense effort to the project, he was not disheartened. He felt that he had learned a great deal about a major issue which was not being considered by anyone and that his insight would be useful in the future. Bucky's intuition proved correct, when less than six years later he was invited to tackle the sanitation problem again, and he completed the development and production of his mass-produced Dymaxion Bathroom under the auspices of Phelps Dodge Corporation.[71]

Bucky did experience one unfortunate consequence of his work with Standard Sanitary. Always the genial host, he had invited some of his fellow employees to the ongoing parties at his rooftop apartment.[72] His boss, the head of the research department, was particularly impressed with the residence, and in 1931 Bucky returned from a lecture tour to find himself evicted. The research manager had secretly contacted Bucky's landlord and offered him a much higher rent for the space, but the two conspirators had to wait for Bucky to break his lease obligations. The 1931 lecture trip provided the opportunity as, with his usual disregard for financial matters, Bucky had forgotten to pay the rent in advance as promised, and upon returning he found his ideal space had been rented to his former boss.[73] Once again Bucky was without a home and dependent on his wits to negotiate the perils of daily life on the street.

With typical Fuller timing, Bucky had embarked upon his newest endeavor the previous year while he still enjoyed the luxury of a somewhat permanent residence and a regular salary. Although he had no experience in the field of publishing, Bucky felt that he needed a vehicle in which to formalize his ideas for the public, especially those individuals involved in architecture and construction. Never one to start off small, his first endeavor at publishing was the taking over of an entire magazine.[74]

The two Levenson brothers of Philadelphia had been publishing a relatively new architectural magazine named *T Square*, which was subsidized by George Howe, a wealthy architect. The publishing firm of Charles Scribner's Sons also produced an architectural magazine with that same name, and they held the copyright to the title. When Scribner's learned about the Levensons' periodical, they immediately demanded that the Levensons stop using the name *T Square*. Unwilling to become involved in legal issues, Howe removed his financial support, and the magazine was about to discontinue publication when the Levensons contacted Fuller.[75]

In a typical Fulleresque flash of idealism, Bucky decided that the magazine would provide an excellent forum for his latest ideas on construction and would help him to prod the architectural community out of their archaic resistance to technology. Without the slightest hesitation, Bucky resolved to cash in all his insurance policies and become publisher of the magazine, which he renamed *Shelter*. His first act as publisher would be regarded as suicidal by any businessperson. In the midst of the Great Depression, he canceled all the advertising being run in *Shelter*.[76]

Bucky later made the following comment, reflecting on the integrity of his action as well as his total lack of financial concern:

> I felt compromised because your advertising contracts required that you come out on a certain fixed date and I said, "Anybody publishing on a deadline is obviously being forced; he is doing it to make money, and he is not coming out only when he needs to. I put my magazine on a spontaneous basis. It would come out when we had something to say, and when we were ready to say it and say it right."[77]

Bucky's second major publishing decision was even more controversial than banning advertising. To cover his expenses, he raised the price of a single copy of *Shelter* to an outrageous two dollars and promised subscribers that, although he would not publish on a regular basis, they would receive as many issues as they paid for in advance. Two dollars may not appear expensive today, but during the Depression, when a few cents would buy a loaf of bread, it was an enormous sum, akin to charging twenty dollars for a copy of a modern magazine.[78]

Still, *Shelter* succeeded. That success was based largely on Bucky's dedicated enthusiasm and unique editorial content. By November of 1932, the magazine's average circulation had reached 2500 copies, and it was producing adequate receipts to sustain itself. Fuller was able to create that miracle by editing the entire publication himself, while receiving little or no financial compensation. He also persuaded other bright young men who were eager to use the magazine as a forum for new ideas to write articles with little or no financial recompense. For example, the first issue featured the aspiring young architect Phillip Johnson as the "guest editor."[79]

Each issue focused on a single theme which Fuller felt was significant, and true to his word, Bucky would not publish any issue until he felt that it encompassed the latest thinking on a particular subject. For instance, *Shelter*'s second issue concentrated on environmental controls and contained graphic photographs of black smoke pouring out of East Coast factories to emphasize the danger that pollution was creating.[80] Even though most readers were confused by the strange new word *ecology* presented in that issue, some people realized that they were being provided with insight into the future, and readers began to seek more from *Shelter*'s mysterious editor.[81]

The magazine's editor was mysterious because, even though he edited *Shelter* and wrote many of its articles, Fuller's name never appeared in print. Having learned just how jealousy and competition tend to hinder the thinking and openness to new ideas of creative people, including architects, designers, and artists, Bucky decided to remain anonymous. Rather than claim any of the ideas as his, he would simply refer to himself as "4D," an unnamed person or group who presented the ideas.[82]

Although unorthodox, his strategy succeeded until Bucky encountered one factor which would continue to dictate the course of his life: his love of change and action. Writing about ideas was certainly one mode of contributing to humanity; however, given a choice, Bucky would invariably select a more ambitious, active mode of operation which allowed him to participate directly with his fellow human beings.[83]

Following that passion, in the autumn of 1932, he decided to close down *Shelter*. His primary motivation stemmed from his belief that Franklin Roosevelt was going to be elected President of the United States and that his New Deal administration would provide a rejuvenated atmosphere in which liberal ideas and the people who embraced them would flourish.[84]

After making that decision, Fuller informed his subscribers that he was going to put all his ideas and information which had not yet been published into one final gigantic issue and cease publication.[85] Since *Shelter* was on the verge of becoming a very profitable operation, Bucky's decision surprised most people. Those who knew him, however, realized that he was not motivated by financial gain, and within that context such a decision was logical. Although Bucky was able to publish that final issue and close *Shelter* with no outstanding debts, he did not recover any of the money he had invested in the project, and as 1933 dawned, he found himself again homeless and without financial resources.[86] This time he was also without the cushion of his insurance policies.

Bucky continued to develop and employ the characteristics which had supported his constant exploration and ability to "thrive" throughout the previous several years. His mental dexterity and creativity, as well as his controversial but charismatic personality, were his primary assets, and he continued to utilize them during his frequent visits to Romany Marie's, where he would talk for hours on end. It was there that he perfected a speaking technique which he had begun to formulate during his years of silence. The one person with whom he had continued to speak at length during his self-imposed silence was his infant daughter Allegra.[87]

As noted earlier, Fuller was convinced that infants are born with a great deal of intellect. In fact, Bucky believed that infants naturally know more than adults, and he decided to honor that intelligence by always speaking spontaneously to Allegra as if she were a wise adult. The result of that resolution was Fuller's developing a form of consciously "thinking out loud" for his daughter during the sporadic times he was with her. No matter how complex the thought, Bucky was constantly sharing his ideas with Allegra without censorship or concern for her apparent ability or inability to understand. Later, when he began to speak to others, Fuller maintained his discipline of "thinking out loud" rather than delivering a set of preconceived ideas and well-planned words.[88] At Romany Marie's, those spurts of spontaneous thought were well received by a fascinated clientele.

Romany Marie herself was descended from a long line of Russian Gypsy mystics, and a group of mystical types gathered around her. One of the most well known of those individuals was the guru Gurdjieff, who was a frequent visitor to the café, where he and Bucky were known to get into some fascinating discussions.[89] After observing the operation of Gurdjieff and other guru-leaders, Fuller noticed that he himself was beginning to attract a coterie of "followers" who accepted his words as gospel. As he would later recount, "I found myself being followed by an increasing number of human beings, particularly women, who were beginning to make me into some kind of messiah. I became a cult, and that was exactly what I did not want to be."[90]

Bucky wanted each individual to think for himself or herself and not to blindly accept the beliefs of others. In order to dissuade those who sought to unquestioningly accept his ideas and who refused to think for themselves, he made a New Year's resolution that as of January 1, 1933, he would go out of his way to discourage followers. He felt that the easiest way to keep such people from deifying him was to become as unappealing as possible.[91]

Although that rationale was logical in Bucky's mind, it also concealed some of the rougher facets of his personality. He loved outrageous behavior, be it taking a position on behalf of all humanity or standing out as the loudest person at a party. Claiming that his rowdy behavior was predetermined to keep people from following him may well have been his way of justifying behavior he felt compelled to exhibit but of which he was often later ashamed.

Having lived in a raucous freewheeling manner for a good portion of his life, outlandish behavior, whether planned or not, was not a difficult assignment for Fuller. Not everyone avoided him, however. The insights he expounded in his talks and in *Shelter* were recognized as unique and deserving of further evaluation by many people. Accordingly, he continued to receive offers constantly. One such offer would consume the next few years of his life as he shifted his focus to an idea he had first considered in detail during his years of silence: the "zoommobile."[92]

CHAPTER 7

Vehicle of the Future

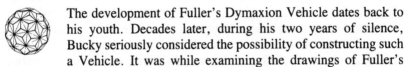 The development of Fuller's Dymaxion Vehicle dates back to his youth. Decades later, during his two years of silence, Bucky seriously considered the possibility of constructing such a Vehicle. It was while examining the drawings of Fuller's unique 4D House that his infant daughter, Allegra, referred to the porpoise-shaped object situated at the house's base as a "zoommobile."[1] The Vehicle was, in fact, more a product of Bucky's vivid imagination than practicality, and it was designed to travel on land or water as well as in the air. Its first appearance was at the base of the Dymaxion House model that Bucky transported from lecture to lecture in an enormous suitcase.[2]

Fuller's idea of delivering Dymaxion Houses by air and installing them throughout the World, including in remote rural locations, presented several unique problems. Chief among those problems was transportation. Though the House was designed to support its residents in an autonomous lifestyle, a means of personal transportation was needed to provide residents with the self-sufficient status Fuller sought for them.[3]

Because Dymaxion Houses were to be installed at any site, including ones serviced by poor roads or no roads at all, the Vehicle which would accompany the House had to fly and make spot landings without runways, just as a bird does.[4] Fuller felt his ultimate Vehicle should also be able to travel over rough terrain and on highways as well as in the water. That concept was so radical that, even though the yet-to-be-invented technology he required to complete the project is now available, such a vehicle has never been developed for use by the general public.

Years after his initial work on the Dymaxion Vehicle, Fuller would theorize that the core of the problem obstructing progress on such a vehicle was a condition he had faced many times during his career and which he perceived as a continuing impediment to humanity's advancement. That single issue is the bureaucratic drive for power and control that pervades any organization, especially governments. Both the Dymaxion House and the Dymaxion Vehicle were conceived to function autonomously. Once built and installed, they would require no roads, airports, electrical power lines, plumbing and sewage connections, or any of the other links through which corporations and governments control individual human beings' lives.[5]

Bucky eventually realized that if such autonomous innovations were publicly available, independent-thinking humans would quickly discover that the bureaucracies dominating their lives were no longer necessary or useful.[6] At that point, governments, corporations, and most other archaic bureaucratic structures would lose public support, and in the natural order of evolution and development, they would collapse. Fuller believed that such a scenario was inevitable and that the production of autonomous inventions would simply hasten its arrival. Within such an environment, resistance from the establishment was inevitable.

Because of the formidable opposition he received from power structures, Fuller also realized that the officials who dominate other individuals through bureaucracies and who prosper from their power would resort to extremes in blocking any innovation which would provide individuals with more autonomy and freedom.[7] Still, he envisioned a new era in which the responsibility for maintaining basic human needs would shift from governments and other institutions to individuals themselves and their autonomous devices such as the Dymaxion House. Fuller's vision of that new era was characterized by a peaceful design science revolution in which changes are predicated upon peaceful innovations that benefit all humankind rather than the wishes of conquering military powers. Design science is predicated on concentrating humanity's resources on life-supporting tasks rather than weaponry. By employing such a strategy, Fuller felt that we could successfully support every human being on Earth.[8]

When Bucky examined his design science revolution idea in detail, he determined that it could greatly contribute to the development and evolution of humanity and that it was, in fact, a necessary element in the success of the human species. Following that logic, Fuller believed that no politician, government, or corporation would be able to entirely obstruct the momentum of a design science revolution which would eventually furnish every person on Spaceship Earth with resources and a freedom never before known to modern humankind.[9]

Although the Dymaxion Vehicle was conceived of as an accessory of the autonomous Dymaxion House, it became more well known and supported than the House because of people's fascination with moving vehicles and the immediate freedom they afford. The human obsession with automobiles and speed was especially prevalent during the late 1920s and early 1930s when Fuller designed and developed the Dymaxion Vehicle. Even though the Dymaxion House would have proved far more beneficial to humanity than the Vehicle, its immobility held far less appeal for the American public than the Dymaxion Vehicle. Consequently, the full-size Vehicle was financed and manufactured well before the House.

Although not truly aware of the possible ultimate outcome of his efforts, Fuller had, in fact, been preparing himself for his work on the Dymaxion Vehicle since boyhood. Both flight and sailing had always fascinated him. Those interests were obvious during a series of 1917 discussions Fuller had with his Newport News naval commanding officer, Commander Pat Bellinger. The boom mast Bucky had invented to rescue overturned seaplanes had provided him with a great deal of attention, and Bellinger was extremely interested in listening to Bucky expound on his seemingly impossible ideas.[10]

During one such conversation, Fuller explained his radical notion of utilizing liquid oxygen in a turbine engine to produce enormous air pressure, an idea which captivated Bellinger. At that time, turbine engines were large, immobile machines used chiefly for generating electricity, and no one considered them adaptable to smaller jobs such as powering aircraft. Fuller himself realized that, even if his ideal turbine engine had been constructed in 1917, it would have failed because no material had yet been developed which could contain the enormous heat and pressure that the ignited liquid oxygen would produce.[11]

Although Bucky had no models or resources for testing his propulsion theories, he intently examined all the potential problems he might encounter in creating a small turbine engine. When he had planned each aspect of his theoretical engine as completely as he felt was possible, he discussed that phase with Bellinger. Bucky thus became one of the first to conceive of a practical jet engine, during a period when the technology which would make such an innovation possible was just beginning to emerge.[12]

At that time, Fuller did not believe that his radical ideas would be approved or even considered by the naval authorities. Still, as a young man in his early twenties, he was thrilled to have an opportunity to discuss them with the experienced aviator and engineer Pat Bellinger. As the two men debated the pros and cons of jet turbine propulsion in greater detail, Fuller realized that such an engine was exactly what was needed for an autonomous flying vehicle which could make pinpoint landings and take-offs. Years later, he would again theorize on the practicality of such a method of propulsion while developing the Dymaxion Vehicle, and in the 1960s and 1970s, he was elated to witness his radical 1917 idea being produced in jet aircraft and rocket technology.[13]

When Fuller began designing his Dymaxion House in 1927 and realized that the primary method of reaching remote locations would have to be flight which employed spot landings and takeoffs, he turned his attention back to the jet turbine concept.[14] He also began studying the different methods of flight employed by birds in an attempt to find the most efficient flying technique for solving his problem.

From his observations and studies of scientific experiments, Bucky learned that all the components of Universe are in motion, and that all motion tends to be in the direction of least resistance. He applied that knowledge directly to his Vehicle's design. Rather than using excessive propulsion energy in an effort to fight or tame Nature as many engineers were doing, Fuller attempted to create a Vehicle which would travel smoothly in a preferred direction because it was confronted by less re-sistance in that direction. In other words, he decreased the resistance in the direction he wanted the vehicle to travel instead of increasing the amount of energy required to propel it.[15] The extremely streamlined body of the Dymaxion Vehicle provides the most conspicuous illustration of Fuller employing the principle of least resistance, but that principle was also used in other, more subtle aspects of the design, such as the use of high-strength, lightweight materials.[16]

Through his extensive observation of Nature, Fuller came to appreci-ate the impeccable streamlining of birds and fish, as well as the design of those creatures which results in maximum efficiency and low resistance in motion. Because of that understanding, he was amazed to discover that designers of land vehicles had made little or no effort to adapt Nature's unmistakably successful, aerodynamic designs.[17]

Over the course of thousands of years, the people who designed ships of the sea had, out of necessity, incorporated streamlining into their work. Weight and speed were major considerations when a group of people was

Fig. 7–1 The second Dymaxion Vehicle. Although each of the three Vehicles included new innovations, the basic design and appearance remained unchanged.

isolated at sea for months at a time. In the early 1930s, engineers working in the fledgling aircraft industry, which was also very weight- and performance-conscious, were also beginning to discover the enormous benefits of streamlining. However, automobiles were still regarded as horseless carriages, and they maintained the box-like shape of carriages well into the 1940s.

Because his intention was to design a vehicle which would eventually travel on water or land or in the air, Fuller was not constrained by popular opinion or traditional modes of operation. Consequently, his Vehicle was designed in the streamlined configuration he determined to be most efficient. The earliest shape of the Dymaxion Vehicle was a direct adaptation of the streamlined design Bucky had observed in fish.[18] In fact, he was so captivated by the simple elegance of the fish's streamlined form that Bucky combined it with his love of flight to create a flying-fish drawing as the Vehicle's logo, a design which adorned all his workers' white coats.[19]

In his comprehensive investigation of streamlining, Fuller found that the phenomenon was more prevalent than he had first realized and, in fact, is a fundamental component in all of Nature's designs. Because streamlining results in the most efficient motion and everything is in motion,

Nature employs it to some extent everywhere. Fuller even discovered streamlining in objects which move only slightly during their entire existence.[20]

For instance, fruits and vegetables move very little, but when movement occurs, they are designed to move in the direction of least resistance with maximum efficiency. Fruit falling from trees contains seeds streamlined to travel in a preferred direction. In the case of fruit seeds, that preferred direction is usually deep into the ground.

When a ripe apple drops from the tree, it splatters, sending its perfectly aimed, teardrop-shaped seeds into the ground. Rain and frost then aid the seeds in continuing to employ their streamlined design to burrow deeper into the earth.

Since Nature's arrangements are the product of thousands of years of adaptation and evolution, Fuller reasoned that it was the duty of thinking human beings to incorporate Nature's proven designs and techniques into artifacts and inventions. Hence, he adapted a strategy of imitating Nature whenever possible in the design of his Vehicle.[21]

Because the Vehicle was ultimately to fly, Bucky studied birds and determined that, based on their primary method of flight, they could be divided into two distinct categories. The first grouping contains long-winged soaring birds, such as the gulls and eagles. The other category contains smaller-winged birds like the duck, which cannot soar but can fly much faster for short distances. In examining the flight pattern of the duck and other members of that class, Fuller found that they begin flying by building momentum, and he began formulating a flight theory incorporating that technique into the future evolution of his Vehicle.[22]

Fuller often utilized the following analogy of a pole vaulter to explain how ducks fly and thereby clarify the flight theory which he hoped to employ in his Vehicle. Although a pole adds weight to an athlete, it also establishes a synergistic advantage when utilized as a tool. An unaided athlete can high-jump only a short distance above his own height, but that same athlete can use a pole to exploit his momentum in vaulting over a bar three times his height.[23]

In equating the vaulting athlete with a duck taking off, Fuller recounted a hypothetical scenario in which, just as the pole-vaulting athlete is about to clear the bar, another, longer vaulting pole magically appears. The athlete swings onto the second vaulting pole, building on his momentum to achieve a greater height as seen in Fig. 7–2. If new vaulting poles continue to appear, the athlete can persist in achieving greater and greater momentum and in moving higher and farther with each successive pole and action. That phenomenon mirrors what occurs when a duck begins a flight.[24]

In taking off, a duck employs its webbed feet to scoop itself up onto the surface of the water in a vigorous run. At the same time, it flaps its short wings very powerfully to achieve additional speed. That process is enhanced by the duck's highly streamlined body design, which augments forward momentum. The duck's momentum increases to a point where the air-pressure differential created by the construction and flapping of its wings provides it with a slight vertical lift similar to the lift of the vaulter's pole. The duck then begins falling forward more rapidly in the direction of least resistance and, at the same time, uses its wings to give itself another short vault upward. By executing this maneuver continuously, the duck employs its increasing forward momentum and speed to provide a gradual lift, just as the vaulter might use pole after pole to build momentum and reach greater heights.[25]

In his ongoing examination of such flight, Fuller also discovered that pilots on aircraft carriers are aware of the benefits provided by the duck's system of flight, and that they emulate that maneuver when taking off on the short runways of aircraft carriers.[26] The first thing that a pilot does upon clearing a carrier's runway is direct the nose of the plane down

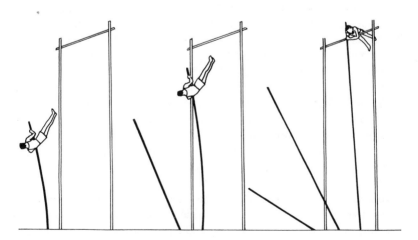

Fig. 7–2 A pole vaulter magically encounters longer poles which allow him to build upon his momentum in a scenario which Fuller employed to explain how a duck takes off and how the Dymaxion Vehicle would eventually fly.

toward the water to build up speed. By executing that maneuver, the plane gains a quick burst of momentum from the pull of gravity, and that burst aids in the plane's becoming airborne. A duck performs an identical procedure. It synergistically employs a combination of forward momentum, streamlined design, short wings, and gravity to travel in the direction of least resistance while increasing its speed and gaining altitude.

Those elements of flight design combine to produce a phenomenon Fuller dubbed "jet stilts." Jet stilts allow ducks to take off from small bodies of water and to rapidly gain altitude. Fuller referred to the duck's technique as a jet because a duck actually pushes air out from under its wings in short spurts in a manner imitated by modern jet engines.[27]

Unlike the force of most modern aircraft jets, the force of the duck's jet stilts can be precisely directed. Initially, the duck thrusts its wings downward to generate altitude. However, as it begins to build momentum and altitude, the jet stilts are shifted to pushing backward in a horizontal direction, and that action increases the duck's forward movement.[28]

Fuller understood that everything in Nature supports a condition of maximum efficiency, and that the design of a duck's head is no exception.[29] That simple configuration assists the duck's flight just as its wings do. The duck's head shape also demonstrates the subtle mannerisms inherent in Nature's design.

In silhouette, a duck's head looks very much like an airplane wing, with a greater length over the curved top than along the straighter bottom. That aerodynamic form is known as an airfoil, and it induces the air traveling a greater distance over the top to be stretched thinner than the air traveling straight across the bottom. As a result of that configuration, more air and greater pressure build up on the lower side of the airfoil than on the top, causing the wing (or the duck's head) to lift. That same aerodynamic principle is employed by blimp pilots to increase the speed and altitude of their ships. They raise the blimp's nose and create a greater pressure under the blimp than above it, producing additional altitude and momentum as well as increasing the blimp's speed.

Fuller's study of the design and operation of short-winged birds provoked him to envision a flying vehicle with small jet engines replacing large wings. Those engines would provide a constant thrust, similar to the pulsating of the bird's short wings, and that thrust would maintain the vehicle's flight.[30]

During Fuller's early 1930s examination of flight, an engine able to provide the immense thrust needed for takeoff without a runway (similar to a helicopter) was relegated to science fiction. Large wings and long runways were still critical to successful airplane takeoffs. At the time, Bucky predicted that a jet technology capable of producing instant lift would be developed in the near future and that such jet technology would be capable of sustaining flight and rendering wings unnecessary.[31] His prognosticated jet engine materialized within twenty years.

He also recognized that by removing the wings from a well-designed flying vehicle, he could reduce its weight significantly. Without wings, a lightweight, streamlined vehicle would be a natural candidate for jet-stilt flight when the predicted jet engines were developed. Bucky imagined an extremely lightweight, superbly streamlined vehicle propelled by jet tur-bine engines on both sides.[32] With such a vehicle at our disposal, he felt that human travel, like that of birds, would no longer be confined to the constraints of airports, roads, and other bureaucratic boundaries, and that autonomous, freethinking individuals could live and prosper wherever they chose.

Bucky's ultimate Dymaxion Vehicle was to have wheels for ground travel and jet stilts for instant takeoff and flight.[33] The two jet-stilt en-gines were to be directed slightly outward so that the effect of their combined thrust would converge just above the Vehicle's center of grav-ity. Because stabilization requires a minimum of three supports, as is modeled in three-legged tripods, the Vehicle needed a third support in addition to the two engines for stable flight. Fuller found that support within the Vehicle's natural tendency to fall forward into gravity. That falling motion would be offset by the forward and upward momentum of the Vehicle in flight.[34]

The tripod balancing phenomenon that Fuller theorized for use in the Vehicle can be modeled by a person walking on stilts as demonstrated in Fig. 7–3. A stilt walker actually balances on a hinge formed by the two stilts, just as the Vehicle would have only two engines. The stilt walker can fall only one of two ways, backward or forward. He chooses to fall forward, and as he does, he shifts the right stilt, moves it forward, and creates a temporary new third leg that breaks his fall.

That new leg also establishes a fresh hinge angled slightly to the left, and the walker begins to fall in that direction. To break the shifted fall, the left stilt is advanced and the forward-falling momentum is again broken. That action initiates a continuing cycle of left, right, left, right, stilt movement. Thus, the process of forward movement on stilts is actually a

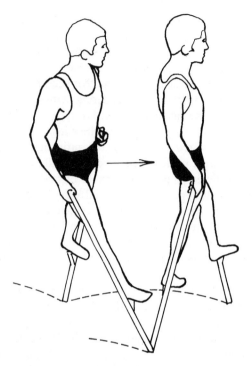

Fig. 7–3   A stilt walker does not move forward in a straight line but advances in a zigzag right, left, right fashion which models the tripodal forward motion Fuller envisioned for the flight of his Dymaxion Vehicle.

constant, delicate falling forward and from side to side, which relies upon an invisible third compression member (invisible stilt).[35]

   An important element of Fuller's design was his requirement that the vector thrust from the two jets on the sides of the Vehicle and the third "invisible gravity" jet in front converge above the Vehicle. That configuration would produce the effect of suspending the Vehicle from a moving tripod of jets and would keep it from becoming top-heavy and rolling over in flight. Because of its jet configuration, Fuller designated the Dymaxion Vehicle as an "Omni-Medium, Twin Jet Stilt Transport."[36]

   He envisioned the Vehicle's jet stilts as being very adaptable. They could be employed to generate altitude or build speed, and they would be flexible enough to be aimed in a wide or narrow mode to compensate for ground terrain and altitude.[37]

From the moment that radical idea came to him, Bucky was aware that developing such a unique method of flight would require the expenditure of millions of dollars and years of intense work. He also believed that his idea would be developed over time by governments and corporations with large budgets and workforces. Hence, he focused his energy and attention on producing scale models which did not fly. He also began writing and speaking about that unique method of flight to stimulate interest and activity among those who possessed the time and resources to develop jet-stilt flight.[38]

Eventually those endeavors did result in funding for a full-scale project which employed the newest technology available. The prototypes Bucky constructed were planned to accommodate jet-stilt flight but were built principally to test the ground-traveling capabilities of the Vehicle, which he believed would someday be adapted to fly without wings.

When Bucky realized that his lack of resources and time would render him ineffective in attempting to develop the jet propulsion technology needed for his Vehicle, he resolved to build a less complex prototype.[39] The purpose of that Vehicle was studying the ground-taxiing qualities of an Omni-Medium Transport which would eventually fly as well as travel on land and water. Fuller's first step in that process was studying the large-pattern relationships between vehicles such as airplanes, boats, trucks, cars, and trains and the primary substance in which they operated (water, air, or land).[40]

In his examination, Fuller found that all vehicles were most vulnerable to damage at the critical moment when they moved from a load-distributing element such as air or water into contact with a crystalline (commonly referred to as solid) element such as land or a rock.[41] In other words, an airplane or a boat is most susceptible to damage when it comes into contact with another airplane or boat or with land. That vulnerability is a consequence of the fact that crystalline elements do not distribute weight, as is the case with liquids and gases. When a vehicle comes into contact with gases or liquids (i.e., air or water), those elements distribute the force of the impact.

In the case of water, we see the splash when a solid object impacts the surface, and that same phenomenon occurs when a solid object collides with an invisible gas such as air. If the object is large enough and the observer is close enough to the occurrence, he or she will feel the offset wave of air as the object passes.[42]

When a vehicle is suspended in air or water, the forces which hold its weight up are evenly distributed hydraulically and pneumatically over the surface contacting the liquid or the gas. That phenomenon is conspicuously evident in the operation of an airplane, which has its entire weight supported by air pressure on the underside of its body and wings during flight. However, once on the ground, the plane's weight must be supported solely by the crystalline (i.e., solid) wheels and landing gear which touch the ground. That concentrated stress results in airplanes' being slow and unwieldy on the ground, even though they are extremely fast and maneuverable in the air.

Fuller found that when a vehicle (which is crystalline) contacts another crystalline structure, such as a rock or another vehicle, the load and weight are instantly focused at the area of contact, and that concentration often creates a serious problem. The hazard is particularly acute in planes and ships of the sea, which are far more delicate than land vehicles. Ships and planes can be constructed without enormously strong or heavy structures because the medium in which they operate most of the time cushions them and distributes load stresses. However, those same vehicles must be protected from rigorous contact with land by the cushioning of pneumatic landing gears in airplanes or pliable bumper guards on ships.[43]

Because of his familiarity with aviation, Fuller knew that a pilot's most perilous moments occur when the plane contacts the Earth in landing and taxis to its final destination.[44] In scrutinizing that dangerous procedure, Bucky discovered a genuine need for research into the properties of streamlined vehicles and for work on designing vehicles that are able to travel on the ground as safely as they do in the air. Such a vehicle would be easier to land than conventional airplanes and would not be as susceptible to treacherous ground crosswinds, which pose formidable problems for aircraft maneuvering on the ground.[45]

Even though Bucky designed and produced the Dymaxion Vehicle during the infancy of aviation and prior to the advent of major streamlining, he was extremely aware of the benefits that more streamlined designs might provide for aircraft. At that time, bulky planes were apt to be caught up in winds on landing and takeoffs. On the ground, they would swerve violently and sometimes flip over from the force of even a mild breeze. To eliminate that treacherous problem, Bucky streamlined his Vehicle wherever possible.

When the first Dymaxion Vehicle was completed, he wanted to test it at high speeds under the most rigorous wind conditions available. As airports were generally nothing more than flattened and graded farm fields, the best paved areas for such tests were public highways which offered long, clear expanses with unexpected gusty winds. Naturally, Bucky decided to use public highways in examining the performance and maneuverability of his Vehicle.[46]

Although it had been built primarily to test the ground-taxiing capabilities of a future Omni-Medium Transport and not as an automobile, in order to test the Dymaxion Vehicle on public streets and highways, Fuller had to purchase an automobile license from the state of Connecticut. And since it was usually seen traveling on streets and highways just like any other car, people began referring to Fuller's Vehicle as the Dymaxion Car. Even though he was continually explaining its potential for air and water transport, it still became known as a car.[47]

Fuller actually began working on the Dymaxion Vehicle concept during his silent years, when he spent weeks sketching his visions of both advanced housing and advanced transportation. Several years later, his friend the sculptor Noguchi produced clay models of the Vehicle from those sketches. The clay models were then used to construct more durable plaster models, which Fuller painted so they would resemble vehicles rather than futuristic works of art.[48]

In building models, Fuller once again created tangible artifacts which people could examine, so that, in some measure, they could experience his new ideas. Thus, Fuller's vision of a radical new system of transportation advanced one step closer to reality.

Although rudimentary, his models of the Dymaxion Vehicle conformed to and supported Bucky's strategy of speaking only when invited and not discussing artifacts until they were constructed in some form. He would exhibit the Vehicle models in conjunction with the Dymaxion House model whenever possible, and their very presence generated interest, questions, and invitations to speak about his ideas elsewhere. Establishing public awareness of his ideas was one of the first hurdles Fuller encountered in producing a practical prototype of the Vehicle. That obstacle provided him with a pragmatic opportunity to test his self-disciplines of building artifacts and speaking only when asked.

Bucky created his models during the onset of the Great Depression, when most of the population was intensely focused on earning a living and acquiring the necessities of life, not on something as obscure as new transportation possibilities. Because jobs and money were in short supply, people increasingly turned their attention from long-range considerations, such as new technology and futuristic housing, to short-term survival issues, such as immediate food and shelter. Manufacturers, however, were still chiefly concerned with creating a demand for and selling new products, and they were willing to try almost anything which might entice the public into spending the little money available.

One of the largest American manufacturing sectors was the automobile industry, whose management was required to continue supporting many of its large factories and to earn profits for stockholders despite the Depression. People in the automobile industry were perpetually inaugurating new schemes for selling cars, and one of the most successful devices used was the auto show. Auto shows featured displays of the newest model cars along with unique exhibits intended to attract potential customers.

The producers of one major New York auto show were in dire need of unusual attractions for their upcoming event when they saw and were impressed by Fuller's Vehicle and House models displayed at an engineering bookstore on Park Avenue. Because of the Depression, only the largest manufacturers could afford the fee required to display their cars at that show, and the producers invited Fuller to exhibit his models free of charge. He was given a prominent booth for his display as well as for describing the theories behind his ideas. At his booth, Bucky met and impressed many well-known people who were captivated by his ideas.[49]

Among the individuals Fuller met during that exhibition was Bill Stout, the man who designed the Ford tri-motor airplane. Stout was also president of the Society of Automotive Engineers, and he was so impressed with Fuller's work that he wrote an article about the Dymaxion Vehicle for the Society's magazine. Years after Fuller had completed work on his Vehicle, Stout would also design and build a famous car, the Scarab Car. Although Stout's car was also radically different from earlier automobiles, it did utilize some of the principles and ideas Fuller had developed earlier.[50]

As a result of his massive exposure at exhibitions and in the media, Fuller's ideas came to the attention of thousands of people, including a wealthy Philadelphia stock trader, Philip Pearson. Somehow Pearson had foreseen the imminent stock crash and had sold short, making an instant fortune at a time when his fellow traders were broke. In another shrewd maneuver, Pearson had converted much of his wealth into cash which he had tucked securely away in safety deposit boxes.

It was Pearson who provided the final impetus for Fuller's next major engineering initiative, the production of a working prototype Dymaxion Car. The investor was keenly aware of the Depression's economic impact, and after visiting Bucky's Dymaxion exhibition, he was also convinced that he would be more satisfied by giving some of his wealth to Fuller for the development of his ideas than by allowing it to languish in safety deposit boxes or to be lost to the Depression. Pearson felt that with financial support, Fuller might develop a vehicle which would be of value to society and that such an innovation might even help to end the most pressing problem of that era, the Depression. He also believed that there was a slight possibility his investment in Fuller's work would result in a commercially viable product from which he could profit.[51]

Although the offer of money was a welcome relief from the financial shoestring Bucky had been dangling upon for years, he did not immediately accept Pearson's benevolence. Because the benefactor was a stranger and Bucky had experienced trouble with financial backers years earlier, Fuller questioned Pearson's motivation. To alleviate his own doubts, Fuller devised a simple contract, which included a stipulation he often referred to as the "ice-cream-soda clause." That provision stated that Bucky could do whatever he chose with the money, including something as frivolous as spending it all on ice-cream sodas.[52]

Bucky's suspicion was not unjustified or illogical because, with multitudes out of work, many people were searching for quick profit-making schemes. As a result of that possibility, Bucky wanted to be absolutely certain that he was free to do whatever he saw needed to be accomplished without the constraints of profits, survival, sales, or other short-term motives.[53]

Pearson accepted Fuller's terms, but as a businessman, he always believed that he could exert financial domination over Bucky and the Vehicle if he chose to do so in the future. He was, however, surprised two years later when the Vehicle became well known and commercial production appeared possible.[54] Only then was Fuller truly thankful for the unique contract. At that time, many people told Bucky that Pearson should profit from what was perceived as his investment, but because the contract was not based on profit and provided Fuller complete freedom to continue his research under any circumstances, he was free to reject business, design, or engineering "advice" from Pearson or anyone else.[55]

Having accepted financial support, Fuller quickly formed the Dymaxion Corporation with R. Buckminster Fuller as its director and chief engineer. He also located and rented an abandoned automobile factory in Bridgeport, Connecticut, which was ideal for his project. The building had previously been used to manufacture the Locomobile, an automobile which, like so many others, succumbed to the Depression.[56]

Bucky arrived in Bridgeport with his money literally in his pocket on March 4, 1933, the day Franklin Roosevelt was inaugurated as President and immediately declared a bank moratorium, freezing all accounts. Fortunately, Pearson had delivered all the promised financing to Fuller in cash, so Bucky was one of the few people in town with a great deal of negotiable currency.[57] Ironically, Fuller was instantly transformed from an insignificant dreamer to a prestigious member of society. The news that he possessed cash and was opening a manufacturing plant spread quickly throughout Bridgeport, and within hours, he was inundated with people seeking jobs or attempting to sell him goods and services for his venture.[58]

Because of the Depression, Fuller was deluged with over one thousand applications for the twenty-eight jobs he had to fill, and he could pick from some of the area's finest craftsmen.[59] The men he eventually hired were predominantly unemployed mechanics, draftsmen, and other specialized craftsmen who possessed a great deal of expertise.

One of Fuller's most difficult tasks was selecting twenty-eight workers from the flood of men who applied.[60] In his interviewing process, Bucky was constantly confronted by the devastation of the Depression, and that experience greatly supported his growing concern for the needs of the "common man."

Because of the extreme poverty gripping the country, many of the applicants had families who had not eaten regularly for months or years. Since he was operating on a limited budget and a tight schedule, Fuller had to engage his staff quickly and begin production. To expedite that difficult situation, he instituted a policy of determining which of several equally qualified applicants for a position had the largest family that had not eaten for the longest time and hiring the man whose family appeared to be suffering the most.[61] The times were so harsh that most of the applicants broke down and cried when Fuller told them they had a job and could feed their families.[62]

The high quality of the applicants was exemplified by the two bodyworkers Fuller hired. In the booming economy prior to the crash of 1929, the Rolls Royce Company of Britain had begun to open a manufacturing plant in nearby Hartford, Connecticut. When the economy and the stock market crashed, Rolls Royce terminated that operation, leaving many of their fine craftsmen stranded in America. As a result of that closure, two of the first men Fuller interviewed and hired were premier Rolls Royce bodyworkers.[63] They were later joined by other exceptional craftsmen who had immigrated to America, including Italian machine-tool workers and Polish hammer workers.[64]

Since completing the single vehicle he planned to construct required a great many diverse tasks, Fuller hired extremely talented men who each had a primary as well as several secondary skills. That multidimensional characteristic produced a unique mix of workers who could execute tasks other than their specialties and could assist one another. The comprehensive nature of his crew also established a synergistic environment in which men with diverse ethnicity and talents supported one another in a spirit of camaraderie, interaction, and cooperation.[65] That spirit resulted in the emergence of many new ideas and possibilities which would not have been discovered by a team of specialists, each focusing on only one single task or aspect of the operation.

Although he strongly believed in the equality of each individual, the pragmatic Fuller also realized that, like a ship at sea, his project required the vision and guidance of a single leader. Since he was the one with the ideas, Fuller established himself as the ultimate authority. Even though he

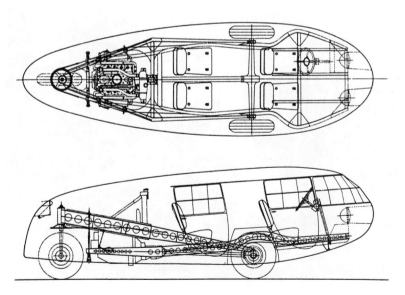

Fig. 7–4 Technical drawings of the Dymaxion Vehicle from above (top) and from the side (bottom).

provided his men with an autonomy which allowed them to work together in harmony and to bring forth ideas for the project, Bucky was practical enough to establish his command quickly.

On the first day of operation, he called a meeting of the entire crew to explain the details of the operation. During that meeting he also re-emphasized the temporary nature of their project and the fact that he intended to complete construction in six months. Bucky reminded every-one that he had only enough money to construct one prototype vehicle and again warned his men that he had no intention of generating a profit or commercially manufacturing vehicles.[66]

Since the men were grateful for any job and had already agreed to the terms of employment, Fuller's restatement of the situation met no chal-lenge, and everyone began working with great excitement. The entire staff clearly understood Bucky's plan was to build a single demonstration vehi-cle for testing the principles he had discovered and was continuing to unearth, and the men accepted his motivation and priorities. Accordingly, they worked harmoniously with Bucky.[67] The fact that he continued to

Fig. 7–5   Some of Fuller's team of workers examining the initial Vehicle's chassis. Starling Burgess is seated at the wheel while Bucky is seen standing at the far right.

drink and party extensively during the limited spare time available also contributed to his dual image as an eccentric genius who was also "one of the guys."

The project moved forward, generally adhering to a tight schedule for the first months. Because Fuller had carefully considered most aspects of few the design for years, the Vehicle began nearing completion within those months.[68] The crew worked on parts throughout the large building, but the Vehicle itself was assembled in a windowless room toward the back of the factory. That assembly room was kept constantly guarded so that no unauthorized personnel could view the Vehicle in a state of partial completion. Bucky did not feel that he was hiding anything from the public, but he simply did not want a partially completed Vehicle creating an image that the project was some harebrained scheme.[69]

After three months of continuous progress, the Vehicle was nearing completion and Bucky was preparing for a series of comprehensive tests when a serious problem emerged. As the workers began to realize that their jobs would soon end and their families might have to return to the misery of the Depression, each of them began to slow his pace, threatening the Vehicle's completion. Fuller understood that the situation was not a conspiracy and that his workers were not reacting out of malice or a conscious desire to condemn the project to failure. He also realized that he had to act expeditiously if he was to save the Dymaxion Vehicle from perpetual incompletion.[70]

To resolve the problem Bucky selected an extremely risky, unorthodox response consistent with his belief in changing environments instead of trying to change people. Rather than talk to his staff about their behavior, he simply initiated work on a second, more advanced vehicle, even though he had barely enough funding to complete the first vehicle. With Vehicle number two in production, the men felt their jobs were secure and quickly finished the first Vehicle so that it could be unveiled to the public on Bucky's birthday, July 12, 1933.[71]

That worker slowdown provided Bucky with several valuable lessons which helped him avoid similar problems in future years. In addition to learning a great deal about production technology, he also gained an enormous amount of insight into basic human behavior. Specifically, Bucky learned just how much the welfare of dependents affects the behavior of individuals, even when those individuals are motivated by the best of intentions. Later, when Bucky finally did close down the operation, that lesson was reinforced. His staff again opposed the imminent closure despite the fact that he had been straightforward with them from the start and clearly had no funds to continue the operation.

Only one of the men Fuller hired for the project did not participate in those actions and had not sought employment on the project. He was the illustrious aeronautical engineer Starling Burgess, whom Fuller persuaded to join the endeavor. Burgess was extraordinarily qualified and accepted the position of chief engineer primarily because he was fascinated by the unusual undertaking and was not involved in any major project at the time.[72]

As a prominent naval architect, Burgess had designed two sailboats which had won the America's Cup, and he was under commission to produce a Bermuda class sloop. Since building that boat was not a large project and Bucky loved sailing, he allowed Burgess to use the Dymaxion Vehicle factory for designing and constructing that boat. The other staff members had no problem with building the ship. In fact, they were thrilled to see another enterprise, which provided additional work, come into the plant, and duties were rearranged to accommodate boatbuilding in the Vehicle's production schedule.[73]

At that time, Burgess was more well known than Bucky. Many people were familiar with his outstanding record as a sailboat designer, but few realized he had also designed airplanes. In 1912, Burgess and the scientist James Dunn had developed the Burgess–Dunn plane for the United States Navy. That plane was the first delta-winged aircraft (i.e., an aircraft with swept-back wings) and the first plane in which pilots could safely remove their hands from the steering mechanism without losing control.[74] Because of his aviation experience, Burgess was particularly eager to work with Fuller in testing the ground-taxiing qualities of the Omni-Medium Transport Vehicle, which Bucky contended would one day fly.

Even though he was far more skeptical of the flight possibilities that Fuller asserted could one day be incorporated into the Vehicle, Burgess's expertise was invaluable to the project. Burgess's previous years of experience in building and testing full-sized prototype boats and airplanes provided the inexperienced Fuller with a great deal of confidence in almost every aspect of production, but Burgess's dynamic personality quickly began to clash with Bucky's similarly large ego.

One classic example of the conflicts which flared between the two men particularly irritated Bucky because he had to confront it every morning when he arrived at the plant. Soon after joining the project, Burgess instructed one of the workers to paint his name beside Bucky's on the sign in front of the building in equally large letters, as if they were partners. Although Bucky quietly tolerated that display in an attempt to maintain harmony throughout the project, it bothered him more than he was willing to admit.[75]

Fuller viewed Burgess, as well as everyone else working on the project, as his employees, an authoritarian position which appears to conflict with his antiauthoritarian stance. However, in an examination of this and other aspects of Fuller's life, the dual facets of Buckminster Fuller emerge. As an ordinary human being, he was susceptible to the same issues as everyone else, including a tendency to dominate others. Although he perceived his ideas as beneficial to all people and the control he asserted as being in the best interest of everyone, his need to dominate could also be seen as egotistical.

The egotistic side of Fuller was quite different from the personality he usually presented. That day-to-day facet was most obvious in Bucky's commitment to his overriding mission and in the talks he presented over the years. He seldom bragged or promoted himself during those talks; rather, he sought to share the urgency surrounding the issues and opportunities which confront all humanity. Thus, although the clash of egos between the personalities of Burgess and Fuller was inevitable, Bucky sublimated that side of himself for the good of the project and the benefits to humankind that he believed it would produce.

Starling Burgess had been raised within a heritage of talent and genius. The Burgess family had a proven record of successful design. Starling's father had designed three sailing ships that each won an America's Cup. His brother was the chief mathematician for the United States Navy's lighter-than-air-structures design program, which produced zeppelin blimps. Starling was a skilled mathematician who loved sailing and designing racing boats for wealthy yachting clients.[76]

Since only the wealthy could afford yacht racing, Burgess was accustomed to demanding the finest materials and the highest quality workmanship on the craft he designed. He required the same high standards for the Dymaxion Vehicle. Both Fuller and Burgess agreed that pioneer inventors and designers, such as themselves, had a responsibility to fabricate prototype artifacts with such quality, precision, and attention to detail that the value of the invention would not be undermined by inferior materials or workmanship. With such superior quality, an invention could be judged only by its performance and not by something relatively insignificant, such as a defect in a part, which could create a minor failure that would shift the spotlight from the benefits of a total project.[77] That insistence on quality resulted in all the Dymaxion Vehicles being built of the finest materials available and being engineered with the same precision used in aircraft and fine racing yachts. Thus, shoddy workmanship did not detract from Fuller's intended objectives or radical design.

The Dymaxion Vehicle project provided Fuller with his first large-scale opportunity to test his conviction that mirroring Nature's designs would be the most effective mode of operation. Bucky had found that the creatures which moved with maximum efficiency had developed on Earth over thousands of years of evolution, and he felt that such successfully evolved designs merited imitation in human engineering projects. He was especially interested in the natural techniques used for the propulsion and the steering techniques of birds and fish.

Observing those creatures, Fuller realized that positioning a steering mechanism—be it the tail of a bird or the fin of a fish—in the front of an animal was ineffective and was never seen in Nature.[78] He reasoned that such a configuration would be analogous to attempting to steer a huge oceanliner by pushing on its front end with a small boat. Similarly, Fuller believed that the front-wheel steering mechanisms, which were fundamental components of automobiles, were just as ineffective. Because he chose to utilize techniques which worked well rather than ones which were popular, Fuller's Vehicle was designed with front-wheel propulsion and a single rear steering wheel, mirroring what he had observed in birds and fish.[79]

The Dymaxion Vehicle's steering and propulsion technology was so revolutionary that even though its conception and development occurred in the early 1930s, only one commercial vehicle has since utilized those innovations together. That vehicle is the large three-wheeled street cleaner which is required to maneuver sharply while cleaning gutters and curbs.[80]

Fuller discovered that front-wheel-drive, pulling propulsion provided an enormous advantage over the commonplace rear-wheel-drive pushing, and he illustrated that advantage using an ordinary wheelbarrow. When passing over obstacles, a wheelbarrow pushed from behind can be dangerous, as its front wheel strikes an obstacle and it lurches backward into the stomach of the person pushing. However, if that same wheelbarrow is turned around and pulled, a person actually lifts it over bumps as it is pulled, and the problem disappears.

Once the simple beauty of that shift became apparent to Bucky, he immediately reconsidered his ideas and designed his Vehicle to include a single, rear steering tire pulled by two front tires so that it would embody a more natural steering pattern as well as the propulsion strategy of the pulled wheelbarrow. In subsequent testing, Fuller also established that such an arrangement resulted in superior traction and reduced skidding.[81]

At the time, most cars employed front-wheel steering and rear-wheel propulsion, which caused the front wheels to be pushed into the ground and produced a great deal of skidding. The characteristic was, and still is, evident in high-speed auto races, where the drivers invariably slide through turns with little steering control. In such instances, the car's front wheels tend to be forced into the ground as the driver maneuvers through turns by allowing the rear wheels to continue spinning rapidly. That act generally causes the rear of the car to slide in the desired direction.

It also results in most auto racing being chiefly a matter of controlling a car's angle of skid. Managing skids for hours at a time is a difficult assignment because once a car begins slipping, the driver has lost most control and a significant part of the car is no longer connected to the ground. In fact, Fuller found that during such skidding a car could be considered somewhat of a flying machine.[82] Only when the car returns to the straightaway does the power of the rear wheels fully engage in propelling it forward again and can the driver regain control as the car again reverts to operating as a land vehicle.

By observing such maneuvers, Fuller was convinced that his final design of a three-wheel vehicle with front-wheel drive and a single rear steering tire provided safety and maximum effectiveness using a minimum of resources. His strategy of always seeking maximum efficiency is also evident in the three-wheel configuration. Bucky understood that three was the minimum number of wheels or contact points required for stability.[83]

Certainly, four wheels provided somewhat more stability, but through experimentation he discovered that the added stability was offset by excess weight and loss of maneuverability. He was also pleased to find that his belief in the fundamental nature of the triangle and triangulation could be incorporated into his Vehicle. Most of Fuller's theories did prove correct, and the final product of his team's effort was a Vehicle which Fuller often contended was more maneuverable and responsive than any car he had ever driven.

The initial public showing of the Dymaxion Vehicle on July 12, 1933, was a great success. Over three thousand people crowded the test track next to the factory to watch as it was put through its paces. That showing catapulted both Fuller and the Vehicle into the public eye.[84] He and his amazing car were prominently featured in newspapers and magazines across the country, while dignitaries began to visit the factory for a look at the future of transportation. Most everyone wanted to drive the Dymaxion Vehicle, and those allowed to were amazed at its performance. Since the steering configuration permitted the back end to move at right angles to the Vehicle's body, it could be parked in a space only a few inches longer than itself.[85]

Fuller often executed that maneuver, to the amazement of onlookers. He would drive the Dymaxion Vehicle into a parking space front end first, maneuvering the rounded nose to within an inch of the car in front of the space. That feat could be performed safely because the driver sat forward of the front wheels, providing an excellent view of everything in front.[86] Modern vans have now embodied similar configurations, but in 1933, it was an unusual arrangement. Once Fuller had positioned the front end of the Vehicle snugly within an inch of the other car, he would aim the rear wheel directly toward the curb, swinging the rear end in perpendicular to the curb.[87]

Whenever the vehicle was driven on public streets, it drew a crowd, but the parking maneuver always captured the most public attention. As a result of that attention, it was often featured in movie newsreels. A newsreel producer would establish a parking space only three inches longer than the Dymaxion Vehicle and call for action. Fuller would then drive down the street very rapidly and pull directly into the space with perfect control and no hesitancy. As long as he was certain he had the excess three inches, Bucky would quickly turn the rear wheel and complete the spectacle with great confidence.[88]

During numerous trial runs, Fuller and his team discovered a great deal about front-wheel drive and the astounding capabilities of the Vehicle they had created. They determined that constant acceleration prevented skids and allowed the execution of very sharp turns. Bucky also found that the Vehicle could perform many unusual maneuvers, and as he practiced them, he continued learning more about the principles behind his design.[89] For example, he found he could slow the Vehicle down to fifteen miles per hour and execute a 180-degree turn while the inside front wheel made a circle only one foot in diameter. Bucky could literally turn the Vehicle and be traveling in the opposite direction within a matter of seconds.[90]

No other car on the road could perform such a maneuver, and Fuller sometimes exhibited it while curious police officers were ogling at the strange car in amazement. His innocent stunt soon backfired, however, as word of the turn spread among the area's police departments. Once officers learned about the amazing car, he was continually being stopped by the police and "requested" to perform the magical 180-degree turn.[91]

In one such instance, Bucky was driving through New York with a group of editors from the *New Yorker* and *Fortune* when he was stopped by a traffic officer at Fifty-seventh Street and Fifth Avenue. While the officer demanded to know exactly what Bucky was driving, Bucky cleverly turned the steering and slowly rotated the entire car around the stationary police officer. That maneuver also backfired because, upon witnessing the traffic-stopping event one block away, the next officer demanded a similar demonstration. The incident was then repeated at almost every successive corner, causing Bucky and his guests to spend nearly an hour traveling one mile down Fifth Avenue.[92]

CHAPTER 8

A Car Is Born

 At nineteen feet in length, the Dymaxion Vehicle was comparable to a Cadillac of the day but longer than an average modern car (fifteen to seventeen feet). The Dymaxion Vehicle was, however, much faster and more efficient than any other car of its era or most modern automobiles, carrying a maximum of eleven passengers comfortably and averaging nearly 30 miles per gallon of gasoline while attaining speeds in excess of 120 miles per hour.[1] Fuller's team achieved those amazing performance records using stock production parts wherever possible and engines from major manufacturers. Most of those parts were, nonetheless, modified for Bucky's purposes.[2]

Only when a component was so radical that no stock component had ever been fabricated did the team forge a part from scratch. That process was evident in the designing and manufacturing of the unique three-structure, hinged chassis with its large holes to reduce weight.[3] During that era, only racing cars utilized such a sophisticated technology to lessen their weight, and the hinged components of Fuller's Vehicle were matchless.

The effectiveness of the Dymaxion crew's work and Fuller's design was exemplified by the Vehicle's speed. The most powerful engine ever installed in it was a modified Ford V-8, which produced only ninety horsepower. Yet, to reach the speed that the Vehicle could attain, a conventional 1933 sedan would have required three hundred horsepower of force.[4]

Because the Vehicle received an enormous amount of publicity, huge crowds formed when it was driven on public streets or displayed. Bucky was also frequently invited to demonstrate it, and he accepted whenever possible.[5] At one such exhibition, he opened the season of a stadium in the Bronx, New York, where midget cars raced. The track's owners invited Bucky to drive a few laps around the track during intermission, and he was delighted to have the luxury of an entire race track to himself. Being a gracious host who loved to show off the wonders of his invention, Bucky returned the favor by inviting several of the stadium officials to accompany him while he circled their track.

Everyone in the Vehicle sensed that Fuller was driving at his favorite pace: fast; however, because of the extremely comfortable ride, they did not feel they were traveling at an unsafe speed. In fact, his passengers commented on how well the Dymaxion Vehicle handled and on the smoothness of the ride. Nonetheless, when the Vehicle came to a stop in front of the grandstand, all the riders were astonished to learn that they had broken the track record by nearly 50 percent.[6]

That accomplishment amazed the officials because their rear-wheel-drive racing cars were always skidding around the track, yet the Dymaxion Vehicle broke the speed record without skidding once. When Fuller explained that the lack of sliding was a result of the Vehicle's advanced design and front-wheel drive, the officials became even more interested in his ideas. The fans were also amazed because, in establishing the new record, everyone in the Vehicle looked and acted like tourists taking a spin in the country rather than serious racers.[7]

Like many other aspects of its design, the Vehicle's steering was also unique. Rather than using conventional rods, internal connections were made with stainless steel cables running from the driver's steering wheel in the front to a post which turned the single steering tire in the rear. Those cables passed through ball-bearing guides built to extremely precise standards. Bucky's and Burgess's sailing experience proved invaluable in devising and fabricating those and other radically experimental components, which were adaptations of the best racing sailboat technology and materials. Burgess had invented similar parts for sailing ships, and his expertise helped Fuller to equip the Dymaxion Vehicle with some of the most remarkably elegant parts ever seen on an automobile.[8]

Among the less obvious innovations that Fuller and his staff introduced were a rearview periscope on the roof to compensate for the lack of a rear window and to provide the driver with a broad perspective of everything behind him. The rear engine compartment was also innovative, providing easy access to every section of the engine as well as the Vehicle's other mechanical components. In fact, when the engine compartment door was opened, the engine appeared to be sitting up on a workbench, ready to be examined and easily repaired.[9]

As with any experimental project, problems had to be worked out, and the entire team was perpetually improving and upgrading the Vehicle. Initially, maneuvering it through strong crosswinds was particularly difficult because the streamlined design caused the Vehicle to be prone to turning into the wind. Although the steering system was extremely tight and would maintain the operator's chosen course, that natural tendency of turning into the wind transferred the impact of wind forces onto other areas, such as the tires. At times that pressure created such a large distortion in the tires that the Vehicle was extremely difficult to steer. In fact, on a windy day, driving the first Dymaxion Vehicle was said to be like flying a light plane.[10] It was during one such windy test drive that Fuller realized the Dymaxion Vehicle was not an invention which could be made available to the general public without significant improvements.[11]

Even though he and his associates continued instituting modifications which diminished that and other problems, Fuller did not completely resolve the issue until years later. Thus, in an effort to prevent accidents, only individuals trained by Fuller or one of his staff were allowed to drive the Vehicle for extensive periods or during inclement weather.[12]

Despite its flaws, once the prototype was successfully demonstrated, the Dymaxion Vehicle was acknowledged as an amazing automobile. Still, most people did not believe that it could one day be transformed into a flying machine. Most conspicuous among the skeptics was Starling Burgess. He believed that Bucky's jet-stilt-propulsion idea could create enough lift for hovering at low altitudes above relatively flat surfaces such as water, but that such jets would never provide the constant power required for high-altitude flight.[13]

With each new ground test, Fuller, on the other hand, became more and more certain that the vehicle could be modified to fly. He felt that future technology would provide the jet flight capabilities he had envisioned, but he realized that the requisite technology would not be available for at least twenty-five years and would probably be developed at great expense by governments for military deployment.[14]

Since that time, governments have produced the jet engines Bucky envisioned during the late 1920s. The United States military even fabricated small, portable rocket jets which can be attached to the wings of planes to assist in takeoff. Those jets operate as stilts by providing additional lift, and they have been instrumental in helping stranded planes to take off from harsh polar regions. Further development has resulted in even smaller jet packs which individuals wear on their backs and fly using hand-held controls.

The United States Marine Corps also conducted experimentation that resulted in a prototype jet flying transport similar to the Dymaxion Vehicle. Unfortunately, they made the mistake of not focusing the thrust of their jets to converge above the center of gravity of the vehicle, and it tended to roll over. Although that was a minor issue that Fuller had anticipated in 1932, it was a major stumbling block for the Marines, and they terminated their project.

Nevertheless, their work and the achievements of other individuals and institutions clearly indicate that Fuller's concept of wingless jet flight is workable. In fact, one futuristic portrait of a flying transport predicted to replace the automobile in the next century hangs at the Air and Space Museum in Washington, DC. That painting depicts a flying vehicle almost identical to Fuller's Dymaxion Vehicle except that it is to be propelled by small jet engines on either side, just as Bucky prophesied.

When the first Dymaxion Vehicle was completed, Fuller hosted a huge celebration attended by numerous dignitaries. Among the guests were the distinguished aviators Amelia Earhart and Sir Kingsford Smith, as well as many of Fuller's artist friends from New York, including Diego Rivera. As a result of the publicity he had received, Earhart and Fuller had met and become friends, and she respected the Dymaxion Vehicle as well as his idea of transforming it into an Omni-Medium transport. Following the unveiling and a test drive, Earhart felt confident enough about the Vehicle's quality and significance to request a favor of Bucky.[15]

She asked to use the Dymaxion Vehicle as her official car during a celebration in which she was to receive the Gold Medal of the National Geographic Society. Although that request might appear audacious, Earhart's motivation was completely supportive of Bucky's work. She sensed that by utilizing his Vehicle at her official functions in Washington, DC, she would generate more interest in and support for Fuller's ideas and work. After listening to the scheme, Bucky agreed, and the Vehicle was soon on its way to Washington.[16]

Earhart's strategy was a resounding success. She had been invited to stay at the White House by Mrs. Roosevelt, and the Dymaxion Vehicle was again showcased in the national spotlight for several days. Fuller was extremely grateful for her support, and as they grew to be even closer

Fig. 8–1 The 1933 public unveiling of the first Vehicle. The car in the background is a contemporary Franklin. The man holding his coat and standing between the Vehicle's doors on the far side is the artist Diego Rivera.

friends, he found that despite her fame she was an extremely humble person who perpetually supported worthwhile causes and people.[17]

Although the Washington, DC spectacle catapulted both Fuller and the Vehicle into greater prominence, its most remembered exposure occurred at the Chicago World's Fair of 1933 and 1934. In 1933, the Fair's producers were eager to exhibit a full-size prototype of the Dymaxion House, which still retained its exotic "House-on-a-Pole" design. Because his ideas were unique, most of the components for such an endeavor would have had to be fabricated from scratch, and Fuller calculated the cost of building the House to be far more than he could afford or garner from investors.

When confronted by the economic reality that Bucky presented, the Fair's producers also declined to finance that construction and suggested he display a scale model of the Dymaxion House. At that time, Fuller had just spent more than a year building working prototypes of the Dymaxion Vehicle and did not feel that simply displaying a small model would be worthwhile or would efficiently utilize his time. He was also obligated to continue working at the Bridgeport plant, where his crew was nearing completion on the second Dymaxion Vehicle, which was progressing on a financial shoestring.[18]

The Fair's producers were so certain that even a model of Bucky's House would support their primary goal of attracting more people that they initiated fabrication of a scale model for exhibition. With a model of his Dymaxion House on display, Fuller could not help becoming involved with the Fair and began speaking there whenever possible.[19]

At the time, a group of Englishmen were considering purchase of the second Dymaxion Vehicle, which was on the verge of being scrapped for lack of funds. The English group dispatched a British aviation expert, Colonel William Francis Forbes-Sempill, to inspect the original Vehicle. Forbes-Sempill traveled from England on the *Graf Zeppelin*'s special voyage to the Chicago World's Fair, and he was accompanied by another expert, the Air Minister of France.[20]

Bucky was asked to provide the original Vehicle for their use and examination at the World's Fair, and he quickly began making preparations. By that time the first Vehicle had been sold to Gulf Oil and was managed by Al Williams, a man who had been the Navy's premier test pilot and had retired to assume the position of sales manager for Gulf.[21] Williams himself vigorously supported Fuller's work and often acknowledged the Dymaxion Vehicle as "aviation's greatest contribution to the automobile industry."[22]

One of Gulf's best-selling products was aviation fuel, and the company produced air shows to promote its fuel. Williams convinced the management at Gulf that purchasing the Dymaxion Vehicle and featuring it as the official race car at all Gulf's air shows would greatly enhance their promotions. The Dymaxion Vehicle was a solid investment for Gulf management, increasing attendance at their events, just as had been true wherever it was displayed.[23]

Learning of the British expert's upcoming visit, Bucky telephoned his friend Al Williams, who immediately assigned a former race-car driver named Turner to drive the Vehicle to Chicago for inspection. Turner arrived in Chicago just as the *Graf Zeppelin* was making its brief stop at the World's Fair to drop off passengers prior to departing for the nearest mooring facilities in Akron, Ohio.[24] He then acted as chauffeur and guide for the two men during their visit.[25]

Their World's Fair tour was an enormous success, and the two men enjoyed traveling in and inspecting the Dymaxion Vehicle. They also found themselves becoming part of the spectacle which the Dymaxion Vehicle created wherever it appeared. Then, a few days after their arrival, an unexpected message arrived late one evening informing all passengers that the *Graf Zeppelin* was returning to England immediately, and that those who wished to return on that voyage should be prepared to leave early the following morning. Thus, Turner and the Vehicle were summoned for a final quick trip to the airport at seven o'clock the following morning.[26]

The next information Fuller received about the Dymaxion Vehicle was in the following day's newspapers. Typical headlines read, "Freak Car Rolls Over, Kills Race Driver, Injures Dignitaries" and "Three-Wheeled Car Kills Driver." *The New York Times* reported, "The machine skidded, turned turtle and rolled over several times. Police say it apparently struck a 'wave' in the road."[27]

At the time, Bucky was in Bridgeport working on the second Dymaxion Vehicle and immediately called a trusted friend and fellow engineer in Chicago, asking the man to begin investigating the accident. Bucky also flew to Chicago to conduct his own inquiry. There he found that the accident had occurred almost directly in front of the main gate of the World's Fair, and that the Dymaxion Vehicle had been quickly removed from the scene and hidden at a remote storage area.[28] Although he was intimately involved with the Vehicle, none of the authorities cooperated with him in even locating the Vehicle. When he finally did find and meticulously examine it with the help of other engineers, Fuller could find no systems or parts which could have malfunctioned to cause the accident.[29] Accordingly, he continued his investigation.

In reading accident reports and speaking with eyewitnesses, he discovered that the Dymaxion Vehicle had rolled over, puncturing the button-down convertible top and killing Turner, who was fastened in his seat by a seat belt. The French Air Minister had not been wearing his seat belt and was thrown through the lacerated top, landing on his feet with barely a scratch. Colonel Forbes-Sempill had been sitting next to the driver and was taken to the hospital in critical condition. When Forbes-Semphill's condition improved, Fuller visited him, hoping that the gentleman could offer some insight into what had occurred. During one of those visits Bucky was asked to wait while the Colonel received a call wishing him a speedy recovery from his good friend, the King of England.[30]

As the accident quickly escalated into a major news item and an international scandal, Fuller felt increasingly guilty that his Vehicle had supposedly killed one person and seriously injured another. Forbes-Sempill did eventually recover so that he could testify at the coroner's inquest, which was required because someone had died.[31] That inquest, however, could not be conducted until the Colonel regained his health two months after the accident, and by that time, the incident was no longer a sensational news item.[32]

Forbes-Sempill and the French Air Minister both testified that the Dymaxion Vehicle had been traveling down the highway when another driver pulled alongside them so his passengers could get a closer look at the strange Vehicle. During Fuller's early public test drives, he had frequently encountered such spectators and had found them to be the most hazardous aspect of driving the Vehicle on public roads.[33] He was often confronted by the frightening situation of cars pulling alongside him, with the driver eyeing the Dymaxion Vehicle rather than the road ahead.

Because Turner was a professional driver and realized that his Vehicle was much faster than anything else on the road, he had accelerated to escape the gawkers. The other driver simply followed alongside, continuing his harassment. When Turner changed lanes, the other car continued its pursuit. Finally, the other driver moved so close that his car hit the tail of the Dymaxion Vehicle, causing Turner to lose control and the Vehicle to roll over. As fate would have it, the other car was owned by an influential South Chicago parks commissioner, and his car had been illegally and expeditiously removed from the scene before any reporters arrived. During the coroner's inquest, the fact that a collision and not a freak rollover had occurred was discovered and exposed.[34]

The coroner ruled mutual responsibility for the accident, and no fault was assigned to either vehicle or driver. He also declared that the accident was not the result of the Dymaxion Vehicle's malfunctioning or being poorly designed. Still, two months had passed, and the Vehicle had received an enormous amount of negative publicity during that period. Since the resulting assignment of no fault was not a sensational story like the accident itself, the final resolution received very little media coverage.[35]

Al Williams spoke candidly to Fuller and expressed his feeling that Bucky had an obligation to inform the general public that the Dymaxion

Vehicle was soundly designed and not at fault in the accident. Williams suggested that since the media were no longer interested in the story, the best method of apprising the public of the truth was the construction of another Vehicle which would generate sufficient positive attention to eliminate the negative image. Thus, public relations became the predominant motivation for building the third Dymaxion Vehicle.[36]

The following year, Bucky did accept a formal invitation to exhibit and demonstrate the latest Dymaxion Vehicle at the 1934 edition of the Chicago World's Fair. Because of its maneuverability, one of his demonstrations consisted of having the Vehicle "dance" a zigzagging maneuver, which was referred to as a "waltz," down the main street of the Fair. Twice each day, Fuller would also drive the Vehicle through the entire fairgrounds, completing his route at a pageant entitled "The Wings of a Century," which portrayed Chicago's history as a great railroad center.[37] During the program, actual railroad locomotives and trains passed directly in front of the grandstand on special tracks, but the grand finale of the show featured one small car. Bucky would enter the arena driving the Dymaxion Vehicle at high speed from one side of the grandstand, quickly slow down to fifteen miles per hour, and execute his famous 180-degree turn.[38]

The stunt was particularly thrilling because the audience had been primed by news stories of the previous year's accident. When they saw the suspect Dymaxion Car spinning around, most of the crowd was certain that it would flip over as it had during the accident. Having performed the maneuver hundreds of times, Fuller knew it to be perfectly safe, but he was willing to give the folks a show in order to demonstrate the safety and effectiveness of his unique Vehicle and to eliminate some of the earlier negative publicity. In fact, the Vehicle's steering was so tight and exacting that Fuller could actually release the steering wheel and the Vehicle would continue the turn without his assistance.[39]

During his second year at the World's Fair, the Vehicle regained its reputation as a masterpiece of modern technology and aroused the interest of some influential individuals, including Henry Ford. Shortly after Ford witnessed Bucky's performance, he called Fuller at the Bridgeport plant to extend a 70 percent discount on anything Ford Motor Company manufactured which could be used on the project. Ford's offer was an enormous aid to the financially burdened Fuller, and his team immediately began using Ford parts and engines whenever possible.[40]

Fuller's original intention had been to build one prototype Dymaxion

Vehicle for testing and then shut down the operation. He had initiated production of the second Vehicle so his workers would finish the first one and because the English group was interested in purchasing it. Following the 1933 accident, no one was interested in purchasing the completed second Vehicle, and Fuller was quickly running out of money.[41] Still, he began work on the third Vehicle to remove the negative stigma spawned by the accident. Because of low funds and the workers' fear of losing their jobs, that proved to be a very difficult undertaking.

Only another unrelated event kept production of that third Vehicle on schedule. When Bucky's mother died that year, her estate was divided among her children, and he received a small inheritance, which he used to cover production costs for the third Vehicle. Each time he was nearly out of cash, Fuller would hastily sell some of the stocks or bonds he had inherited.[42] His brother and sisters were extremely cooperative in supporting those transactions, but the bank, which was coadministrator of his mother's will, invariably caused frustrating delays.[43]

As with almost all his experiences, Fuller attempted to learn from his difficult dealings with that bank, and in those dealings his growing mistrust of profit-oriented corporations was again bolstered. He believed that banks were conspiring to manipulate prices, thereby increasing their profit from the sale of securities at the expense of their customers.

As Fuller perceived that deception and his victimization by banks, such manipulation occurred because banks were continuously receiving large orders to buy and sell stocks and bonds from trust funds. The management of several banks would then manipulate prices by agreeing to withhold those large orders until a predetermined moment, when the price would rise or fall because of their combined transactions. Those banks would also assemble the buy and sell orders from small customers for specific stocks and bonds, and would use those collected orders to increase the strength of their position when they made the large transactions. The action of those bankers resulted in considerable delays when Fuller needed to sell small amounts of stock.[44] Such delays combined with the Depression to create major production problems, and in one instance, Fuller's financial situation was so tenuous that the local sheriff visited him at the plant.[45]

Bucky was in debt to several Bridgeport merchants, and the sheriff warned him that his creditors believed he might leave town or close the plant without satisfying his debts. Such behavior had become routine during the Depression, and businesspeople quickly demanded payment on outstanding debts at the least indication of possible failure. Having no intention of defrauding his creditors, Fuller was very cooperative and allowed the sheriff to examine his financial records so that he could vouch for the fact that funds were forthcoming and bills would be paid.[46]

At that time, the third Dymaxion Vehicle was nearing completion, and Bucky was attempting to gradually lay off his staff and shut down the operation. The workers, however, were not cooperating with that process and were again doing everything they could to sustain production and retain their jobs. Fuller also owned all the inexpensively purchased machinery used in production, and suddenly the plant windows were being broken and the machinery was being vandalized. Bucky had only leased the building, but he had no idea what to do with the equipment, which, because of the Depression, no one was interested in purchasing at any price.[47]

Having taken the initiative to speak with Fuller's staff before visiting him, the sheriff advised Bucky that his crew truly feared for their jobs and were not going to allow the operation to be shut down. He then suggested that the only way to close the operation gracefully was to have him conduct an official sheriff's auction, liquidating all assets and closing the plant. After carefully considering all his options, Bucky agreed to the plan. The sheriff's sale allowed Fuller to complete the project with dignity, having produced three Dymaxion Vehicles, and without declaring bankruptcy. Although that was not the way Bucky had envisioned completion of the project, it was the best alternative available at that moment, and he left Bridgeport with no debt and having learned a great many lessons.[48]

With his work on the Dymaxion Vehicle, Fuller bolstered his belief that cars represent much more than a mode of transportation. He regarded cars, like so much of his environment, as tools. His experience in closing the Dymaxion Vehicle plant caused him to reevaluate his use of all types of tools in relationship to his strategy of acting as an individual and taking initiatives on behalf of all humanity.[49]

Although he was always true to the essence of that strategy, Bucky adjusted it whenever he encountered something not working at maximum efficiency. One such lesson occurred with the Dymaxion Vehicle. After completing production, Fuller realized that purchased equipment was a burden. Accordingly, he resolved to limit his possessions and to avoid owning equipment whenever possible.[50]

During the early 1920s, his company owned most of the machinery in the five Stockade Systems construction factories, but that machinery was simply transferred when the entire operation was sold to another corporation.[51] That experience did, however, forge the foundation of

Fuller's understanding of the complexities and problems inherent in ownership, and that understanding was intensified with the closure of Dymaxion Vehicle production.[52]

Following the Dymaxion Vehicle project, Fuller's first choice was to lease tools and equipment whenever possible. That revised strategy provided him with greater freedom, and he was, consequently, able to shift from one field to another more expeditiously and with greater ease. Following that change in strategy, Bucky was never forced to continue work on a project which he felt had been efficiently taken as far as possible simply because he owned the tools or the facility to produce a particular item. The peril of ownership was a major lesson of the Dymaxion Vehicle operation, a lesson which supported Bucky's success in other enterprises over the years.[53]

Once the Dymaxion plant had closed, a great many people called to console Fuller. When the first of those friends offered their regrets that his automobile operation had failed, Bucky was truly surprised, and he questioned them. When a friend would politely allude to Fuller's inability to mass-produce the car commercially, Bucky would once again explain that his intention had been to develop and construct a prototype Vehicle for testing particular principles and not to enter the commercial automobile business. In his inimitable, candid style, which often shielded some disappointment at not having succeeded in the public eye, he would further recount that, regardless of other people's perceptions, he had succeeded in his mission.[54]

In truth, Fuller had, in a few short years, achieved much more than most individuals could have hoped to accomplish. He had built his Vehicle, and in testing principles with the Dymaxion Vehicle, he had obtained insight into numerous areas, including manufacturing, business, production materials, and aviation. He had also contributed to the whole of humanity by developing and promoting many discoveries which advanced automobile engineering and manufacturing.

For example, shortly after Fuller's work, most auto manufacturers began streamlining their cars, and many years later, American auto manufacturers began implementing front-wheel drive.[55] Fuller felt that those innovations, like all ideas, required a gestation period between the time they were first considered and their commercial implementation, and, by bringing innovations into public awareness, the Dymaxion Vehicle had initiated that gestation process.[56]

*

Although Fuller maintained the idealistic stance that he had never wanted to commercially produce the Dymaxion Car, other sentiments were conspicuous. Bucky did, in fact, receive and respond to overtures from every major American automobile manufacturer. In working with companies which claimed interest in mass production of his Vehicle, Fuller was able to acquire a great deal of insight into the automobile industry, and that insight once again reinforced his belief that businesses were interested in profit at any expense.[57]

The most significant of his interactions were with Chrysler, Packard, and Studebaker. The investment group of Hayden Stone owned Curtiss-Wright Aeronautical as well as both the Studebaker and the Pierce Arrow automobile companies, but Pierce Arrow had succumbed to the Depression, leaving its production facilities in Buffalo, New York, idle. Hayden Stone management believed those facilities would be ideal for production of the Dymaxion Car, which they wanted to rename the Curtiss-Wright Dymaxion Car. Their plan was to use Curtiss-Wright's knowledge of advanced aircraft technology in conjunction with Fuller's designs to produce the most advanced car on the road.[58]

The management at Curtiss-Wright felt that the aircraft industry had adapted and utilized automobile production technology for many years, and that the Dymaxion Car provided a unique opportunity for the automobile industry to exploit the latest aircraft-industry technology. Their plan was to continuously upgrade the Car so it was always revolutionary and to market it as the premier model of the Studebaker line.[59]

Fuller had almost reached an agreement permitting Hayden Stone to produce the Dymaxion Car commercially when a seemingly unconnected and unexpected issue resulted in the termination of their plans. During the early years of aviation, Curtiss-Wright had felt that the creation of airports would stimulate sales of the planes they were manufacturing, and the company began producing both airplanes and airports at the same time. As a result of that effort, at the onset of the Depression, Curtiss-Wright owned small airports across the country. President Roosevelt's New Deal administrators felt that upgrading those airports would be an excellent job-producing government project, and the United States government entered into a tentative agreement to purchase all of Curtiss-Wright's airports as one facet of its effort to stimulate the economy.

Curtiss-Wright planned to invest the capital it had received from that sale in the development and production of the Curtiss-Wright Dymaxion Car, but for an unknown reason, the negotiations became deadlocked, the final agreement was never approved, and Curtiss-Wright terminated the

Dymaxion Car project due to lack of funds. At the time, Curtiss-Wright's justification for terminating the proposed enterprise appeared reasonable and in no way connected to the Dymaxion Car itself. However, when Walter Chrysler explained to Bucky why the Dymaxion Car could never be commercially manufactured, Fuller felt those deadlocked negotiations were probably more than mere coincidence.[60]

Walter Chrysler, founder and president of the Chrysler Corporation as well as an elder statesman of the automobile industry, was a man with interests similar to Fuller's. He was captivated by modern, streamlined cars, and following the Dymaxion Car's successful demonstrations, he too expressed interest in commercially manufacturing it. Chrysler was so fascinated with modern innovation that he had initiated the development of Chrysler Corporation's somewhat streamlined Airflow, which was introduced just one year after Fuller first exhibited the Dymaxion Vehicle. Although no information had been shared by the designers of the two vehicles, detailed comparisons of the Dymaxion Vehicle and the Airflow soon followed.

The most publicized encounter between the two cars occurred in the heart of New York City in 1934. At that time, automobile shows were important events where new models were vigorously sold and promoted. Chrysler Corporation planned to unveil the Airflow at one of the country's largest auto shows, scheduled for the Grand Central Palace in New York City. Because the company was mounting an enormous advertising campaign for the Airflow, they leased the most expensive display position at the show.[61]

With the Depression continuing to dominate the American economy, many smaller manufacturers still could not even afford to rent a space for displaying their automobiles, and the show's producers offered Fuller free exhibition space for his Dymaxion Vehicle. Since he had been provided with display areas and had successfully exhibited the Dymaxion Vehicle at similar events, the promoters were certain that Bucky's Car would not only fill one of their empty booths but would also help attract larger crowds. When the management at Chrysler Corporation learned that Fuller's streamlined Vehicle would be exhibited at no cost, they were furious and threatened to withdraw from the show, leaving the producers with no choice but to withdraw their invitation to Fuller.[62]

Having encountered similar treatment on other occasions, Bucky understood their predicament and simply continued with his work. The event did, however, generate several stories in the New York newspapers, and as fate would have it, the police commissioner of New York City, General Ryan, read about the incident. General Ryan had seen and been impressed with the Dymaxion Vehicle at another exhibit and invited Fuller to park his Vehicle in front of the Grand Central Palace during the show.[63] Thus, with the approval of New York City's police commissioner, Bucky and his Vehicle were prominently displayed, at no charge to spectators, directly in front of the auto show, where they literally stole the spotlight from the exhibition.[64] Throughout the show's entire run, more people were consistently gathered around the Dymaxion Vehicle than at any exhibit inside.[65]

Upon hearing of that incident, Walter Chrysler personally invited Bucky to drive the Dymaxion Vehicle out to his estate on Long Island for complete examination.[66] Bucky accepted Chrysler's invitation, and after a test ride and a detailed inspection, Chrysler confided that the Dymaxion Vehicle was the type of car he had attempted to manufacture years earlier.[67] Chrysler went on to explain that he had envisioned producing a technologically advanced automobile, but that by the time the gigantic bureaucracy at Chrysler Corporation was finished with his ideas, the car barely resembled his initial concept. That car was, however, eventually produced and marketed as the Airflow. Although the Airflow was an exceptional

Fig. 8–2 The 1934 Chrysler Airflow.

vehicle for the time, it was so removed from his original, revolutionary dream that even Walter Chrysler conceded he did not like it.[68]

Mr. Chrysler was delighted to discover that Bucky had produced a car so advanced yet so similar to the one he had attempted to manufacture. During their conversations, he explained the excessive corporate system of checks and balances to Bucky, emphasizing how that system obstructed creative inventors.

Chrysler recounted that even as founder and chief executive he could not persuade his corporation to work on an idea which could not be guaranteed financially successful. He believed that radical concepts, such as the Dymaxion Vehicle, were much too risky for corporate executives, whose primary objective was maintaining a profitable organization rather than producing the best possible product. Chrysler also indicated he felt that if Bucky avoided the burdens of a huge organization such as his, Fuller could actually produce the Dymaxion Car commercially. In fact, Chrysler was so impressed with Fuller's Vehicle and certain of its potential that, following their first meeting, he often asked Bucky to take members of his staff and board of directors for rides in the Dymaxion Vehicle so they could experience the type of car he had attempted to produce.[69]

In subsequent meetings and conversations, Chrysler also asked Bucky to work with him on cost comparisons between the two vehicles. They had both produced prototype cars during the same era, and Chrysler wanted to determine cost and effectiveness differentials between an individual developing a prototype and a large corporation performing that same task. The two men found that Chrysler Corporation had required three times longer and four times more money to produce the Airflow prototype than Fuller had needed to produce the Dymaxion Vehicle prototype. In addition, Walter Chrysler considered the Chrysler Corporation prototype to be an ineffective, compromised car, whereas both men agreed Fuller's Vehicle was an ideal, modern car.[70]

Walter Chrysler was profoundly impressed by Fuller's demonstration of how much more effective one individual could be than an entire gigantic bureaucracy. After receiving that praise, Fuller went on to speak and write about the possibilities of individual achievement with a newfound confidence predicated on his firsthand experience with the Dymaxion Vehicle and the commendation of a highly influential corporate executive.

Fuller's interactions with Walter Chrysler also provided him with many valuable insights into the automobile industry. Chrysler was the person who explained why major manufacturers would never produce the Dymaxion Car commercially, and his prediction was ultimately fulfilled as each of the major automobile companies considered and rejected production of the Dymaxion Car. Walter Chrysler warned Bucky that because most used cars were sold several times and remained in the marketplace for many years, sales of used cars outnumbered sales of new cars by approximately five to one. The abundance of used cars in the marketplace . was especially hard on dealers, who were forced to accept them in trade and were usually inundated with used cars.[71]

That glut proved to be a tremendous burden on the American economy because most unsold cars were purchased and retained under agreements with financial institutions and were, therefore, actually owned by bankers until sold. Bankers believed that if the newest cars were too technologically advanced and desirable, no one would want the inferior used cars they owned. Those cars would then be drastically devalued, and such a devaluation would create massive losses for the financial institutions.[72]

Exceptionally advanced new cars would also generate financial problems for bankers and other businesspeople by remaining in service much longer, wearing out fewer parts and consuming less fuel. A technologically superior, highly desirable vehicle, such as Fuller's Dymaxion Car, simply could not be tolerated within the structure of the automobile market. Consequently, Mr. Chrysler believed that influential individuals worked diligently to ensure that superior vehicles would never be produced commercially.[73]

Walter Chrysler further explained that he felt that automobile company executives may have genuinely wanted to produce the Dymaxion Car, but that they were subject to the influence of bankers and stockholders who would never allow management to jeopardize the automobile market by producing such a superior vehicle. New car models were, at times, improved through engineering developments, but such innovations occurred over the course of many years so that they would not affect the value of used cars. Both Chrysler and Fuller agreed that the major modifications proclaimed by manufacturers from year to year were (as they still are) primarily superficial alterations in appearance which excite customers about apparently new models without costly retooling of manufacturing plants or the devaluation of the cars on the road or in the marketplace.[74]

The two men also agreed that major technological advances developed within other industries had been withheld from immediate implementation in the automobile industry in an effort to maintain a stable economy and to keep powerful individuals secure in their positions.[75] Had employing advanced technology for the benefit of humanity been a priority for those people controlling the economy, the Dymaxion Car would have been perfected and mass-produced and would now be regarded as a relic of the past rather than a vision of the future.

Although only three prototype Dymaxion Vehicles were produced during the early 1930s, the Dymaxion Car concept did emerge again for a short period during 1943. At that time, Fuller was working as Director of Mechanical Engineering for the Board of Economic Warfare in Washington, DC, and he was contacted by the famous industrialist Henry J. Kaiser. The two men had met earlier, and Kaiser had been impressed with Bucky's innovative ideas and design skills.[76]

One of Kaiser's enterprises was producing and marketing the Kaiser automobile. Kaiser's company was one of several mid-sized auto manufacturers which sought to overtake the larger companies by producing a revolutionary new car, and Henry Kaiser invited Bucky to design such a car for him.[77] Like most industrialists, Henry Kaiser was looking forward to the end of World War II and the shift of American industry from munitions to peacetime goods. Kaiser, like so many others, felt that cars would be in great demand when servicemen returned home, and he wanted to be ready with a car that would carry four passengers comfortably, accelerate very rapidly, get excellent mileage, and be much less expensive than anything else on the market.

After discussing the exact requirements with Kaiser, Bucky agreed that he could design such a vehicle, but because of his responsibilities with the Board of Economic Warfare, he could work on the project only during his spare time. Both Kaiser and Fuller's boss agreed to that plan, and Fuller initiated an amazingly rapid design program. Because of the war, competent engineers were scarce, but Bucky persuaded an engineer friend of his, Walter Sanders, to work on the project.[78] Sanders's architectural partner was also recruited to help with the project, but the three men had to work within the confines of the busy schedules and the New York location of Sanders and his friend.[79]

Never one to let circumstances deter him, Bucky convinced the two men to work on the new car design in the evening after putting in a full day at the office. When they agreed to his scheme, Bucky found himself in the awkward position of having a day job in Washington, DC, and evening work scheduled for New York City. Thus, he began taking the 5 P.M. train to New York, working all night with his partners and riding the first morning train back to Washington. With such a hectic schedule, the only opportunity for sleep was during his train rides, and for a solid month, Bucky slept only on the train.[80]

Because Fuller had spent several years producing the original Dymaxion Vehicles and had continued to consider refinements after the project had been concluded, the work progressed quickly. Accordingly, at the end of only one month, the three men had actually completed a full set of engineering drawings as well as a scale model of the new car.[81]

Their design remedied a significant flaw in the design of the original Dymaxion Vehicles by mirroring a phenomenon Bucky had observed within Nature following the completion of the first Vehicles. His new car adapted a movement pattern he found in the swimming of the horseshoe crab. Horseshoe crabs are equipped with a broad, crescent-shaped tail which is assisted in delicate balancing situations by a secondary long tail. Using that dual tail configuration, horseshoe crabs swim easily across streams and travel particularly well in powerful crosscurrents too strong for many other creatures. Their tails are specifically designed for maneuvering in such turbulent situations. In fact, the navigational problems crabs encounter traveling across strong currents are relatively far more treacherous than those found while traveling through wind currents on land or in the air because water is far denser and does not yield as readily as air.[82]

During the initial stages of the first Dymaxion Vehicle project, Fuller had not been extremely concerned about crosscurrents because the Dymaxion Vehicle passed through air, which is less resistant than water. As mentioned earlier, only after numerous experimental outings did he discover that because of its streamlined shape, wind currents influenced his radical design more than they affected traditionally shaped automobiles. Specifically, strong winds sometimes caused the tail to lift slightly. Later, Fuller speculated that the horseshoe crab configuration, with its combined broad and narrow rear sections, could eliminate that instability problem, and he utilized such an innovation in the Kaiser car design.

His new design employed a rear steering tire but was markedly different from earlier models in that the rear tire was mounted on a beam which could be extended from the back to lengthen the Vehicle. That process provided much greater stability at high speeds, and when the Vehicle slowed down, the rear tire beam would retract to provide improved maneuverability.[83]

In addition, Fuller arranged each of the Vehicle's three wheels to be powered by a separate, low-horsepower (fifteen to twenty-five horsepower), lightweight (less than five pounds each) engine, which he was to create later. Each of those three revolutionary engines was to be coupled with one wheel by a variable fluid-drive system, and the Vehicle's speed would be governed by altering the amount of fluid in the coupling assembly rather than by controlling the output of the engines. Fuller estimated that such a system would be so efficient that it would require that only one of the three engines be running to maintain highway cruising speeds while delivering forty to fifty miles per gallon of gasoline.[84]

The updated Dymaxion Car was to be outfitted with a single seven-foot-wide seat which could comfortably accommodate four passengers plus the driver and could be converted into a large bed if a person wanted to sleep in the car. Fuller calculated the weight of the Car to be as little as 620 pounds and no heavier than 960 pounds. It would also have been much more economical to mass-produce than anything on the market.[85]

Once his work was completed, Bucky called Henry Kaiser, who suggested he make a complete presentation at a suite in New York's Waldorf Hotel. At that meeting, Mr. Kaiser and his top engineers examined Fuller's work, and after cautious inspection by all the parties, Kaiser decided to produce Fuller's design. He then asked Bucky how much he wanted to be paid for his services, and Fuller's financial naiveté and innocent trust of his fellow human beings suddenly reappeared. Bucky asked only that the costs which he had incurred be paid, an amount of only $2500. As Bucky did not include anything for himself in that figure, Henry Kaiser generously suggested that Fuller simply be included as a partner in the endeavor and be paid a share of the profits.[86]

The two men agreed to those terms as well as a development strategy in which Kaiser's engineers would begin working within weeks on a prototype at his West Coast plant under Bucky's supervision. Fuller's supervisor at the Board of Economic Welfare also believed the project could be valuable to society and agreed to give Fuller a leave of absence to work on the project.[87]

With everything in place for his trip west, Bucky waited for the details from Kaiser's management; however, days quickly turned into weeks and then months with no calls or letters from Kaiser. Fuller attempted to phone Henry Kaiser several times during that period, but the industrialist would not answer or return his calls. When Bucky began to receive queries from Kaiser engineers concerning various aspects of his design and heard rumors of a new car being developed by Kaiser, he became extremely distraught.

Finally, after months of waiting, one of Henry Kaiser's assistants explained that after the industrialist approved a project, he usually turned it over to his representatives and seldom paid much attention to it. Months after that explanation, Bucky had given up hope that the new car would be produced, when he learned that he had, in fact, been deceived. Rather than bringing Bucky out to supervise the creation of a revolutionary new vehicle, the Kaiser organization had hired a former Chevrolet engineer named Alexander Taub to supervise the project.[88]

Taub did not understand Fuller's radical ideas, and one of his first decisions was to install a six-hundred-pound engine into Fuller's design. That engine weighed more than Fuller had calculated for the entire car although it replaced the three small engines Fuller had designed to weigh only a few pounds each. Following that news, Fuller heard nothing about the project until after the war, when an article about Kaiser's attempt to build a radically new vehicle was published. The article stated that Kaiser had tried to build a Dymaxion Car designed by Fuller, but that the design had proved inferior when a test driver lost control of the car and hit a tree. The article went on to explain that after that accident, Kaiser's engineers had scrapped the revolutionary project and had returned to the production of conventional cars. Although the inaccurate information in that article infuriated Bucky, he was already busy with other endeavors and resigned himself to the fact that nothing more could be done about that project. Thus, it became another valuable learning experience.[89]

Even the termination of the first Dymaxion Car project had been less of a blow to Fuller than his betrayal at the hands of Henry Kaiser. Following closure of the Dymaxion Vehicle manufacturing plant in 1934, Bucky decided it was time to move his family and himself back to New York City. Anne and Allegra had moved from Long Island to an apartment just outside Bridgeport the previous year, and Allegra had been enrolled in a local school. However, in 1934 Bucky insisted that her innate intellect should be protected from the problems of rigid school systems that he had encountered. He felt that his daughter should be enrolled in a more open school in which free thought was the norm. After some discussion, Anne agreed, and they began preparations for a move to New York so that Allegra could enter the progressive Dalton School.[90]

The New York that Fuller returned to was far different from the one which he had left less than two years earlier. The Depression had created great fear among the population, as people sought to maintain a decent lifestyle for themselves and their families. Rather than the openness of a mixing pot of cultural sharing, Bucky found that divisions were the rule as small neighborhood groups banded together for economic survival, and "outsiders" were not welcome. That same mood pervaded the national and international scene as military powers began to muster forces in preparation for war. Among the nations struggling for power were Germany, Italy, Great Britain, and Communist Russia under Stalin. Although no one was certain which side Russia would support in a war, most Americans feared the "communists."

Yet during the unrest of the Depression, a fringe of liberal Americans perceived socialism as a cure and joined the Communist Party. A number of those new Communists had been Bucky's chums at Romany Marie's years earlier, and they were now under government surveillance. Fuller realized that even people like himself might come under suspicion if they were observed in public with the Communist sympathizers, and more importantly, that he would be questioned about his friends if he was seen with them. Not wanting to create trouble for his friends, his family, or himself, Bucky, like so many others, simply decided that Romany Marie's was off limits. That attitude spelled the end of Romany Marie's as his Bohemian headquarters and left him without the camaraderie he so loved.[91]

Although he had changed many of his habits, Bucky still continued to drink, socialize, and party with friends. Not being able to do so at Romany Marie's, he sought a new hangout and soon found it in the "Three Hours for Lunch Club," which met regularly for long luncheon discussions. Bucky felt that the Three Hours for Lunch Club was a much safer place to congregate because it was simply an informal aggregation of famous and aspiring writers.[92]

Fuller had devoted two years to the production of the Dymaxion Vehicle and was in the process of shifting gears. His next undertaking would be a continuation of his years of silence. He was about to write his first bona fide book, and the "Club" provided him with a social as well as a literary environment in which to share ideas. It was also the place where he met a man who would become one of his best friends and supporters, the famous author, Christopher Morley.[93]

CHAPTER 9

Practical Applications

 Bucky adjusted easily to life back in New York City. He had, after all, been traveling throughout the East Coast states displaying his revolutionary Vehicle and lecturing to enormous crowds for the past two years. He had also persisted in examining and questioning all aspects of his environment, both seen and unseen. As his thoughts amassed, Bucky could not help but assemble them in written form, and he soon found his collection of writings growing to the size of a book. With that mass of material confronting him, the next major project requiring his attention was obvious. He thus devoted a great deal of 1935 and 1936 to writing a book he esoterically titled *An Adventure Story of Thought*. Later, in an attempt to entice readers, the title was changed to *Nine Chains to the Moon*.[1]

Without a publisher or an editor to provide a critique of his work in progress, Fuller was fortunate to be a member in good standing of the "Three Hours for Lunch Club." While the Club members were noted for their raucous gatherings and parties, most of them were also experienced writers, and they provided Bucky with literary assistance and encouragement.[2]

Although his book was completed in the summer of 1936, Fuller had no formal experience or training as a writer and could find no publisher who would seriously consider his work. He had, however, become close friends with Christopher Morley, one of the premier authors of the day, and a man who had published over forty books with Doubleday. Morley had just begun working with the publishing firm of J. B. Lippincott and suggested that he champion *Nine Chains to the Moon* at that company.

When their new star writer "suggested" that Lippincott publish Bucky's book, management could hardly contest his wish, and Fuller was offered a contract.[3]

Progress toward publication proceeded smoothly until an editor carefully read the three chapters which focused on Albert Einstein. Fuller had devoted a great deal of time to studying Einstein's work over the years, and he greatly admired the celebrated scientist. Because Einstein's work had significantly influenced him and he felt that it provided a critical new context for all human beings, Bucky dedicated three chapters of his book to explaining and interpreting Einstein's theories.[4]

When the Lippincott editors read those chapters, they immediately reevaluated their commitment to the book. At the time, a common legend claimed that Einstein had stated that only ten people in the World were capable of understanding his ideas. One of the Lippincott editors wrote Fuller and reported that they had examined the lists of people who were supposed to understand Einstein and could not find Fuller's name on any of them. The editor continued by advising Bucky that they had, in fact, not found his name on any list of known scientists, educators, or prominent individuals, and Lippincott's management felt that by publishing *Nine Chains to the Moon* they might be facilitating an enormous public hoax.[5] He further informed Bucky that although they were concerned about his credibility, Lippincott would publish his book on the basis of Morley's recommendation, but it would have to be printed without the three chapters on Einstein.[6]

Being a brash young man with nothing to lose, Fuller wrote the editor and suggested that he solicit an opinion from the one true expert on Einstein, Albert Einstein himself. Bucky did not expect anything to occur as a result of his outlandish proposition, so he was truly surprised when he received a call from a Dr. Morris Fishbein one month later. Dr. Fishbein told Bucky that his friend Albert Einstein would be visiting New York City the following weekend, had read Fuller's manuscript, and wanted to know if he would be available to discuss it during the visit. Bucky instantly made himself available and arrived at Dr. Fishbein's apartment precisely as directed. He was shown into a long drawing room with a high ceiling and tall windows overlooking the river. At the far end of that room sat a small man with dozens of people surrounding him.[7]

Fuller was escorted through the crowd and introduced to that man, Albert Einstein. He would later recount that he felt some sort of aura or energy field surrounding Einstein, and he was certain that he was in the presence of greatness. After their introduction, Einstein immediately rose, telling Bucky that they had to speak in private. He then led the awestruck Fuller into a smaller study where the manuscript of *Nine Chains* was neatly stacked on a table underneath a reading lamp.[8]

After sitting down in front of the lamp with Bucky on the opposite side of the table, Einstein relieved Fuller's anxiety by telling him that he had read the manuscript and thoroughly approved of Bucky's analysis and explanations. He also said that he would notify Lippincott of his approval so the book could be printed in its entirety. Einstein continued by revealing that he was particularly impressed with the chapter "$E = MC^2 = $ Mrs. Murphy's Horsepower," which envisioned what life would be like for an average person if Einstein's theories were correct. He told Bucky that he never imagined that his ideas had any practical application, and he thanked Fuller for demonstrating pragmatic uses of his theories.[9]

Even years later Fuller would claim to remember Einstein's exact words: "Young man, you amaze me. I cannot conceive anything I have ever done as having the slightest practical application. I evolved all this in the hope that it might be of use to cosmogonists and to astrophysicists in gaining a better understanding of the universe, but you appear to have found practical applications for it."[10]

Fuller left that momentous meeting with a renewed sense of purpose. In the three contested chapters, he had explained that Einstein's work was a key element necessary to understanding Universe and the evolution of humanity, and Einstein had agreed with his assessment.[11] From that day on, Bucky wrote about and discussed Einstein's awareness of Universe and human evolution with greater confidence, and he became a master at simplifying Einstein's complex discoveries for ordinary people.

A fundamental component in Fuller's interpretation of Einstein's work was the importance of understanding that change was normal and natural. Prior to Einstein's theories, most people accepted as truth Newton's laws, which state that a body at rest remains at rest unless acted upon by an outside force. In contrast, Fuller recognized that Einstein had provided the foundation for an entirely new interpretation of Universe in which ceaseless movement is the norm. Years earlier, physicists had discovered that everything in Universe is composed of energy, and Einstein calculated that energy equals mass times the speed of light to the second power.[12]

Fuller further inferred that if everything is composed of energy, and if energy is the result of mass perpetually moving at the speed of light (i.e., approximately 186,000 miles per second), all aspects of Universe are therefore moving and changing very rapidly. Such insight clearly establishes change, rather than rest, as normal.[13]

Although Fuller's book was finally accepted in its totality by Lippincott in 1936, it was not printed and distributed until 1938 and was not acknowledged as a classic examination of the human condition until the 1960s when the magnitude of Fuller's vision became widely appreciated. The book did, in fact, have flaws other than Fuller's labyrinthine style of writing. Most notable was his series of twenty-two predictions which were to occur prior to 1948.

That section, which was published only in the introduction of the original 1938 edition, contained outlandish statements such as: "Beamed Radio transmission of power employing gold as the reflecting surface. The main system of general education to go on the air and screen."[14] It was not that Fuller's prognostications completely missed the mark, but that he misjudged the timing of many of them.

When considering the broad context of the evolution of all humanity, as Fuller did, a matter of decades became almost imperceptible. Still, the showman in Bucky seldom resisted the opportunity to wake people to impending issues which he felt would escalate into major problems by making spectacular predictions that became even more astounding because of the time frames he proposed.

Needless to say, *Nine Chains to the Moon* was no best-seller. Few people purchased the book until it was reprinted decades later. Most critics, unable to decipher the intricacies of Fuller's thought, simply panned the book. In fact, even Lippincott's management had trouble following what Bucky was attempting to portray.[15] Even the cover of that original edition reflects that lack of understanding. Despite the fact that the book had nothing to do with copper or Bucky's job, because Fuller was working for the huge copper-mining company Phelps Dodge when *Nine Chains* was published, the book was released with a strange copper-colored cover.[16]

Fuller's association with Phelps Dodge began after another low period of his life. Having devoted nearly two years to writing *Nine Chains to the Moon,* Bucky once again found himself financially destitute during the summer of 1936. Still, he was able to employ his dynamic personality to earn enough to make ends meet. Bucky's outgoing personality also resulted in his becoming friends with William Osborn, a man with family ties to and a great deal of influence in the gigantic Phelps Dodge Corporation.[17]

Like so many others, Osborn was impressed as well as inspired by Fuller's ideas, and when Phelps Dodge management decided to expand from simply mining copper to manufacturing copper products, Osborn "suggested" that Bucky be employed as a source of new ideas. Thus, in the fall of 1936, Fuller was offered a position as director of research with Phelps Dodge at a large salary.[18]

Although he had consciously resolved to operate as an individual whenever possible, Bucky's concern for the financial welfare of his young daughter and his wife helped him to justify accepting that job. Prior to making the decision, the management at Phelps Dodge also agreed that he would be allowed a great deal of freedom and latitude in choosing exactly where to focus his attention.[19]

With his vast experience in the automobile industry, it was natural that Fuller's first project at Phelps Dodge was a new brake system for cars and trucks. Bucky's design consisted of a revolutionary solid bronze drum fitted with rubber insets to dissipate heat very rapidly, thereby solving the problems of overheating, grabbing, and fade, which had plagued automobile and truck braking systems for decades.[20] His new brakes also cut stopping times by nearly 50 percent and were the forerunner of the now-popular disc brakes.

Bucky also developed other innovations, but his most publicized work at Phelps Dodge focused on the production of an invention he had conceived and begun working on years earlier, the single-unit Dymaxion Bathroom. Bucky first envisioned the concept as a result of his idea to compartmentalize most of the "rooms" in his Dymaxion House. By designing them in that manner, he felt rooms could be inexpensively mass-produced and "installed" within the framework of the House just as a stove or furnace would be installed. Such an innovation also provided Dymaxion Houses with the potential for being instantly upgraded, with more advanced rooms replacing outdated models.[21]

The concept of modular rooms offered a great deal of potential benefit in the area of human sanitation because, at that time, private indoor bathrooms were primarily limited to the homes of upper-class, urban residents. The predominant rural sanitary facilities were still traditional outhouses, and most low- or middle-income people living in apartments were forced to share a single bathroom with several families. Hence, a self-contained, private bathroom which could be easily installed,

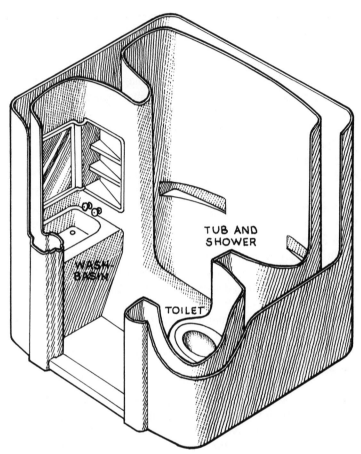

Fig. 9–1   Cutaway illustration of the single-unit, Dymaxion Bathroom.

owned by an individual, and moved just like any other appliance seemed
to be an easily marketable product.

The Dymaxion Bathroom was designed to be manufactured on assembly lines in four sections which were assembled at the installation site.
Those easy-to-transport sections could then be put together and connected
to water, electricity, and drainage wherever needed.[22] Bucky's innovative
design permitted an ultramodern bathroom with toilet, sink, shower, and
tub to be installed in any house or apartment within hours. The entire unit
occupied a floor space only five feet by five feet, weighed only 420

Fig. 9–2   Lower section of the Dymaxion Bathroom.

pounds (the weight of a conventional bathtub), and could be manufactured for about three hundred dollars. It could also be disassembled and moved with ease. Fuller even incorporated within the unit electrical strips for independent heating and a fan for cooling as well as small innovations such as a side hole rather than a spigot in the sink so babies could be bathed in greater safety.[23]

Within months of beginning, Bucky and his team had constructed prototype units using copper coated with either other metals or a colored synthetic resin. Although he utilized copper because Phelps Dodge's primary concern was mining and refining copper, Fuller speculated that the bathroom would be far more efficient in a lightweight plastic-like material that he predicted would be invented sometime in the near future. That material is the Fiberglas now used in the fabrication of modern tub and shower sections.[24] Fiberglas was also the material employed in the construction of the modern, self-contained bathroom unit similar to the Dymaxion Bathroom which became commercially available in the early 1980s.[25]

Having witnessed the undermining of his earlier bathroom project by

Fig. 9–3   The lightweight Dymaxion Bathroom sections being assembled.

the plumbers' union, Fuller paid a great deal of attention to the concerns of that group, and he marketed his idea to them splendidly. In fact, the plumbers' union was so impressed with the Dymaxion Bathroom that it was enthusiastically endorsed in the union's official magazine, *The Ladle*. Their rationale for endorsing an invention which appeared to eliminate plumbers' jobs was purely financial.[26]

Using his pragmatic logic, Fuller convinced union officials that mass production of such a device would result in many more people's buying private bathrooms and having them installed. He then showed that every time bathroom owners moved, they would naturally want their personal bathroom disassembled and reinstalled at their new home.[27] Thus, an entire new market of renters who had not previously used the services of plumbers would open up.

Bucky's public relations effort was, however, in vain. One of Phelps Dodge's largest customers was his former employer Standard Sanitary, and when they learned that the Dymaxion Bathroom was being resurrected, an uproar ensued. Because of their negative experience with Fuller and his Bathroom years earlier, the management of Standard Sanitary was resolute in its opposition to rejuvenating the project, and as a result of that resolve, Phelps Dodge's management had no choice but to discontinue the project.[28]

Although the Dymaxion Bathroom, like so many of his inventions, did not become commercially successful, Bucky was again satisfied with his work and the fact that he had advanced the project to a new level of development. Under the auspices of Phelps Dodge, he had produced thirteen prototype Dymaxion Bathrooms, most of which were sold to private parties and installed in homes when the project was terminated. Those units were so durable and well designed that at least half of them were known to be satisfactorily functioning as late as the mid-1970s.[29]

Always the showman who wanted others to "experience" his inventions as tactilely as possible, Bucky personally acquired one of the Dymaxion Bathrooms for exhibition. While still working for Phelps Dodge, he mounted that unit on an old pickup truck and drove up and down the East Coast with his new display. Not all those exhibitions were, however, solemn occasions, and more than once, he was seen driving the pickup through the sedate streets near the Hewlett home in Lawrence, Long Island, with assorted Hewletts riding in the Bathroom streaming toilet paper behind them.[30]

Because he always kept one eye on the future, Fuller regarded his revolutionary Dymaxion Bathroom as a rudimentary step in applying modern technology to resolving the issue of human sanitation. Bucky believed that human beings had not even scratched the surface of that issue, and accordingly, he continued his quest for more effective systems throughout the balance of his life. During that probe, the single invention he created which he felt could replace much of the Dymaxion Bathroom's functions and would revolutionize human cleanliness resulted from remembrances of his days in the navy.[31] Bucky recalled that when he and his shipmates would go on deck during a storm, even the thickest engine room grease would be quickly washed off their skin by the combination of wind and water.[32]

Having observed that phenomenon on numerous occasions, Bucky decided to investigate what had occurred, and he soon came to appreciate the cleansing power of ordinary water when it was atomized with air under high pressure. Employing that insight and with the assistance of students in his design classes during the late 1940s, Fuller designed and produced another invention.

That artifact was the "fog gun," which utilized atomized water under pressure to clean human skin far more efficiently than anything previously known. With the fog gun, a person could take a complete shower using less than one pint of water, and since no drainage was required for the tiny amount of water used for washing face and hands, much of the human cleansing process would require minimal plumbing.[33] In Fuller's futuristic scenario, the act of cleansing was to become a simple "in-the-bedroom" process akin to dressing.[34]

Having previously experienced the wrath and power of the plumbing industry and being extremely busy with several other projects during the period he developed the fog gun, Bucky decided to simply build and test his idea with the help of his college students. After months of fabrication and careful testing, his students found the gun to be a completely satisfactory system of cleansing, which, in fact, caused less damage to skin than ordinary soap and water.[35] Thus, another significant artifact was created and left until a time when future generations would require it.

Although Phelps Dodge was forced to scrap the Dymaxion Bathroom, the company's management did not attempt to use Fuller as a scapegoat. They realized that Bucky had delivered the product they had requested, and he remained with the company, studying the future of copper. It was during that examination that he truly began to appreciate and understand the flow and use of raw materials and the patterns to which those materials conform.

Fuller was particularly interested in the primary industrial material of the era, metals. As he was working for a copper company, Bucky devoted most of his time to the scrutiny of copper and soon found it to be a bellwether metal.[36] Being a bellwether metal meant that the use and distribution of all other metals followed the pattern of copper. In other words, if a shortage of copper occurred, a shortage of other metals was certain to follow. Because of that phenomenon, Fuller's examination of copper flow provided him with a broad understanding of many other metals as well.

During his scrutiny of the copper industry, Bucky was amazed to find that he could actually calculate the amount of that metal which had been mined as well as the probable amount of copper which remained in the Earth and could be economically removed in the future. Thus, he was able to determine the total amount of copper available to humanity. That exercise suggested the possibility of similar studies for other materials and the ultimate potential to formally document all the Earth's natural resources and how they were being employed. Eventually, Fuller would expand that idea into his World Resources Inventory, which was the first comprehensive assessment of all available resources as well as how they could be more effectively utilized on behalf of all humanity.[37]

Bucky began his metal research by observing the fundamental fact that metal resources, like most elements on our Planet, rarely leave the Earth, and that any elements removed from the Planet are in such small quantities that they are not statistically significant. He then examined the fate of Earth's fixed body of metals after they had been mined and incorporated into products, and those products had been used up or had become obsolete.[38]

For example, during World War I, a great deal of copper was mined and employed in the production of armaments and electrical generating equipment. When more efficient, alloy metals began to replace copper in those applications, engineers found new applications for copper. One of the largest such uses was in the rapidly expanding communications industry. By studying the cycles of metals, Fuller found that mined copper was regularly recycled every 22½ years. He also ascertained that because of World War I, more copper had been taken from the ground in 1917 than had been mined during the entire previous history of humankind.[39] Hence, he predicted that a great volume of recycled copper would be available 22½ years later in 1939.[40]

Few people paid any attention to that forecast. Even in July of 1939, when the docks of New York harbor were clogged with barges full of scrap copper, Fuller was not acknowledged for that insight. He was also ignored when he suggested that scrap metal was a valuable resource rather than discarded trash, the common contention among profit-oriented businesspeople who sought to rid themselves of an unwanted commodity.[41]

Because of such attitudes the enormous inventory of American scrap copper was sold to Germany and Japan in the late 1930s. Shortly after that massive series of sales, those two countries formally became enemies of the United States and, ironically, returned the copper to the United States in the form of munitions which were manufactured from melted-down American scrap and fired at American troops.[42]

Fuller learned a great deal about large-scale, global patterns as well as resource flow and management from his study of copper. His most significant discovery was the fact that each time metals are recirculated, an important nonphysical element is added to them. That component is knowledge, and it enables humans to accomplish much more with the same amount of resource metal. For example, a modern communications satellite weighs approximately five hundred pounds, and some of that weight is copper. However, the satellite outperforms and makes obsolete the 175,000 tons of transatlantic copper cable previously used to carry messages across the Atlantic Ocean.[43]

Both the satellite and the cable are fabricated from resources which have existed on Earth since the Planet was formed epochs ago, but when advanced knowledge was added to the materials used in the satellite, those basic substances were employed far more efficiently than ever before. By adding knowledge to materials, humans are continually able to do much more with fewer and fewer resources. Fuller perceived that phenomenon of added value through applied knowledge to be a significant demonstration of humanity using mind rather than muscle to grow. He also believed that the survival and flourishing of all humankind was predicated upon more efficient utilization of resources, resulting from the application of newly discovered knowledge.[44]

Bucky himself was continuously attempting to increase the knowledge of the management at Phelps Dodge with his constant flow of reports. Those reports were, however, written in a style *Time* magazine would later christen "Fullerese," and Phelps Dodge executives could rarely decipher his communications. One executive did return a report which was to be forwarded to the board of directors, telling Fuller that something should be done because he couldn't understand a word of it. The executive was certain that the board would not be able to understand Fuller's message until Bucky read it aloud to him, punctuated by short pauses. It was then that both Fuller and the executive realized that the report was easily understood when read broken into short phrases. The executive suggested that Fuller revise the report into a similar style, but he was completely surprised when he received the new version.[45]

He called Fuller into his office and said, "This is lucid. But it is poetry, and I cannot possibly hand it to the president of the corporation for submission to the board of directors."[46]

When Bucky insisted that it was not poetry but simply a broken-up rendering of his previous work, the executive said he would solicit the opinion of two poet friends whom he was having dinner with that evening. The following morning he reported to Fuller, "It's too bad; they say it's poetry." Hence, the report was changed back into its original form, but with a great many dots, dashes, and asterisks to separate the enormous sentences and paragraphs into the more easily understood poetic phrases.[47]

Thus began Bucky's career as a poet. Over the years, he would often turn to that format for sharing his ideas in a more easily understood form. Fuller would eventually publish several volumes of that writing as poetry and would be acknowledged in that field when, in 1962, he was awarded the Charles Eliot Norton Chair of Poetry at Harvard.

In 1938, Fuller was presented with another opportunity to develop ideas which would result in additional new inventions and would also more directly influence public opinion. Because his work at Phelps Dodge had once again catapulted Bucky into the public spotlight, Ralph Davenport, managing editor of *Fortune* magazine, felt Fuller would be an asset to the magazine's staff. Accordingly, Davenport offered Bucky a position as "Science and Technology Consultant" at a generous salary of $15,000 per year.[48]

Davenport's offer was simply too lucrative, both financially and in its potential to influence the course of human action, to be refused, and Bucky became an employee of *Fortune*. Although most of *Fortune's* management was skeptical that the radical Fuller could fit into their somewhat conservative and deadline-oriented environment, Bucky surprised everyone. He was, in fact, an ideal employee, always arriving at his office on time and remaining until closing. What he did during those hours, however, was somewhat controversial.[49]

While at *Fortune*, Fuller wrote very few articles. Instead, he devoted most of his time to the single pursuit at which he excelled, thinking. He would then share his thoughts with anyone who cared to listen, and almost everyone at the magazine seemed interested in his outlandish ideas. Editors and writers were constantly dropping into his office to garner the most recent workings of his mind, and many of those thoughts became the subjects of *Fortune* articles. As was generally the case wherever Fuller hung his hat, the one group who did not appreciate his contribution to the operation were the accountants, who, during one of their periodic examinations, found his editorial output to be well below that of almost everyone else.[50]

When the accountants questioned Bucky's efficiency and productivity, Davenport brusquely set them straight and concisely clarified Bucky's true value to the magazine as well as to humanity. "Look, you," he replied, "as managing editor, I have to have new ideas every single, solitary day, and sometimes I just dry up. So I send for Bucky, and Bucky comes in and talks for about two hours. By the time he leaves, I have more ideas than I can publish in the magazine in ten years. He recharges my batteries, and he's worth twice what he's being paid." With that, the inquiry was closed.[51]

During his tenure with the magazine, Bucky continued developing techniques for more easily conveying the complexities of his ideas experientially. In one such effort, he began displaying huge charts and graphs more and more. He also continued refining his "One Town World" mapping representation of the Earth so that it would more clearly display the entire Planet without visible distortion. Fuller believed that he could reveal the true deployment of global resources and influences of technological advancements much more easily on such a map. Thus, the endpapers of *Nine Chains to the Moon* were printed with a rudimentary representation of what Fuller claimed was the most accurate global mapping system known to humankind: the Dymaxion Map. Displays of resources using that early Dymaxion Map also appeared in *Fortune* during Fuller's tenure at the magazine, and those illustrations began to generate interest in the Map itself.[52]

Always working on several ideas at the same time, Fuller also continued refining and expanding his mathematical theories in search of Nature's coordinate system and Nature's geometry. By 1939, he had studied the history of mathematics in great detail and was certain that several of the fundamental concepts of plane geometry were erroneous.[53] Clearly, elements such as infinity and the intangible plane did not conform to the experiential environment in which Bucky found himself operating.

Still, Bucky's continual outpouring of ideas and projects did not inhibit him from his habitual lifestyle of frivolity. In fact, with a steady income, he tended to drink and attend parties more than ever.[54]

Not all of Bucky's habits appeared so contrary to his commitment to the well-being of humanity. In fact, one peculiar practice he developed during 1939 was actually a precursor of the future of exercise. It was during that year that Bucky suddenly decided he needed more exercise and began jogging. At the time, few people other than athletes were seen running in public, and the stocky figure of Bucky in shorts and a sweatshirt trudging through Central Park must have raised some eyebrows among the pedestrians. With short legs, Bucky's physique was not very well suited to his chosen sport, and he began referring to his form of jogging as "dog trotting."[55]

Even in an activity as individualized as running, Fuller's belief that he knew what was best for everyone and his love of companionship were evident, and some bizarre situations resulted. For instance, on weekend evenings at the Hewlett's Martin's Lane estate, Bucky would often persuade a crowd of the young men to join him for a late-night run. After having devoted most of the night to conversation and drinking, Bucky would suddenly stand up and shout, "Come on, Dog-Trot Club, time for a run!" Jackets and pants would then fly off, as the group prepared to join Bucky in a rowdy jaunt through the tastefully sedate neighborhood. Having witnessed that scene numerous other times, most of the neighbors would simply stare out at the midnight runners, commenting, "It's that crazy Bucky and his friends, again."[56]

Crazy or not, Bucky's friends always seemed to recognize his genius and to support his projects. One of the most palpable evidences of that support occurred in the early summer of 1940, when he and Christopher Morley were driving through the Midwest. During that trip Fuller noticed hundreds of identical grain-storage bins scattered on farms alongside the road.[57]

After stopping to question a farmer, Bucky discovered that most of the bins were identical because they were all manufactured by the Butler Manufacturing Company of Kansas City. Because Butler's grain bins were superior in both design and quality to those manufactured by other companies, they had become extremely popular, with news of the exceptional product spreading from farmer to farmer by word of mouth over the years.[58]

Always the designer and inventor, Bucky examined the Butler bins and quickly decided that they could easily be converted into shells for small prefabricated houses. As that concept formulated in his mind, Bucky explained to Morley that with the use of mass-production techniques, the cost of manufacturing such a structure would probably be in the astounding range of one dollar per square foot. Although Morley found Fuller's concepts fascinating, he was not prepared to become a captive audience for Bucky's incessant praise of Butler's grain bin. As the

Fig. 9–4   A Dymaxion Deployment Unit installed as a residence.

Figs. 9–5 and 9–6   The furnished interior of a Dymaxion Deployment Unit. Curtains hanging from the center of the roof were used to divide the single ''room'' when necessary.

miles passed, the praise persisted, and Bucky began explaining the details of modifying Butler's design to create inexpensive homes and other forms of shelter.[59]

At the time, Morley had just completed a novel entitled *Kitty Foyle*. After listening to Fuller's ideas for hours, Morley suddenly declared that the fictional heroine, Kitty Foyle, wanted to see the housing idea pursued. He continued by explaining that if the novel was a success, he would finance Fuller's designing such a structure and traveling to the Butler Manufacturing headquarters in Kansas City.[60] As fate would have it, *Kitty Foyle* was an instant success. In fact, the book was so popular that it was made into a movie, and Bucky was soon journeying to Kansas City with house plans in hand.[61]

In Kansas City, he conferred with the president and the top management of the Butler Manufacturing Company, and in 1940, Butler and Bucky's newly formed Dymaxion Company embarked on a joint venture to develop mass-production housing units christened Dymaxion Deployment Units or DDUs.[62] DDUs employed the standard grain-bin design and manufacturing techniques in conjunction with Fuller's innovations and modern mass-production technology.

Twenty feet in diameter, the structures proved to be an ideal size for the primary unit of a small dwelling. Bucky created a more humane interior by lining the inside with Fiberglas-backed wallboard for both aesthetics and insulation. He also added a translucent vent in the center of the conical shaped roof to allow in both light and fresh air.[63]

The structures were designed to be installed on a circular brick floor which was covered by a combination of several materials and topped by Masonite flooring.[64] Inside, the large round room could be divided into three smaller pie-shaped rooms by heavy curtains strung on wires. To bring in additional fresh air and light, Fuller installed circular porthole windows which were glazed with acrylic plastic, the first use of that material for windows.[65]

In addition to design innovations, Bucky also developed unique new construction techniques which later became important in the production of geodesic domes. For example, because he was concerned with worker safety and ease of construction, Fuller devised a system of erecting the building's prefabricated sections from the top down rather than from the bottom up, as is normally the custom. That innovation eliminated dangerous work on high scaffolding and allowed a structure to be erected in a much shorter period of time.

Fig. 9-7   The first Dymaxion Deployment Unit suspended on its mast prior to its final installation at the time its "natural air-condition" was discovered.

Fuller's unique system required a large metal ring several feet in diameter attached to a pulley and hanging from a mast in the center of the building site. Workers fastened individual roof sections to the ring, and when the entire roof was complete, it was hoisted up. The construction workers then attached the upper wall sections to the finished roof, and the structure was again hoisted by means of the pulley. Next, the lower wall sections were fastened in place, and the finished structure was gently lowered onto its foundation.[66]

One extremely hot afternoon during the construction of the first prototype DDU, Fuller chanced upon an amazing characteristic which he would later apply in many of his buildings, including geodesic domes. The completed Unit was still suspended from the central mast slightly above its foundation, when E. E. Norquist, president of Butler Manufacturing, and his top management came out to inspect the work in progress.[67]

Being an inquisitive as well as an adventurous individual, Norquist entered the metal building even though the outside temperature was over one hundred degrees and the Unit had no ventilation or electricity for fans.

Naturally, everyone, including Fuller, expected Norquist to come running back out of the scalding-hot metal structure, which was receiving the direct rays of the sun, and they thought he was joking when he exclaimed that the structure was, in fact, air-conditioned. Since Norquist was their boss, the other staff members played along and cautiously entered the DDU, expecting to encounter an inferno of heat. They were, however, amazed to discover that the interior was extremely comfortable, being at least twenty degrees cooler than outside.[68]

Fuller and the other engineers immediately began working to unravel the mystery and discovered cool air currents flowing down from the hole in the center of the conical roof and out the lower edges of the suspended building. Bucky quickly uncovered the principles which had created that cooling effect. He determined that the sun's rays heated the air nearest the outer shell of the DDU to temperatures even higher than the outside air, and that that scorching air rose to create a low-pressure vacuum effect around the bottom edge of the structure. Because of that lower pressure around the structure's base, which was suspended a few inches above the ground, air from inside the structure rushed out the bottom, producing lower pressure inside. That lower pressure could be balanced only by air rushing in through the other opening, the small ventilator aperture at the top of the DDU.[69]

Since that opening had a much smaller total area than the entire lower edge, more air per square inch had to enter through the top than was leaving through the bottom, so that the entering air was compressed. The resulting cooling could then be easily explained by the Bernoulli principle which states that the more a substance is condensed, the slower its molecules move and the colder it becomes. Thus, the air entering through the roof aperture was actually cooling the entire DDU. Remarkably, the warmer the outside temperature, the faster the air was forced to enter, and that phenomenon resulted in even lower inside temperatures.[70]

Once Fuller understood what had occurred, he became an instant proponent of that form of "natural air-conditioning." He also discovered that the phenomenon was particularly efficient in spherical or circular structures.[71] Years later, Bucky would become even more supportive of that method of environmental control, as he determined that a small opening in the center of a geodesic dome combined with vents around its lower edge produces a similar cooling effect with no additional expense.[72]

Like so many of Fuller's inventions, the Dymaxion Deployment Unit suffered several setbacks even though it was sponsored by a major corporation. Fuller's original idea was to produce an inexpensive home which could be mass-produced at a factory, delivered anywhere in sections, and erected in a matter of days by almost anyone. The structure was to provide an ideal home for small families and to be especially beneficial to those who had been rendered homeless by the Depression.[73]

Weighing only thirty-two hundred pounds (the average weight of an automobile at that time) without the brick floor, it conformed to Fuller's modular intent. Because of the center mast pole, an instruction sheet, and a simple tool kit, the DDU could be easily assembled by a single person if necessary.[74] Realizing that the simple structure would have to accommodate the necessities of life if it was to be accepted as a home, Bucky also designed a second, smaller grain-bin unit which could be attached to the main Unit and which contained a small kitchen, a bathroom, and an additional bedroom.[75]

While working on the design, the comprehensivist Fuller could not help but consider every aspect of living in the radical new structure. Thus, he calculated that a completely furnished two-bin home, which included a stove and refrigerator, could be sold for approximately $1250.[76]

Because of World War II, the DDU was not, however, destined to become known as the family home Fuller envisioned. In fact, the first stages of the American war effort nearly terminated DDU production before it began in earnest. With industrial giants gearing up for war production, many materials were scarce and not allocated for something as curious and nonessential as a house made from grain bins.

Although World War II nearly finished the DDU, it also made the Unit famous. The British were fighting Germany, and several British military leaders had read about Bucky's latest project. While searching for an inexpensive method of housing their troops, they examined the DDU and were impressed enough to place a large order. However, before manufacturing could begin, the war escalated, weapons became a priority for England, and the order was canceled.[77]

Following that setback, the United States became heavily involved in supporting Great Britain with armaments, and steel became a rationed commodity. Naturally, top priority was given to military applications, and inexpensive experimental housing for civilians was relegated to an even more insignificant priority.[78]

Then, just as it seemed that the project had been condemned, Fuller's strategy of creating an invention and waiting for an emergency proved itself. At the time, the United States was expanding its support of the Allied forces. As the United States defense effort increased, small aggregates of American troops and technicians were being dispatched to remote locations where housing was unavailable.[79] Most important among those troops were the groups being assigned to work with a new military device, radar. Radar stations were being established at extremely isolated sites where they could be maintained secretly, and the DDU seemed like an excellent air-deliverable means of quickly housing troops and protecting equipment at such sites.[80]

After examining Fuller's DDU, several military officers felt that the invention could alleviate many of their remote housing and storage problems, and they requested that a prototype unit be installed at the Washington, DC, Haynes Point Park for further inspection. Following a positive response, the military leaders submitted a massive order for the Units, and Butler Manufacturing was catapulted to the top of the government's steel priority list.[81] Within weeks, Butler's mass-production operation was established and running twenty-four hours a day, and both the Company and Bucky began doing something to which he was not accustomed: earning substantial profits.[82] Soon small cities of DDUs were springing up in faraway locations such as the South Pacific islands and the Persian Gulf.[83]

Always the consummate designer, Bucky made certain that in addition to a complete instruction packet, every shipment contained the portable mast and the tool kit which facilitated expeditious construction. Because of such attention to details and the quality of his design, both Fuller and Butler began receiving praise from both the personnel for whom the DDUs provided quality living accommodations far superior to anything ever experienced by remote troops and from those who had to erect the DDUs.[84]

The DDU's rapid success was, however, just as suddenly terminated by World War II when Japan attacked Pearl Harbor on December 7, 1941, and the United States officially entered the war. Steel then became the sole property of armament manufacturers, and with no steel, DDU manufacturing abruptly ground to a halt, leaving Bucky again searching for the next project requiring his attention.[85] This time, however, he could do so without his usual financial predicament, since the DDU project had resulted in profits for both himself and Butler.

Working at Butler had also resulted in another positive benefit. During that period Bucky had chosen to again end one of his most deleterious habits, drinking. In a moment of insightfulness, he suddenly realized that both he and his ideas were, at times, dismissed as inconsequential because of his copious drinking. He also felt that if he was to make a truly significant contribution to humanity through a continually advancing design-science revolution, he had to be taken seriously at all times, and so he simply decided to stop drinking and smoking.[86]

Prior to that moment in 1941, Bucky had relied on drink to soothe rough times and build friendships. However, he gradually came to realize just how much alcohol and the image it created of him interfered with his work. Fuller truly wanted his ideas to be considered seriously and, thus, without any special treatment program or pressure from other people, he simply gave up alcohol and smoking for the balance of his life so that he could be of greater service to humanity.[87]

CHAPTER 10

Searching for Perspective

 One of the issues Fuller focused on following the termination of Dymaxion Deployment Unit production was the problem of accurately displaying the entire Earth on a single map. He had been surveying that issue for years and had published some interpretations of global resource usage and distribution on rudimentary ''Dymaxion Maps''; however, he had yet to perfect his idea.[1]

Bucky's concept of displaying the entire Planet on a single practicable-sized sheet was by no means new. Human curiosity has always motivated people to learn as much as possible about their environment and in particular their immediate surroundings, but for most of human existence, that nearby environment had amounted to the distance an individual could walk.

By examining history, Fuller found that people's range of physical environment was continually expanding to include more territory as their knowledge grew. He also discovered that once human beings determined the Earth was spherical, individuals promptly began searching for satisfying methods of viewing the entire Planet.[2]

Maps flourished, but mapmakers discovered enormous problems in attempting to transfer even limited information from a sphere to a flat surface. The primary problem they encountered can be demonstrated by peeling an orange and flattening the skin, invariably tearing much of the peel's surface. Similarly, in the process of shifting from globe to flat map, splits occur.[3] Those breaks represent distorted data, an unsolved problem which has plagued humanity for centuries.

With the onset of easily accessible worldwide communication and transportation, most people's primary environment has expanded to include all of what Fuller later referred to as Spaceship Earth, and Fuller found that the innate curiosity of a majority of individuals had burgeoned in a similar manner.[4] Since scientists and explorers have determined the Earth to be spherical, that curiosity has generally been satisfied by accurate globes, which do not embody the extreme distortion inherent in most flat maps.

Globes are precise because they are spherical like the Earth itself, but that very characteristic limits their usefulness. Because of the spherical shape of a globe, an individual can examine only a portion of the Earth's surface at one time, and such a limited perspective does not promote understanding of the true relationships inherent on our Planet.

Fuller had learned a great deal about maps and the Earth in general from his extensive naval experience, both in the military and as a civilian. Because their livelihood and lives depend on navigating without the landmarks available on dry land, sailors have always demanded the best navigational tools, and accurate maps are one of the most important such tools.

For centuries, maritime societies have been confronted by and have seriously examined the problems inherent in mapping a spherical surface upon a flat surface such as a piece of paper. More recently, modern technology has been applied to that problem in an effort to depict the spherical Earth in a manner suitable for maritime navigation. Bucky found that engineers and mathematicians who had no knowledge of sailing even went so far as to propose using globes for marine navigation. Such a notion amused pragmatic sailors, who understood that a globe containing enough detail to be of any use in navigation would have to be much larger than the largest ship at sea.[5]

Because of his knowledge of and experience with both mathematics and the sea, Bucky was able to examine the problem from the more realistic stance of a sailor. He understood that modern sailors rely on two types of maps: comprehensive ones displaying large portions of the Earth and more precise sectional charts of specific areas which provide great detail and are the primary instruments of navigation.[6]

In his quest for a better map, Fuller studied the history of marine navigation and found that sailors had attempted to employ the same Mercator map of Earth (which was developed in the 16th century and which continues to be displayed and used in most schools and institutions) for navigation. Those sailors discovered that map to be adequate in many instances, but they ultimately established that their success was due to the travel routes they used rather than the accuracy of the maps.

Sailors found that the Mercator map worked well because they sailed primarily in the warmer, less hazardous waters near the equator, where the Mercator map exhibited the least distortion. Their Mercator projection maps, however, became less and less accurate as they ventured further from the equator because a Mercator map, like most other maps, is generated by mechanically projecting the Earth's features from a globe to a sheet of paper.[7]

In the instance of the familiar Mercator map, the paper which will become the map is initially rolled into a cylinder that touches an Earth globe at the equator. The Mercator projection technique can be envisioned by imagining a single light bulb at the center of a clear globe which has the Earth's landmasses etched upon it. The cylinder of paper is then positioned to touch the globe only at the equator, and, when the central light is turned on, a shadow is projected onto the paper. Tracing that shadow on the paper creates a classic Mercator map projection.

Because the equator circle and the paper are in direct contact, that line touches the globe and contains no shadow distortion.[8] However, just as the size of a shadow increases and its clarity decreases with increasing distance from a light source, greater distortion occurs as the map shadow is projected farther from the equator.

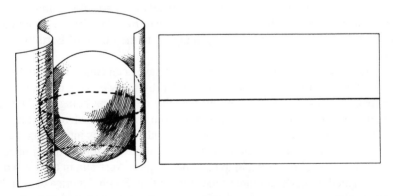

Fig. 10-1 The Mercator map projection method. A flat sheet of paper is wrapped around a globe (left), and no distortion occurs only on the equator line, which runs down the center of the sheet when it is laid flat (right).

The problem with such a technique is extremely evident near the poles, where the shadow expands to such an extent that the exact North and South Poles cannot be projected onto a paper cylinder of any size. That phenomenon is the result of the fact that light passing through the globe's two poles travels parallel to the cylinder of paper and, consequently, never comes into contact with it.

By studying mapping techniques, Fuller found that mapmakers had historically redrawn the highly distorted areas of the Mercator map to compensate for the gross inaccuracies farther from the equator and to produce a more acceptable representation. Because those alterations tend to be aesthetic rather than mathematically calculated, such work was seldom precise and continued to promote a grossly distorted view of Earth.[9]

For example, on the redrawn, commonly used version of the Mercator map, Greenland is displayed as three times the size of South America when South America is actually three times the size of Greenland. Still, with no better alternatives, people have accepted and utilized the warped Mercator representation of Earth for centuries.

In his examination of cartographic history, Fuller found that, like most knowledge, humanity's quest for a true graphic representation of Earth developed gradually as each generation expanded on the work of its ancestors. He also discovered that during the twentieth century the search for more accurate representations of Earth escalated dramatically.[10] Within the past few decades, researchers examined thousands of possibilities in an effort to uncover a better method of transferring data from a spherical globe to a flat surface and bringing integrity to the image of Earth we all share.

Because the majority of human beings have been conditioned to think of the Earth as the static object represented by the Mercator map seen in classrooms, popular opinion holds that the new maps being produced, which frequently appear very different from traditional Mercator maps, are novelty items. Although some people find such maps interesting, few believe that those unconventional representations portray a true image. After all, how valid could a map be which depicts Australia in the form of an elongated kidney bean three times longer than South America or which depicts entire continents as seeming to have been stretched from their most northerly and southerly points?

When Fuller first became an explorative cartographer seeking an accurate global perspective during his 1927 period of solitude, his motives were not academic as was true of so many other investigators.[11] Instead, Bucky initiated his quest as a means of helping bring attention to and solving worldwide problems at a time when few people considered, much less sought, answers to global issues.

He believed that the emerging globally oriented society would be confronted with serious problems that could be resolved, or at least limited in intensity, by an accurate World map and that perpetual exposure to misleading World maps, as has been the case throughout history, only complicates and discourages such solutions. As a result of that conviction, Fuller set out to examine the World map problem and to provide the best solution possible.[12] That quest continued throughout his life, and it produced some uniquely comprehensive inventions as well as options to significant yet often unconsidered issues of our emerging global society.

One such issue was technology's influence on military combat. With the advent of modern aircraft and bombing techniques, humans have witnessed entire cities being demolished in a matter of hours or minutes. People are also aware of the fact that intercontinental nuclear missiles provide an opportunity for destruction far beyond the scope of anything previously known. However, our perception of that massive capacity for devastation is frequently concealed by conventional World maps, which portray "the enemy" as situated a great distance from "us." Such maps conflict with most people's individual experience of a "shrinking" Earth on which human beings feel they are moving closer to one another as they learn more about the people and cultures of lands previously considered distant.

Fuller felt that every human being could trust the single attribute which has always provided individuals with an authentic perspective and could be applied to the issue of war. That characteristic is an individual's personal experience, and it has guided many people to understand that "the enemy" is, in reality, much closer than institutions lead us to believe or than children are taught in school.[13]

Increasingly rapid air transportation and communication has moved every citizen of Spaceship Earth nearer to his or her neighbors, regardless of the countries involved. Fuller believed that displaying the true physical relationships between landmasses, nations, and individual human beings on a map would promote a much more honest impression of the human condition on Earth at this time.[14] He also felt that such a view could provide individuals with a new perspective of "the enemy."[15]

Fuller realized that once individuals begin to recognize their true relationship to tiny Spaceship Earth and the other inhabitants of our

Planet, the issue of an enemy's position would lose its significance. He felt that people who are provided with a trustworthy perspective of Earth on a regular basis would begin questioning the validity of labeling human beings living at distant sites or adhering to other religions or forms of government as "the enemy."[16]

A comprehensive global perspective supplies each individual with an opportunity to personally experience and understand, as Fuller did, that all the inhabitants of Spaceship Earth form a single aggregation living on the small segment of our Planet's land which is inhabitable. When the facts that all the land on Earth covers less than one third of its surface and that 90 percent of the human population lives north of the equator are portrayed on an undistorted World map, an individual can begin to comprehend the tightly packed nature of humankind's living arrangement.[17] Within such a context, destroying a neighbor's resources is more easily appreciated as the destruction of one's own life support.

Early in his mission on behalf of all humanity, Fuller realized that the global scope of his operation necessitated graphically portraying World data such as population and resources. His first attempt at such a display was presented on a standard Mercator map, but he quickly discovered that the map itself transformed information into misinformation.

Bucky often commented that his frustration with ordinary maps was analogous to an individual's seriously examining him or herself in an extremely warped mirror. Although such mirrors entertain us at amusement parks, when a person needs to be certain he or she looks just right for an important meeting, a severely deformed mirror becomes a real problem.[18]

Fuller's attempts to deploy his precise global data on an ordinary map were monumental failures, but they provided him with valuable learning experiences. Following those episodes, he understood that even the most reliable information would be deformed and misrepresented by a distorted map, just as a person's image is distorted by a warped mirror. Those experiences did not hinder his examination of World issues, but in each instance of portraying global data on ordinary maps, his presentation was marred by a lengthy explanation of the problems inherent in the maps he used. Growing frustration led Bucky to realize that emerging global humanity needed a map with no visible distortion, and time and time again he was spurred to satisfy that requirement.[19]

One of the initial tasks he undertook in studying maps and searching for a superior World map was to painstakingly produce detailed freehand drawings of the entire Planet. Although that job was laborious and time-consuming, it provided him with a global familiarity and appreciation experienced by few other people.[20] His intimate knowledge would later prove to be extremely beneficial on numerous occasions, not the least of which was during the years he spent perfecting the Dymaxion Map.

Through his studies, Fuller determined that despite the absence of undistorted World maps, our innate curiosity and technology have created a unique modern appreciation of geographic, and consequently human, relationships.[21] For instance, individuals are beginning to understand that the shortest distance between two points on the Earth is represented by the arc of a great circle (i.e., the largest circle which can be inscribed on the surface of a sphere) rather than a straight line drawn on conventional maps.

That phenomenon can be observed by plotting a course between New York and Paris on a traditional World map and then on a globe. The difference between those two routes clearly demonstrates the problems inherent in traditional global maps.

Bucky felt that if individuals studied the more comprehensive field of spherical geometry before they studied limited plane geometry, they might notice a relationship between their classroom studies and the reality of modern travel—including the possible travel of destructive intercontinental missiles.[22]

During his later years, he was pleased to witness that, as a result of humanity's need to comprehend our spherical Planet, spherical geometry (and specifically the concept of great circles) was entering the realm of common knowledge. The most conspicuous example of a great circle is the Earth's equator, but the number of possible great circles on any sphere is infinite. In essence, the arc of any great circle is analogous to a straight line on a flat surface because it is the shortest surface distance between any two points on a sphere.

Knowledge of great circles and spherical geometry has historically remained within the domain of scientists and mathematicians. It was considered interesting but seldom relevant to "practical" matters for most of human history when an individual's maximum daily travel was restricted to the speed of a pack animal and the shortest distance between two points was usually a visual "straight line."

With the advent of air travel, however, the entire Earth has become a single transit system, and its natural barriers are of little concern to most travelers. Consequently, the "as-the-crow-flies" technique of navigation,

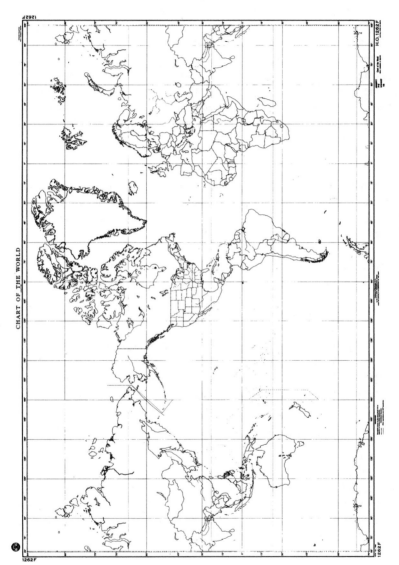

Fig. 10–2   A standard Mercator map published by the United States government.

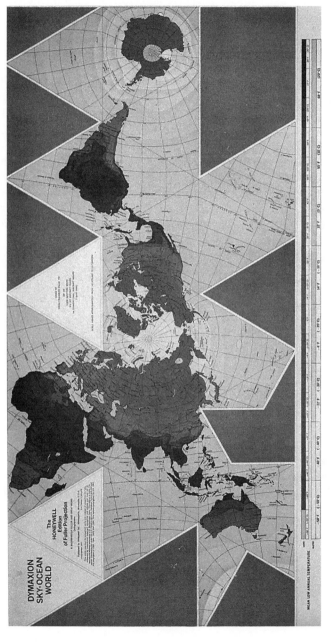

Fig. 10–3 The Dymaxion Map in its most familiar configuration. Note the difference in the size and the relationship of the landmasses when displayed with minimal distortion.

which relies upon great circles whenever large distances are involved, has become extremely important.

By understanding great circles and being exposed to ever more sophisticated displays of information, today's globally oriented individuals are beginning to be more aware of the major distortions inherent in nearly every large map. Fuller first discovered those same distortions in the 1920s. The most conspicuous of them appears in the Southern Hemisphere, where the entire continent of Antarctica is frequently eliminated or is portrayed as a jagged fringe along the lower edge of the map.

A century ago, when it was believed that the northern and southern polar areas were merely blanketed with masses of floating ice and when Antarctica was unexplored, such distortion was insignificant. However, modern technology has drastically altered the significance of that seemingly innocent misrepresentation. Today when a great deal of human activity affects the totality of the Earth and its inhabitants, as witnessed in the recently discovered hole in the ozone layer over the Antarctic region and fallout from nuclear accidents, we can no longer afford such deceptions.

Early in his experiment to determine what one individual could accomplish on behalf of all humans, Bucky realized that as people expanded

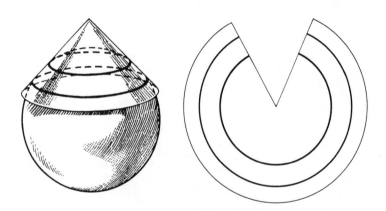

Fig. 10–4   The polyconic projection method. Paper is positioned so that two circular lines touch the globe (left), creating two lines of no distortion when the paper is laid flat (right).

their exploration and influence into every area of the Planet, humanity could no longer relegate any location to an insignificant status. Yet, he found that the maps which greatly influence people in creating their personal interpretations of Earth—including the relationships between landmasses, nations, resources, and individual human beings—continued to grossly distort many areas.[23] Frequently, such maps even lend credence to the erroneous belief that the misrepresented areas are not important or are so far removed that they are irrelevant to the daily life of most people.[24]

Fuller felt that such maps thoroughly alter the truth and consequently influence all global activities, including news and education. He also believed that the emerging global human required an accurate, undistorted representation of our relatively small Planet so that individuals could formulate accurate decisions beneficial to all humankind.[25] The Dymaxion Map provided that representation.

Before even attempting to develop his new map, Fuller devoted a great deal of time to studying mapping techniques and discovering the weaknesses and strengths inherent within each of them. Because distortion is intrinsic in any transfer of information from a spherical (i.e., globe) to flat (i.e., map) surface, Fuller found it to be the critical problem in the creation of a World map.[26]

Since all traditional processes of conversion require that the globe touch the map's flat surface at one or more points, that point or series of points on a map contains no distortion. The Mercator map provided Fuller with a classic example of the distortion pattern which occurs in all mechanical map projections. More deformities always appear the farther a site is located from the point or points of contact between the sphere and the flat surface, and because the handful of methods employed in the manufacture of most maps all utilize a similar projection procedure, each of them generates a similar pattern of misrepresentation radiating from that point or points of contact.

Although the Mercator map projection is by far the most commonly used, maps which require greater accuracy are generated by a technique known as *polyconic projection*. In that method, a shadow image of landmasses is projected from a globe onto a cone rather than a cylinder of paper, and because the cone intersects the globe twice, it creates both a large and a small circle where no distortion occurs. The polyconic map projection is most advantageous when accuracy over a specific area is necessary because the tract between the two circles of intersection is far less deformed than the remainder of the map.

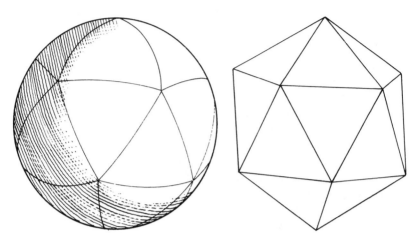

Fig. 10–5   The Dymaxion projection method. A spherical icosahedron (left) is inscribed upon a
World globe which is then transformed into a planar icosahedron (right). That planar icosahedron
is then cut along its edges to create the Map.

Polyconic projections reduced deformities more than any other projection technique prior to Fuller's invention of the Dymaxion Map. The
uniqueness of the Dymaxion Map lies in its lack of visible distortion. In
fact, the Dymaxion Map is the first and only map which provides a view
of the entire Earth's geography with no visible distortion in shape or
relative size of continental masses and no fabricated rifts in those landmasses.[27] With Fuller's Map, the entire surface of our Planet can be
perceived as what he christened a "One-World Island in a One-World
Ocean."[28]

That view is the direct result of Fuller's patented system of map
production. Rather than relying on traditional shadow projection techniques with their inevitable distortion, Bucky applied his knowledge of
mathematics to developing and fabricating the Map, creating it by mathematically transferring data from a spherical globe to a flat map surface.[29]
Although many people today would find such a task rigorous but not
impossible with the assistance of computers, Bucky confronted the job in
the 1920s and 1930s without even a calculator, and he was forced to make
all his calculations manually.[30]

Because of his explorations of spherical and plane geometry, Fuller
understood that even if done with total accuracy, such a transformation
would still produce some distortion, but he felt that if he relegated that

minor distortion to the three quarters of the Earth covered by water, it would not be apparent or meaningful.[31] He also determined that because his method—like the Mercator technique, which produces no distortion along the equator—would produce no deformities along the edges of the structure, he would be most effective if he employed a geometric form that provided a maximum number of regular sides (i.e., identical in length) to transfer data.[32]

Fuller found that the single structure which best suited his needs was a regular icosahedron, composed of twenty equilateral triangles. And after years of experimentation, he devised a mathematical method of transforming a regular spherical icosahedron into a regular plane icosahedron in order to reproduce the spherical information accurately on a flat surface.[33]

A spherical icosahedron is simply an icosahedron in which the edges have been curved to such a degree that the structure is spherical rather than planar. It can be visualized as a volleyball which is constructed of three-sided, regular (i.e., equilateral) triangles rather than five-sided regular pentagons.

Fuller's transformation can be visualized as removing just enough of the triangulated volleyball's air so that the arcs of the triangles can be straightened out, thereby creating a planar icosahedron. That structure contains edges identical in length to the arcs of the arcs in the spherical version, but the inner areas of the triangles are flattened and thereby produce a structure which can be "cut apart" along any of its edges to produce a flat map.

As those arcs are transmuted into straight lines, the areas within the triangles shift slightly inward creating a small amount of distortion, but the arcs themselves, which become the edges of the Dymaxion Map triangles, remain intact. If the distance between two locations situated on one arc is two thousand miles on the globe, it remains exactly two thousand miles on the Dymaxion Map.

To further reduce deformities, Fuller dissected each of the twenty large triangles into smaller triangles and performed the same process on the smaller shapes. And by repeating that procedure on an ever smaller scale, he was able to create a Map in which the distortion is so minute and evenly spread among all the landmasses that it cannot be detected by the human eye.[34]

The fact that Fuller's methodology altered shapes inwardly and reduced them in size rather than expanding them is also crucial.[35] Deformities cannot be avoided when one is transforming data from spherical to

planar surfaces. However, just as the enlargement of a photograph produces a much greater distortion than its reduction, so projecting an image
outward, as is the case with map projections, creates much greater distortion than transforming it inward in a reduction process, as is the case with
the Dymaxion Map transformation.

Once he had determined a method of transformation, Fuller was
confronted with the problem of how to position his spherical icosahedron so
that after the transformation to a planar structure he could cut the structure
apart without creating unnatural breaks in the surface of any landmasses.
Because he sought to display the Earth with no map-generated breaks in
continents and an accurate representation of the relationship between landmasses, Bucky again initiated a system of patient trial-and-error experimentation to determine the best configuration.[36] That procedure escalated into a
two-year process of methodically turning and examining a World globe
encased within a spherical icosahedron grid representing the triangles of the
evolving Dymaxion Map.[37]

Eventually, Fuller did uncover the configuration now recognized as
the Dymaxion Map. Upon close inspection of that Map, the difficulty of
his task becomes apparent.

For example, on the eastern coast of both South America and China
and on the western coast of Australia, the edges of the triangular sections
come extremely close to the landmasses. Another instance of his careful
work can be seen in the area surrounding the continent of Africa, which is
positioned very near the edges of the triangles that constitute the outer
border of the Map.

Fuller's patented method of inward map transformation can be likened to his work with geodesic domes because in both instances he relied
on the same geometric principles and mathematical knowledge. However,
when developing the geodesic dome, Fuller projected outward from the
icosahedron to the sphere, and when developing the Dymaxion Map, he
projected inward from the sphere to the basic icosahedron Map structure.[38]

Utilizing that unique method of transformation, Fuller was able to
invent the first World map which provides a ''cosmic perspective'' of
every location in a single glance. That perspective has been important to
nautical navigators for centuries but impossible to display on the distorted
World maps of the past.

The cosmic view is one in which the center of the Earth and the
astronomical zenith (i.e., the point in the sky immediately overhead at any

location) are represented as directly above and below every point on the surface of the Map.[39] In other words, if an individual were to stand at any location on an oversized Dymaxion Map, his or her relationship to the Map would be identical to his or her relationship to the Earth itself if he or she were to stand at that same location on the Planet.

Fuller's intimate knowledge of our Planet and the importance of his map became even more conspicuous during the 1960s, when the great circle he had discovered and labeled the "Dymaxion Equator" years earlier was shown to be a significant component of the American space program. He always contended that when anything, including the Earth, is viewed comprehensively, facts which were not previously noticed become blatantly apparent. The Dymaxion Equator is one such overlooked phenomenon.

Fuller initially felt that the Dymaxion Equator's great circle was important because it circumvented the entire Planet without crossing any continent except a tiny portion of North America. Although the Dymaxion Equator follows a path far from the Earth's equator, it became extremely important when America began testing rockets and missiles from Cape Canaveral.[40]

The section of Florida in which Cape Canaveral is located is, in fact, one of the few landmasses along the Dymaxion Equator. Consequently, any rocket launched from that site and following that route can circumnavigate the Earth almost entirely over water, a characteristic particularly useful during the early years of testing when many of the unmanned rockets crashed while traveling on that route.[41]

The specific course of the Dymaxion Equator is almost identical to the one used by the United States for Earth-orbiting rockets and missiles. Traveling easterly from Cape Canaveral, this Equator passes just north of South America and just south of Africa. The route then proceeds past the northwest coast of Australia and across the neck of New Guinea, and it reenters North America in Southern California. Utilizing the Dymaxion Equator route, scientists are able to safely test experimental rockets and missiles with little fear of destroying ground-based life or property in the process.[42]

Fuller realized that, just as is true of the Earth's equator, his Dymaxion Equator's great circle must possess an axis. The ends of the Earth's equatorial axis are clearly delineated by the North and South Pole, and Bucky's curiosity led him to search for similar Dymaxion equator sites. He found that the Dymaxion Equator divided the Earth's surface into two distinct halves with less than 4 percent of the Planet's population residing in one half, which includes only South America, Australia, the Antarctic, and numerous small islands, and over 96 percent living in the other half.[43]

Because he realized that the site of the Dymaxion Equator's pole located in the populated hemisphere would be located at the single location with easiest access to the most inhabitants of Earth, Bucky devoted a great deal of time to precisely calculating its location. He discovered that site to be just south of Volgograd in Russia. Although, he was unable to determine if the significance of that location was known to the Russians, Fuller was later amazed to find it was the location of their primary rocket-launching base.[44] In other words, the Russians had chosen to launch their rockets and missiles from the single site nearest to most humanity, a little-known fact which could ultimately influence the course of human history.

Although the United States did not become formally involved in World War II until 1941, the American public was interested in that conflict years earlier, and that interest prompted the management of *Life* magazine to devote an entire 1939 issue to the war. They also decided that since Fuller was a fairly well-known global prognosticator and futurist, he might provide some pointed insights for their war issue.[45]

At the time, Bucky was still working for *Fortune,* a magazine under the same corporate management as *Life,* and he was certain that the United States would eventually enter World War II on the side of the British. When he was asked to work on an article about the war for *Life* magazine, Fuller was obliged to comply with management, but he was also eager to research the future of that conflict and to share his global perspective with its readers.[46]

Life's editors specifically requested that Fuller formulate a global, grand strategy for defeating the Germans, and he quickly set to work on the task.[47] Although he was in the process of establishing and elaborating his strategy for supporting the success of all human beings, Fuller also believed that true global cooperation would not be possible until 1970.

He predicted that during 1970, technology would have advanced to the point of doing so much more with fewer resources that all humanity could be supported with the Earth's resources. Fuller also theorized that at the time of that shift, war as well as politics would become obsolete.[48] While working on the *Life* article, however, he patriotically believed that supporting the Allied cause was of prime importance.[49]

At the time he began the article, most observers realized that German military forces tended to be pushing outward from their centrally located inland empire and defeating other armies by driving them toward the sea where they would be trapped. The majority opinion held that effectively stopping Germany's ceaseless initiative was a momentous task at best, and most likely impossible. Although he believed humanity was evolving toward a global society in which war would be obsolete, Fuller realized that humankind had not reached that point, and he was intrigued by the problems of the war. He also believed that defeating the German forces was essential to the freedom and development of humanity.[50]

After the war's completion, Fuller would learn that the Allies' great military leaders had also been working out a strategy for defeating Germany at the same time. Their plans were, however, maintained in great secrecy for years.[51] The strategy which was eventually implemented by the Allies was called the *soft-belly strategy* and was fabricated by Winston Churchill and his staff during the late 1930s.

That plan required American troops to first enter Africa and cross that continent over land before attacking the German troops in North Africa. When that maneuver was complete, the American troops were to continue to advance across the Mediterranean Sea into what was regarded as the Germans' vulnerable underside through Sicily and Italy.

With a great many of the German forces diverted to defending that southern front, other Allied forces could then successfully invade Europe via the more direct route across the English Channel. Although the Allies ultimately used that strategy successfully, Bucky proposed an alternate plan designed to employ fewer troops and less time.

In his report to the editors of *Life,* Bucky brazenly proposed a sweeping American air attack on Germany across the seldom-considered North Polar route. His scheme involved no ships, but an armada of aircraft launched from northern airfields throughout the United States and Canada and towing gliders packed with troops and equipment. The gliders themselves were designed by Fuller so that they could be quickly dismantled into sections upon landing in Europe and reassembled for other uses. Those reassembled sections were to serve as readily available shelter for command posts, storage, and other vital functions.[52]

Fuller analyzed all the elements of his plan in great detail and designed gliders which were simple yet strong. Because metal was in short supply and the gliders would fly only once, they were to be constructed of plywood. The sections were to be standardized and mass-produced at large factories before being shipped to airfields for assembly. Always the consummate designer, Bucky validated the feasibility of his glider design by submitting it to several aircraft engineers who corroborated its practicality and flightworthiness. The experts further pronounced the craft's design extremely sound and functional.[53]

Fuller's plan called for American forces traveling in airplanes and gliders to rendezvous on the Eastern Front with advancing Russian ground troops and to undertake a joint effort driving the Germans south and west. With that maneuver, the joint Allied troops could force the Germans to retreat south and west toward the sea, just as the Germans had done in conquering neighboring nations.[54]

Life's management was pleased and excited about Bucky's initial idea, and they assigned several people to support him in creating an extensive article describing all aspects of the strategy. Although Fuller and his team worked diligently over the subsequent two months, when the article was finally published he was astounded to find he was not listed as the author. In fact, his name was not mentioned anywhere in the issue.

Despite his never being told exactly why *Life*'s management chose not to publicly acknowledge his authorship, Fuller did receive a letter from the magazine's senior editor thanking him for his outstanding work and attesting to the fact Bucky alone had devised the strategy proposed in the article. Since a great many of the initial meetings had taken place at that editor's home, he was obviously aware of Fuller's contribution to the article.[55]

Following the principles of his operating strategy, Fuller did not exhaust much effort attempting to understand *Life*'s failure to publicly acknowledge his authorship of the article and the plan. Instead, he simply proceeded to the next task he felt required his attention.[56]

Then, years later, when he was working as Director of Mechanical Engineering at the United States Board of Economic Warfare, Bucky was shown just how practical his plan was. While working at that Washington, DC, organization during World War II, Bucky was contacted by Gary Underhill, a man who had been a strategic military adviser at *Life* when the article was written. Underhill had assisted Fuller on the project and was clearly aware of the plan's origin. Having grown up in a military family, he had become an expert in military strategies, especially German warfare. When the United States entered World War II, Underhill was summoned to the Department of War where he was serving when he contacted Bucky.[57]

Underhill invited Bucky to join him for lunch, and at that meeting he produced a copy of a German magazine. As a respected member of the

War Department's planning team, Underhill had access to a great deal of intelligence including ordinary German magazines and newspapers, but Bucky had no idea why he was being handed what appeared to be a German military magazine.[58]

Underhill explained that the periodical, a 1943 edition of the official German military magazine *Wermacht*, contained a story which might be of interest to Fuller. He then produced an English translation of a major article which contained a complete German translation of Bucky's earlier *Life* article on defeating the Germans. The piece was followed by a series of commentaries in which top German military leaders candidly stated that Fuller's strategy was the best possible method of defeating their forces. The leaders further recounted that they were not worried about the article because they felt American leaders were too naive to grasp the potential value of such an innovative strategy and to utilize it against them.[59]

Although Bucky also felt that most political leaders lacked long-range global vision and never expected his plan to be implemented, he made certain every aspect of it was, in fact, practicable. Thus, while most Americans viewed his strategy as interesting but unequivocally impossible, the German military acknowledgment provided unmistakable proof that his global perception and comprehensive thinking ability could be successfully applied to "practical" matters.

Fuller had learned that when he sought and utilized a broad perspective, his pragmatic solutions could be far more successful than those offered by individuals with less comprehensive points of view. He realized that because most people focused their attention on limited issues of basic survival, he was discovering phenomena and relationships which few of his peers could even imagine, much less develop into practical solutions. However, he also believed that his vision and solutions would prove useful to future generations.

Accordingly, Fuller solved problems which would confront future generations of human beings before those issues had emerged and before those who would require the solutions had even been born. A small percentage of Bucky's insight, such as the geodesic dome and his strategy to defeat Germany, were practically utilized or recognized during his lifetime, and he was acknowledged as a visionary because of them. Fuller, however, often recounted that he was merely observing large-scale patterns and reporting on the next steps in the evolution of human beings and our environment.[60]

It was during World War II that Fuller first witnessed the appearance

of one such large-scale pattern: many people actively seeking a comprehensive representation of Earth. He also found that Americans were the
population most attracted to developing such a map because, while most
other countries were fighting World War II within their borders or a
relatively short distance from those borders, America was being forced to
dispatch troops across both the Atlantic and the Pacific Oceans.[61] Because
of that situation, United States citizens remaining at home were extremely
interested in learning where their family members and friends were being
assigned. A similar concern had arisen during World War I, but the
modern communication technology of the World War II era, especially
the radio, established a public intimacy with the battlefront never before
experienced by civilians.

Global maps also became more prominent during that period. Since
such maps made enemy nations appear extremely distant from the United
States, Americans felt protected by the vastness of the oceans. American
citizens did, however, believe that more advanced submarines, warships,
and airplanes exposed the United States coastline as a vulnerable target,
and they became concerned about the actual distances between those
coastlines and enemy bases. Having studied the history of war, Fuller felt
that World War II was unique in that it represented the first instance in
which nations that had previously felt protected from one another by
distance began to experience just how close "enemy" countries had
become as well as how vulnerable they were to attack.[62]

The German ability to exploit air power during World War II in
conquering entire nations within a matter of days unmistakably demonstrated that technology had established a condition in which our Planet
could be circumnavigated in an infinite number of directions. Since his
period of examination during the 1920s, Fuller had been aware of that
situation and its emerging effects. Prior to the practical application of
flight, east–west ocean trade routes had been the principle paths of traffic,
and most attacks could be resisted by defending those few passages.[63]

Explorers and soldiers had, however, always sought "back-door"
routes for surprising other nations, and the airplane opened up an infinite
number of those passages. One of the most obvious such routes is the
North Polar passageway used by airplanes as well as submarines traveling
under the ice. Because the majority of humanity resides in the Northern
Hemisphere, the opening of that route greatly reduced the distance (i.e.,
travel time) between most concentrations of population.

Bucky understood that the "new" travel routes which were changing
the scope of human life had always been available, and that they belonged

to no individual or nation. He also felt that as modern technology had provided a growing number of individuals with access to global travel, more and more people were interested in discovering accurate methods of viewing the Earth and uncovering more effective routes of transportation.[64]

While examining global development, Fuller found that humanity was also beginning to appreciate the fact that our Planet is a tiny sphere floating in a gigantic cosmos.[65] Earlier societies had perceived Earth as inconceivably immense and their fellow Earthlings as very distant, but with governmental leaders possessing the ability to destroy not just the "enemy" but all life on Earth in a matter of minutes, that axiom of protection through distance had become invalid.

Although Bucky believed that a comprehensive understanding of our Planet was a critical requirement for all Earth's inhabitants, he found that that perspective was not supported by the officials of most governments and other institutions responsible for the creation and distribution of maps. Instead, those organizations developed maps which graphically fostered the interests of the individuals and institutions in power. Rather than displaying information as truthfully and completely as possible, these maps frequently contain consciously designed-in distortions, such as exhibiting the country in which the map was published as proportionally larger than its neighbors or as positioned in the center of the map.

Less innocuous examples of purposeful distortion can be found throughout history. Bucky discovered one such representation while working in Washington, DC, during World War II. At that time, the United States government was printing a version of the Mercator world map to display the progress of United States troops, especially those in the South Pacific, where Allied forces were meeting extreme resistance.

To foster the impression that the Japanese forces were a great threat to the security of America and that the United States forces were rapidly advancing toward Japan, the United States government published what Fuller considered purposefully altered World maps on which the distance between the West Coast of America and Japan was portrayed as much smaller than was actually the case. The United States government then utilized those falsified propaganda maps to contend that Allied troops were making great progress in their mission to defeat Japan when they conquered the tiny South Pacific island of Tarawa, located in the Gilbert Island chain. Although Tarawa is situated in the general direction of Japan, its proximity to Japan is nowhere near what was displayed on those distorted maps. In fact, Tarawa is actually an insignificant island located so far south and east of Tokyo that it is farther from that city than Chicago is from London.[66]

Had people studied globes rather than the government's contorted maps in search of information about Tarawa and the American troops, they might have acquired some of the insights which Bucky discovered and which are now so crucial to air transportation. A globe clearly displays the fact that the shortest route between Japan and the United States follows a northerly course near the North Pole, and that the Allied troops would actually have been closer to Tokyo had they captured the North Pole rather than the island of Tarawa.

As Director of Mechanical Engineering for the Board of Economic Warfare during World War II, Fuller spent a great deal of time examining such troop-movement strategies. He also increased his appreciation of the fact that in an environment increasingly dominated by the airplane, great circle routes were far more critical than the previously utilized ocean trade routes. Thus, he devoted most of his examination of transportation routes to those shortest-distance, great-circle routes rather than to the commonly used more indirect ocean routes.[67]

Because he was a comprehensivist, Fuller recognized that his initial great-circle calculations of distance and flight duration did not take the Earth's dynamic atmosphere into account. He paid particular attention to the motion of air currents, which provide head and tail winds for aircraft. When he included the prevailing influence of air currents in his calculations, Bucky discovered that in air travel time, the North Pole is, in fact, more than 30 percent closer to Tokyo than the island of Tarawa.[68]

Being an individual who attempted to utilize every new piece of information and experience which came his way, Fuller was quick to incorporate his understanding of air currents into his development of the Dymaxion Map. He realized that even though some maps had attempted to integrate the great circle concept of shortest distance travel, they were static and could not be rearranged to comprehensively display the atmospheric conditions, such as air currents, which affect air travel.[69]

To solve that problem, Fuller began to envision his Dymaxion Map as an artifact which could be dissected and shifted to accurately display different factors and data.[70] The result of that idea was the first and only World map which provides a comprehensive sectional view of our Planet. Each triangular segment of the Dymaxion Map can be shifted in relation to its neighboring sections to create unique arrangements which provide distinct views of the dynamic interrelationships operating on the Earth's surface. Those interrelationships include wind and ocean current patterns as well as more static distances between locations.

Fuller did not believe that the Dymaxion Map was the ultimate portrayal of Earth's surface, but he and others did envision it as a great leap forward in displaying a true picture of our Planet. Because he wanted future generations to have a true global perspective available to them, Bucky went so far as to perfect the Dymaxion Map prior to the practical advent of computers, even though he predicted their coming and knew that they would make the creation and use of the Dymaxion Map much easier.

Although Fuller developed early versions of the Dymaxion Map during the 1930s, its first major publication occurred in a February 1943 issue of *Life* magazine. Having been victimized in dealing with *Life*, Bucky was far more cautious when the management of the magazine approached him with another proposition, and he made certain that he was involved in the entire process from concept to final production.

Because the article *Life*'s management wanted was conceived and scheduled during the height of World War II, when major news items were regularly given priority over other articles and Fuller's article was to include a complicated, full-color cutout version of the Map, preparation stretched for more than a year. Having worked with the Time-Life organization earlier, Bucky was not bothered by the editors' policy of slow development for expensive, nonnews projects, especially since he was contentedly working at the Board of Economic Warfare during that period. He did, however, take advantage of his previous association with that organization and his friends in the magazine industry to learn just how thoroughly the Dymaxion Map was being scrutinized prior to investing a great deal of time and effort in the article.[71]

The editors wanted the Map examined for two possible problems. First, they sought certification that it was as accurate a representation as Fuller claimed. And second, they wanted to be certain that it was, in fact, a new discovery and not something which had been known to geographers and which had simply never risen to public attention. To protect themselves and their reputation as a leading news magazine, the *Life* editors hired a panel of "experts" to study every facet of the Dymaxion Map. Members of that panel included the United States State Department's chief geographer, the American Geographic Society's chief cartographer, and two mathematicians.[72]

After months of study, those experts agreed the Dymaxion Map was, in their words, ''pure invention,'' a term which at the time was filled with negative connotations.[73] Although not specifically expressing their true conviction, the term implied a somewhat dubious nature which was the product of a vivid imagination rather than of research and scientific facts. In their simple statement, the team of experts attempted to convey their emotional reaction to an upstart like Fuller contending that he had invented a superior map predicated on principles unfamiliar to such noted specialists. The team was, however, forced to communicate that response to *Life*'s management without discounting the logic of the undeniable evidence that Bucky had presented or without risking their reputations as learned men. Because of those factors, the experts were unanimous in reporting that, although they could find no geographic, mathematical, or scientific flaws in Fuller's work, it did not conform to any known geometry, mathematics, or geography. Accordingly, they acknowledged that they could not fully explain why Fuller's Map was so accurate or how it had been created.[74]

In reality, the team of experts believed Fuller had somehow tricked them, but with no concrete evidence of a hoax, they were unwilling to risk their reputations and challenge him. Hence, they were compelled to agree that the Dymaxion Map appeared to be a valid new concept and that it had not been copied from earlier cartographic or mathematical work. The ''pure invention'' comment was included in their report merely to embody their hopeless feeling of having been duped, but it would prove far more useful than damaging in later years. After a great deal of deliberation, *Life*'s management decided to accept the experts' verification and to publish an eighteen-page article about the Dymaxion Map which included full acknowledgment of Bucky.[75]

By that time, Fuller had already entered into a process of acquiring a patent for the unique cartographic process he had invented and used in creating the Map. Within days of initiating that process, Bucky's patent attorney learned that no patents of cartographic innovations had been granted since the turn of the century when the chief patent examiner had made a ruling that all possible mathematics and projection methods used in mapmaking had been exhausted. Because of that ruling, the patent office followed a strict policy of simply rejecting any such applications.[76]

Fuller's patent attorney found new hope when he read the ''experts''' report to *Life* because, without incurring any expense, he and Bucky could produce indisputable testimony that the Dymaxion Map was, in fact, ''pure invention.'' When the experts' conclusions were presented to and considered by the United States Patent Office, those in charge had no choice but to grant Fuller the first patent on a cartographic innovation (No. 2,393,676; January 29, 1946) of the twentieth century.[77]

Despite the experts' belief that the Dymaxion Map was nothing more than an elaborate hoax, it was enormously popular with the general public. At that time, magazines were a primary communication and entertainment medium, and they were pretested for public reaction, just as is done with many consumer products today. The standard procedure was to print a small trial run of a magazine issue, to expeditiously test it for public response, and to base the size of the actual printing on that reaction. In the case of the Dymaxion Map, the public responded so positively that *Life* produced and sold a record printing of over three million copies.[78]

The dominant element of the eighteen-page Dymaxion Map article was a cutout Dymaxion Map and instructions for assembling it into a globe-like structure. When the issue was published in February of 1943, the United States was deeply entrenched in the bitter fighting of World War II, and little reason for joy or excitement was to be found anywhere. Yet, years later, Fuller would vividly recall his experience of the Sunday when the magazine first hit the newsstand.[79]

He was returning to his job at the Board of Economic Welfare in Washington, DC, after spending the weekend with his family in New York and was amazed to see young children all along his route of travel proudly parading through the streets carrying their Dymaxion Map "globes." At that moment, Bucky felt great pride in having created an instant of excitement and happiness for both children and adults in America with his "global toy."[80] Even decades later, many older people's initial response on being questioned about Buckminster Fuller would be a recollection of the joy and exhilaration they experienced in assembling and examining the small paper structure which provides humanity with a truly global perspective.

Life magazine's publisher, Henry Luce, was also excited about the possibilities spawned by the Dymaxion Map. Luce and Fuller had become acquainted years earlier and had grown to be good friends. Several weeks prior to the Map's publication, Luce invited Fuller to spend the weekend at his Westchester, New York, home and to demonstrate the benefits of the Dymaxion Map to a party of visiting guests. The actual display took place following dinner, when Bucky filled the parlor with several different copies of the Map and other drawings illustrating its usefulness. After the group convened, Bucky began his lengthy explanation and was in the process of describing how the Map could greatly alter both global and local operating strategies when Henry Luce interrupted his discourse.[81]

"Bucky, I believe every human being has an exact opposite, and you are mine," Luce remarked. Bucky replied that he was honored to be selected for such a lofty position and asked why he had been chosen. Henry Luce then explained that he thought Bucky believed "something" far more meaningful than humans governed Universe, while Luce, on the other hand, believed that humans (and Henry Luce, in particular) possessed free will and could do anything they chose.[82]

In considering that comment following the presentation, Fuller began to understand why Luce was always upset when Bucky explained and demonstrated how the large patterns operational throughout Universe are far more influential than individual humans. Bucky also believed that although Luce was impressed with his work as an individual, he was clearly agitated by artifacts such as the Dymaxion Map, which could be interpreted as evidence that large patterns controlled all Universe.[83]

Fuller felt that because Henry Luce was a World leader, he believed anything of significance was the result of human beings—especially powerful ones like himself. Luce's interaction that evening and at other times fueled Bucky's insight into the mentality of leadership and power still prevalent among influential individuals and reinforced his beliefs in the vulnerability of such people of power. He sensed that people like Henry Luce, with his magazine empire, felt they could greatly influence the outcome of events almost anywhere in the World and, consequently, believed they were actually controlling Universe or at least the portion of Universe familiar to them.[84]

One of the events Henry Luce was attempting to influence at the time of the Dymaxion Map publication was popular American support for England. In fact, he had accepted the task of managing all British public relations in the United States, and because of that responsibility, he was constantly meeting with diplomats.

Two such diplomats from Australia had stopped in New York for a brief rest while on a journey to meet with Winston Churchill in England just prior to the *Life* magazine publication. Luce decided those men would be ideal candidates for presenting the Dymaxion Map to Winston Churchill before the magazine hit the newsstands. Consequently, he asked Bucky to meet with the diplomats and explain the Map to them.[85]

Having demonstrated his inventions before, Bucky was very aware of possible pitfalls and quickly assembled two separate information packets containing Dymaxion Maps. One packet was for use in instructing the diplomats so that they could explain the Map, while the other was for them to use in their presentation to Churchill. Although both packets contained Maps which Fuller had dissected into their triangular sections, they were different. The Australian's Map was reassembled with Australia in the center while the other Map was formed with England in the center. Because they were native Australians, the Australia-centered arrangement appeared perfectly natural to the diplomats. Once they were comfortable with the strange Map, Bucky explained that it should always be arranged with the viewer's country in the center. Only then did he show the diplomats his Churchill version of the Map with the triangular section containing England in the center, and that particular rendition was, in fact, presented to Churchill, who was extremely impressed with it.[86]

Fuller averted a great deal of rejection when presenting the Map because he had learned that individuals have an innate perception of the World predicated upon their regional heritage and that violating that individual legacy only generates more problems, especially when one is attempting to communicate information. With that insight, Bucky instituted a policy of assembling the Dymaxion Map with the viewer's country in the center whenever possible.[87]

Since any sides of the Map's adjoining triangular sections can be joined or separated without violation of its integrity, one arrangement is as valid as another. However, Fuller invariably discovered that displaying the Map with the observer's country in the center neutralized a person's natural resistance to looking at his or her country relegated to the inferior "far reaches" on the edges of a map, as is usually true of World maps produced in "foreign" countries.[88]

Fuller had first encountered that phenomenon two years earlier when he spoke to the mathematics department of Haverford College in Pennsylvania and used a sectional Dymaxion Map to demonstrate his projection method. Following the lecture, Bucky was invited to the home of the head of the department, where the two men decided to conduct an experiment with the Map and the professor's young children. The men placed the triangular Map sections on the floor and invited the children to play with it. Although the sections are unlike conventional puzzle pieces in that they do not have unique grooves and can therefore be assembled in many different ways, the children soon realized that by matching the continental shapes and the colors printed on the triangular sections they could construct interesting shapes which often correctly represented landmasses. They also appeared to relish assembling and reassembling the World in different and sometimes unusual formations.[89]

During one such experiment, the "helpful" father and college professor intervened to exhibit his wisdom when he explained that because the United States was not positioned with north pointing away from them, the children had assembled the World upside-down. Having spent a great deal of time contemplating the concepts of *up* and *down,* Fuller realized they were relative terms and, in fact, were not a part of his universal cosmic perspective. He also felt that with no "formal" educational indoctrination, the children were perfectly comfortable with the Map configured in any number of ways, and that the "learned" professor's attempt to educate his children was actually an exhibition of his conditioned ignorance and narrow perspective.[90]

That event reinforced Bucky's appreciation of the importance of "natural" truth experienced by young children as opposed to commonly accepted dogma presented at educational institutions. Again, he was made aware that formal education tends to discount experience in favor of authoritarian doctrines which restrict and divide individuals rather than expand perspectives and support the development of a global society.

CHAPTER 11

Expanding the View

 Because so many modern human activities affect not only local communities but the entire Planet, Fuller's Dymaxion Map has proved especially beneficial to individuals involved in planning. People who forecast future trends and devise strategies involving global issues find the Dymaxion Map particularly helpful because it allows them to observe all Earth's landmasses portrayed in accurate relationship to one another. It also provides them with a framework on which to display and examine their findings. By plotting information such as population or agricultural production on Fuller's Map, discerning planners who use the Dymaxion Map can more easily observe large-scale patterns and trends and, consequently, formulate more informed, intelligent decisions. They can also arrange their ideas to operate within what Fuller felt is the "ocean of modern transportation": the air, domain of aircraft.[1]

As a result of his study of and work within the maturing field of aeronautics, Bucky began to appreciate the aeronautical perspective that both the land and water areas of the surface of the Earth provide the inner boundary of an always available, yet only recently discovered and utilized, unbroken "ocean of air." The outer limit of that ocean of air is the threshold of outer space where airplanes cannot fly.[2] Even in the 1930s, Fuller understood that during the first half of this century, the leading edge of transportation was gradually shifting from water to air in a process unnoticed by most people. With that transformation, the leading edge of innovation also moved from the nautical to the aerospace industry.[3]

The reasons for that change are more apparent when one examines and compares the characteristics of water and air, as Fuller did. He found that because gases such as air are much more fluid than liquids like water, vehicles can travel much faster in air. Accordingly, the aeronautical industry was able to build vehicles which moved much faster than ships of the sea, and those vehicles required technology responsive to constantly increasing speed and rapid change. As airplane speeds increased by leaps and bounds, new technology was required, and that technology had to handle the even more rapidly arising problems inherent in greater acceleration. Because lack of atmosphere in space provides an environment of even less resistance and accordingly greater acceleration, Fuller felt that space represented the next frontier for exploring increased speed, more sophisticated technology, and increasingly rapid resolution of problems.[4]

Bucky was also aware of the shift to the air in the area of communication, where satellites and other wireless devices began replacing cumbersome cables and increasing individual access to information. With the implementation of radio and television signals, which can be received inexpensively, people are no longer limited to the news presented by a few magazines and newspapers.[5] Fuller felt that as individuals acquired more information, they would begin to understand the constantly shifting character of Nature, which supports rather than restricts expansion and rapid changes. With such an understanding, humankind could then begin operating in greater harmony with the ever evolving and changing capacity of Nature, rather than resisting it.[6]

He noticed that within the framework of human development and natural change, the intervals between significant historical events have now decreased from centuries, decades, and years to days, hours, and even minutes. Such a headlong acceleration of history allows each individual to witness and experience many more alterations within his or her environment than had been possible only a decade before. Those experiences of change are also becoming global rather than local phenomena.[7] As a result of a new universality of early life experiences, Bucky felt that people from all parts of the World were beginning to realize that they were in the same boat and must work cooperatively to find comprehensive, holistic solutions to the issues which would affect the lives of everyone on Earth.

Fuller also found that humanity's shift from land and sea transportation to the much quicker and unobstructed air was, and continues to be, evident within social issues. He felt that the rapidly escalating crossbreeding of human beings into a single, dynamic World citizenry was particularly significant because he believed that the successful development of a unified global society is crucial to the future success of humankind.[8] Consequently, he supported that trend whenever possible. The Dymaxion Map was only one of his inventions which aided the development of the global society Fuller envisioned. It did so by helping to expand people's perspective on and vision of themselves and their environment.

Fuller understood that the Earth can be viewed from an infinite number of different perspectives. He therefore had the foresight to organize the Dymaxion Map into sections which can be disassembled and reconstructed in a variety of ways to portray different characteristics and outlooks.[9] Even though each of those unique arrangements satisfies a special need, it simultaneously retains all the positive characteristics of the original Map. In displaying different configurations of the Map, Fuller found that an observer's attention invariably focused upon the sections

Fig. 11–1 The One-Ocean World version of the Dymaxion Map uses the South Pole as its center.

which were positioned in the center, and with that knowledge, he set about devising unique arrangements of the Map for specific purposes.[10]

For example, he laid it out in the "One-Ocean World" configuration for use by sailors and those involved in traveling the oceans. That arrangement centrally positions the Earth's oceans, which constitute the majority of the Southern Hemisphere, and relegates the shorelines, which provide navigational markings for sailors, to the outer sections. With Antarctica in the center of the One-Ocean World configuration, land becomes a border for the Earth's vast water area.[11]

The "One-Ocean World" Map graphically displays many characteristics which frequently go unnoticed. For instance, it clearly portrays the vastness of the Pacific Ocean and emphasizes the fact that its longest axis does not run north and south as most people believe, but extends from Alaska's Aleutian Islands southeast to Cape Horn on the southern tip of South America. The One-Ocean World arrangement also vividly displays the fact that the Indian Ocean and the Atlantic Ocean are, in reality, secondary sections of a single immense ocean.

In experimenting with different configurations of the Dymaxion Map, Fuller came to better understand the psychological power of maps and in particular how the common Mercator map greatly affects the thinking and perception of most people.[12] To more clearly understand that phenomenon, he conducted a simple experiment with the One-Ocean arrangement of the Dymaxion Map and the more common Air–Ocean configuration, which most people simply call the Dymaxion Map.

Fuller displayed those two distinct arrangements of the same twenty triangles for less than a minute to individuals who were unfamiliar with the Dymaxion Map, and then asked each person to describe what he or she had seen. Although both Maps contained identical sections, the majority of the participants reported that the Air–Ocean version (i.e., with the landmasses in the center) contained an average of 75 percent land, and that the One-Ocean version was composed of 90 percent water.[13] In actuality, both Maps display Earth's true surface proportion of approximately 30 percent land and 70 percent water. However, by a simple rearrangment of the sections with either water or land in the center, people were easily misled to perceive two very different notions of reality.

Fuller understood that one arrangement of the Map was no more significant than another but merely displayed a different viewpoint. He also realized that the Map's transformability was an important feature because it allowed any individual to create a configuration in which the geographical location identified as the person's "home" was centrally positioned.[14] With a center triangle displaying the familiar "home," the Map seems more ordinary and familiar, so an observer is less likely to object to its radical appearance. Once such a tolerance has been established, motivating a person to consider shifting the sections and examining the Earth from another perspective becomes a much smoother process.

Because the Dymaxion Map can be presented in this less intimidating manner, Fuller felt it was a valuable tool for assisting people in attaining the global perspective so vital to our expansive modern society.[15] Rather than invalidating people's experience by proving their previous perception of Earth to be wrong, Bucky and his Dymaxion Map provided a method of first corroborating an individual's experience and then demonstrating the feasibility of other possibilities. Bucky believed his strategy would be effective because it gently eased individuals away from the dogma of tradition. He understood that every human being is deeply conditioned by the limited perspective of his or her heritage and education but that such a narrow view no longer benefits an individual or society.[16]

Fuller found that by positioning different continents and nations in the center of the Dymaxion Map and examining relationships, individuals could also begin to understand the motivation and the logic behind the actions of individuals and national leaders in "foreign" countries.[17] Although any size of Dymaxion Map will suffice for seeking such insight, larger Maps provide the most intimate geographic details of the events and beliefs which most influence all the travelers on Spaceship Earth. Such information and insight are crucial to the personal understanding which Fuller felt each of us must achieve if we are to successfully survive the technologically induced movement of individuals into more intimate contact with our Earthly neighbors.[18]

For example, with Australia in the center of the Dymaxion Map, an observer can begin to appreciate just why the inhabitants of that country remain rather isolated despite the advent of high-speed modern transportation and communication. Examining the geographic perspective of other societies provides people with a graphic experience of the influence of location on a culture and a better understanding of exactly where other people are "coming from" as the human species advances toward a unified global society.

Fuller himself gained great insight into solving several global issues by examining different arrangements of the Dymaxion Map in great detail. He also found that in gaining new perspectives, he developed a stronger compassion for the conditions experienced by his fellow beings around the World.[19]

Because the Earth's surface is nearly three-quarters water, Fuller felt that one of the most significant World perspectives has always been that of the sailor-navigator, and the Dymaxion Map is the first map to accurately display that view of the entire globe on a single sheet.[20] A sailor examines maps from the perspective of being at the center of the surface of a water-covered sphere. He or she views the entire Planet as blanketed by a single ocean, 30 percent of which has peaks tall enough to crop up through the ocean's surface. Although most people perceive those outcroppings as the land on which we humans live, the sailor-navigator regards them as high mountains around which to guide his or her ship.

Such a reckoning of the Earth's surface may appear strange, but Fuller felt that understanding it was important because it provides another interpretation and that, like all interpretations, it presents unique insights. Through his examination of history, Bucky found that the sailor's perspective has been especially beneficial to cultures in which humans have sought new territories that could be reached only via water routes. His Dymaxion Map permits any person to easily display and view features of the Earth's environment which have been understood and utilized by sailors and navigators for thousands of years, yet that remain relatively obscure to most people.[21] Fundamental among the characteristics Fuller himself examined are the movement and influence of the Earth's water and air atmosphere.[22]

The Earth spins in an easterly direction and influences all the atmosphere's water and air to travel in that same direction. Since water and air are more fluid than solid land, they move much faster, and the momentum generated is greater near the poles, where fewer landmasses obstruct rapidly moving air and water. Every location on Earth makes one revolution each day, but the distance traveled at the equator is much greater than that closer to the poles. Consequently, equator locations move much faster, in terms of distance traveled over a specific amount of time, than elsewhere. That relative speed decreases proportionally as the distance from the equator increases, and it reaches the extreme of no spinning motion at the poles.

Although the rapidly traveling atmospheres (water and air) at the equator produce enormous momentum, they do not rival the speed of atmospheres near the poles because of the equator's land obstructions. When eastbound equatorial water and air strike slower moving land-masses, the collision produces swirling eddies which spiral in the direction of least resistance, toward the poles.

That same phenomenon also occurs near the poles, but with less frequency because of fewer obstructions in those regions. When the polar air and water do strike a land obstruction, they also spiral in the direction of least resistance, toward the equator.

Bucky discovered that the combination of these two fundamental phenomena produces most of Earth's weather patterns and storms. He also concluded that storms generally arise over water when the fast-moving air traveling east strikes the western coast of landmasses and its momentum is redirected westward. Following that shift, the traveling air encounters the continuing momentum of other air currents carried with the Earth's easterly spin, thereby producing storms.[23]

The phenomena of moving air and water have been known to and used by sailors for thousands of years. In studying ancient marine cultures, Fuller found that early sailing societies realized that the Earth's movement and changes result from natural forces which can be analyzed, charted, and exploited. He also found that modern "sailors of the air" were examining and charting the Earth's volatile air and water environments just as their nautical predecessors had done for centuries.[24]

Until recently, neither group of global navigators possessed the means of accurately charting its findings on a comprehensive World map. Fuller's Dymaxion Map resolved that issue by providing a framework for accurately plotting global patterns of both air and water currents as well as a means of uncovering new discoveries predicated upon those patterns.

One such discovery of modern air adventurers is that the same forces which move the Earth, its water, and its lower level air environments produce rapid air currents at higher elevations. Because they encounter fewer land obstructions, high-elevation currents are much stronger than ones nearer the Earth's surface. Knowledge of that phenomenon has allowed pilots to save fuel and travel eastward far more rapidly by ascending to higher elevations and taking advantage of prevailing wind currents.

From his examination of global patterns, Fuller discovered that high-elevation currents afford the greatest benefit around the polar regions where winds are rarely obstructed by landmasses.[25] For example, traveling by jet from northern Russia to Alaska requires only one third the time required for traveling in the opposite direction. Furthermore, the flying time from Alaska to Russia is usually much shorter when a plane follows the longer-distance easterly route via Greenland. It is also generally quicker to fly on an easterly route from Norway to Minnesota via Siberia and Alaska than to take the seemingly "direct" westerly route.[26]

Less than a century ago, such information was speculative and of little relevance to practical issues; however, it is now as important as the railroad schedule was at the turn of the century. Still, invisible atmospheric conditions and major new air-travel routes cannot be accurately displayed on conventional World maps. They can, however, be precisely exhibited on the Dymaxion Map.

Fuller felt that the Dymaxion Map and the polar air routes are even more meaningful in light of the potential for destruction which modern intercontinental nuclear missiles possess. Conventional maps portray Russia and the United States as being separated by enormous distance, but the Dymaxion Map vividly illustrates the proximity of the two countries. Using it, an individual can clearly see that the superpower enemies are actually neighbors separated only by short polar travel routes which can be traversed in a matter of minutes by modern missiles and aircraft.[27]

One of Bucky's earliest and most controversial global discoveries was his theory that humanity has perennially migrated in a northwestern direction since human beings first walked the Earth.[28] Fuller initially detected that migration when he began plotting historical population centers on globes and the Dymaxion Map and found that humanity's major population centers have continually advanced along a northwest path originating in the South Pacific islands.[29]

Fuller believed that migration began in earnest when early maritime cultures learned to sail westward into the wind, and then he conjured up the following hypothesis.[30] Since human migration began, the dominant flow has followed the Sun on its mysterious westward journey. After that major movement reached Europe, it continued northwest to North America rather than southwest to South America.[31] Bucky uncovered a plausible explanation for that trend by studying multicolored versions of the Dymaxion Map. His explanation is still discernible on his most recent multicolored Map rendition, on which the colors delineate average yearly low temperature and the border between green and yellow areas represents the "freezing line." Along that edge, the average yearly low temperature is thirty-two degrees Fahrenheit, the point at which water freezes.

North of the freezing line, a person without proper clothing, shelter, and food would most likely freeze to death during the winter. Yet, if early settlers were to stray too far south of that line, they would be without ice, which was often stored underground and proved vital to food preservation during the summer. Remaining as near as possible to the freezing line also proved beneficial in that diseases tended to be less prevalent in colder regions. That advantage was, however, counterbalanced by the fact that a person roaming too far into the colder northern region was susceptible to sudden death from winter storms. In other words, living directly on the freezing line provided the best possible environment for human beings who did not possess modern climate-controlling artifacts. Hence, westwardly migrating pioneers seldom erected major settlements far from that boundary. This was particularly true during the early evolution of the human species, when clothing had not yet been developed and most people roamed naked or nearly naked.[32]

By carefully plotting data on his map, Fuller determined that humanity's northwest migration followed the thirty-two-degree freezing line from the South Pacific, across Asia and India, and into Europe. That migration pattern continued to advance into the United States where, since 1790, it has been formally documented through accurate census records.[33]

In studying the American aspect of global human migration, Fuller found that the first census and all those following it provide clear records of the way in which America's advancing population followed the thirty-two-degree freezing line. At the time of the initial census in 1790, the center of American population was concentrated near Philadelphia, and it advanced slightly south and west along the freezing line until the Civil War. The first noticeable deviation from that migration along the thirty-two-degree border occurred in 1851, when the center of America's population shifted from slightly south of the line to slightly north of it.

That incident coincided with the introduction of mass-produced steel and the flourishing of the industrial revolution in the United States. With those innovations, humanity, and Americans in particular, entered an era of transformation from an agrarian to an industrial society. At that time, scientists and engineers began developing inventions which permitted environments to be controlled by humans and which created year-round comfort in the most inhospitable climates. For example, practical structural steel permitted the construction of larger and larger buildings which could be created only by use of the steel skeletons that eventually resulted in skyscrapers. As architects designed larger structures employing steel frameworks, engineers began discovering how to heat those buildings with another new invention, central heating. With the advent of such innovations, people could move north of the freezing line without fear of being at the mercy of the elements. Still, the general flow of human migration continued to follow the northwestern advance which had originated centuries earlier.

When Bucky recounted his theory of human migration to his friend Henry Luce during the late 1930s, he also explained that global power centers had followed the population centers along the northwestern route. As a strong advocate of the British Empire's war effort in the United States, Luce was particularly interested in how such a migration had affected England, and Fuller explained that the British Empire, which Fuller regarded as a front for British economic interests, had also expanded along that northwest course.[34] Using available data, the Dymaxion Map, and his understanding of large-scale patterns, Fuller predicted that the economic hub of the British Empire would soon follow global migration patterns and move into North America. He also suggested that the logical site for such a new center was Canada.[35]

At that time, the sophisticated Henry Luce considered Fuller's prediction to be an unfounded but interesting observation. After all, Luce was in constant communication with top British government and business officers, and he knew nothing of such a shift. Yet, just six months later, Fuller received a note from Luce describing how the British government had shipped their secret archives to Ottawa, Canada. Those archives contained major components of the British economy, and Luce acknowledged Fuller's prediction of shifting power as having been correct. He did, however, moderate his recognition of Bucky's theory by arguing that the move was related not to the migration of economic power, but to a need for additional security as a result of World War II.[36]

Being a person who presented only the complete truth as he perceived it at a given moment, Fuller responded to Luce's note by saying that, like most people, he was aware of the security problems resulting from World War II, but that those factors had not influenced his prediction. Bucky explained that he felt that the apparently significant events of any given moment, which most people believed dictate human action, are, in fact, rarely the motivational force, and that in the majority of instances, the actual cause of a specific event can be perceived only within the context of major patterns.[37]

The migration of humanity over the course of thousands of years is such a pattern, and it constituted the foundation of Bucky's prediction. He understood that most minor events, which are perceived as significant when they occur and lead to major events such as the transfer of the secret archives, are almost impossible to predict with certainty. Large patterns such as human migration are far more obvious, and when perceived, they support an individual's understanding and prognostication of lesser circumstances and phenomena.[38] Because the Dymaxion Map does not present great distortions like the conventional Mercator map, many large-scale patterns are far more evident when displayed on it.

Prior to the advent of air travel, water was the primary mode of global human transportation and communication. At that time, the Mercator map adequately displayed a "sailing-ship World" in which the majority of goods, individuals, and long-distance communications were transported via sailing ship. The primary directions of the lengthy voyages those ships made were east and west. Within that environment, the limited number of cities with good harbors became the most important locations, and the primary routes of commerce extended east or west from those ports.

While examining history, Fuller found that in America, New York had the best natural harbor on the East Coast. Consequently, it prospered to become the largest city. Philadelphia and Boston also had good harbors, but because theirs were not as large as New York's, those cities did not develop to the size of New York. Still, any port city enjoyed a tremendous advantage over inland locations, which had to rely solely on slower, more costly modes of transportation and communication such as the horse and wagon.[39]

If a town was fortunate, it was blessed with a connection to the "ships of the land," railways, but few cities had such connections. Wherever a railroad station was erected, a new city sprouted and grew, just as was the case with ocean ports. In fact, Fuller felt that the growth of the American railway system had simply been a continuation of the large-scale patterns of human migration.[40] Railroads were generally constructed along east and west routes, and because they were designed to serve the areas populated by the most people, they also tended to follow humanity's northwestern migration route along the freezing line.

The settling of the "Wild West" and the expansion of the railroad system provided Fuller with other examples of individuals believing they were operating with some sort of divinely ordained free will while, in fact, they were merely following a large-scale pattern initiated thousands of years earlier. Certainly, each individual has the right to make choices, but Fuller believed that those choices would be much more successful if they

conformed to large-scale patterns.[41] The Dymaxion Map provides individuals with a tool for exploring and understanding many of those patterns
so that they may also learn to operate in greater harmony with other
individuals, the environment, and all Universe.

Fuller's work with globes and the Dymaxion Map provided him with
a uniquely intimate appreciation of the Earth's geography and humanity's
relationship to the Planet. Because he so clearly documented that perspective, his work also offers each of us, as well a future generations, insights
which could support the discovery of critical solutions to global problems.
Bucky appreciated the significance of the fact that 85 percent of the 30
percent of the Earth's surface covered by dry land is north of the equator
and that nearly 90 percent of Earth's inhabitants reside north of the equator. He utilized those figures to conclude that global humanity is, in fact,
concentrated in the Northern Hemisphere and that the Southern Hemisphere is predominantly water.[42] To better examine and understand the
significance of that fact, Fuller developed the "One-Ocean" rendition of
the Dymaxion Map shown in Fig. 11–1. With the South Pole positioned
in the center of that arrangement, he studied the Planet's great oceans in
their true form, as a single interconnected body of water.

He found that most maps portray the World's oceans as isolated
entities and came to feel that such interpretations encourage beliefs which
contradict inborn human instincts. After years of examination and personal experience, Bucky arrived at the conclusion that human beings, and
especially young children, have an innate feeling for the fluid wholeness
of liquids.[43] People naturally understand that all the water within a container is connected regardless of the size or shape of that vessel, and that
the Earth's surface is in actuality a single area covered with water. In
some locations the bedrock beneath our Planet's water juts above the
surface to create "land," but all the Earth's oceans actually constitute a
single interconnected whole, just as is true of water in a container.

Through study, Fuller found that humanity's maps and educational
systems indoctrinate individuals into erroneously believing oceans are
separate, isolated entities. However, with the One-Ocean arrangement of
the Dymaxion Map, Bucky clearly established and demonstrated the reality of Earth's integrated system of water in a convincing, undistorted
display available to everyone.

In his examination of history, Fuller had discovered that the most
advanced societies have always utilized the Earth's natural currents for

commerce, and that the majority of that trading has been inextricably linked to the World's oceans.[44] He also found that national leaders have invariably succeeded in convincing their citizens that those oceans, like nations themselves, are divided into isolated entities. By focusing on such a divisive strategy, leaders could maintain a "divide-and-conquer" mentality among their people, could assert the need for continued war with neighboring nations, could maintain a national pride, and, accordingly, could remain in power.[45]

After some study, Fuller came to understand that although humans relish the idea that they have the power to divide land and water masses arbitrarily and to possess small parcels, from a cosmic perspective such apportioning is absurd. He discovered that early in their history, sailing cultures had envisioned the fallacy of such isolationist beliefs and had initiated an ongoing tradition of unbounded global travel, linking themselves with their fellow human beings by water.[46] Bucky found that marine cultures had been freely sailing between locations on the unbroken highway of the World's oceans for thousands of years; however, in the last centuries, sailing cultures had, ironically, educated their children using the same Mercator map used by land-based cultures.[47]

Fuller discovered that because of such conditioning, the secrets of the Water-Ocean World (i.e., Earth during the period when most commerce and travel was conducted over water routes) were known only to a few individuals. Those people were the ones who regularly traveled the seas and the leaders who exploited nautical secrets for greater wealth. As leaders, they had acquired the clandestine knowledge of the sea from their ancestors, who had passed information and power onto their children as a royal or noble heritage. That secret understanding of the sea is exemplified in knowledge about the primary interconnecting waterways of the extreme Southern Hemisphere, which were essential to the control of trade, and therefore wealth, throughout the Water-Ocean World.[48] Even today, those routes are not visible on the Mercator map because the South Polar regions are not clearly displayed.

Individuals and nations with knowledge of the southerly trade routes became even wealthier by transporting goods from areas of inexpensive abundance to locations where the items were scarce. In that way, inexpensive commodities were transformed into rare, extremely expensive goods during a single voyage. Bucky concluded that such exploitation of the distance and isolation between groups of people had been a common practice among human beings since the dawn of humanity. He saw that such a pattern recurring over and over again, dominating cultures until the last few decades and, in some instances, continuing to persist.[49]

In every instance he examined, a small group of people had purposefully exploited the remoteness of most people from one another by controlling the majority of trade. Fuller believed that such a process, although seemingly unsupportive of humanity, had inadvertently assisted the development of human cultures. For example, Bucky discovered that during the domination of the sea by the British Empire, metals were transported to England from around the World.[50] In England, the metals were combined to produce crude alloys for use in items as varied as weapons and household utensils. To reach England, most of those metals were shipped great distances using the secret trade routes which employed the strong currents flowing eastward around the Antarctic.

Those Antarctic routes constituted a surreptitious superhighway of the seas, and they were available to any ship which could withstand the turbulent conditions created by tenacious currents, numerous storms, and a frigid climate. Still, the Antarctic routes remained fairly well concealed because the people who knew of them were aware that any ship sent into the Antarctic currents had to be extremely sturdy or it would not return. Because of that problem, powerful leaders had exceptionally stalwart ships constructed for Antarctic travel. Less well-constructed ships which attempted such a passage usually turned back and reported that the Antarctic routes were impassible. As a result, only a small armada of ships was able to exploit the most efficient means of commerce between distant ports for most of human history. Bucky felt that since sailing had become commercially viable, the secret of the Antarctic passage had been a significant economic asset which was guarded by any nation or individual possessing it.[51]

Great Britain had been the last country to control that route and, therefore, the Water-Ocean World. Following World War I, global air transportation began to unfold, and by the onset of World War II, air transport had become a major factor in World domination and transit. With the advent of the modern airplane and regularly scheduled flights along an infinite number of routes following World War II, individuals were no longer restricted to water routes and the domination of those who controlled the seas. Instead, a growing number of individuals were now free to travel to almost any location on Earth whenever they chose.

Bucky discovered that although the port cities which had grown into the major centers of population and commerce during the past few centuries still retained a cultural and economic prominence, their prosperity was no longer ensured by location. Instead, those cities were finding themselves in competition for dominance as global hubs of activity with more obscure, inland locations.[52] With the advent of the airplane, the most efficient trade and commerce routes began shifting from the east–west sea-lanes to the most geographically direct route between two sites, and the significance of sites joined by air networks was not necessarily their location. Fuller found that in the new era of air transportation, any city or nation could compete for prominence as a global nerve center, and that the ones which grew in population tended to become dominant. However, he also speculated that as individuals became increasingly autonomous in all areas including thought, economic considerations, and housing, they would feel less and less compelled to live in large cities. By examining census data he found that with the coming of such changes, the

Fig. 11–2   A large Dymaxion Map displayed as part of a 1973 traveling exhibition.

dominant population centers were shifting from seaports to locations along major flight paths, and he felt that that trend would continue.[53]

From Fuller's perspective, such sites will become even more significant in the future as airplane flight paths invariably follow the shortest distance between locations. In our Northern Hemisphere–oriented global society, the majority of those air routes will be established over the North Polar area, causing locations in Canada, Greenland, and Russia to become far more important than ever before.

That newly emerging north–south travel orientation and commerce is extremely evident on the most common arrangement of the Dymaxion Map. That version, which Fuller called the Air–Ocean or Spaceship Earth arrangement, is generally referred to simply as the Dymaxion Map, and it displays all the Earth's landmasses in the central triangular areas with no breaks in continents or visible distortion in size or relationship to one another. It also reveals the efficient north–south North Polar travel routes, which are not visible on most World maps.

Bucky witnessed that north–south Earth orientation becoming increasingly important as air travel grew during the 1950s and 1960s, and he was quick to note the significance demonstrated by a 1961 transportation event which went virtually unnoticed by the general public.[54] At that time, three jet airplanes transported more passengers across the Atlantic than a single passage of the ocean liner *Queen Mary* in a fraction of the time and far more economically. Because they were not at the mercy of ocean currents and conditions, those airplanes traveled along a more direct, northerly route than the one utilized by ships of the sea.

Fuller was one of the first to predict that with the inception of practical ocean passage by planes, sailing ships would become obsolete as an efficient means of human transportation.[55] Certainly, people continue to take leisurely cruises on ships, but he foresaw the jet plane replacing all other means of transportation for traveling between two locations as expeditiously as possible.

Bucky felt that the rapidly increasing transformation occurring within all areas of society was far more significant than most people realized. He frequently reminded his audiences that the dramatic shift from water to air travel represented another large change not popularly predicted even as recently as World War II. Being a observer of major patterns, he had first prognosticated a global air-transportation network in the late 1920s when he was labeled as a wild dreamer at best.[56]

He also realized that the practical onset of air travel would provide individuals with opportunities to live at almost any location on Earth, and he therefore began developing the "autonomous dwelling machine" which later became the Dymaxion House and further evolved into the development of the geodesic dome.[57] Fuller predicted that with such liberty, a great many individuals would elect to reside in rural areas and small remote towns rather than the gigantic cities into which people had migrated in search of work.[58] In the Fuller scenario, the shift to smaller, less centralized communities would combine with the growing importance of inland cities and result in the decline of the large port cities.

As early as the 1970s, Fuller noted that such deterioration had already begun. He contended that the demise of large port cities such as New York and London was (as it continues to be) generally being obstructed only by the corporate and industrial power brokers with heavy investments in port-city real estate. Bucky felt those individuals were unsuccessfully attempting to obstruct the natural trend of human development toward a north–south-oriented, yet ultimately omnidirectional, pattern of travel. They witnessed business and industry leaving the cities and realized that populations would soon follow suit. Those in power also understood that such a migration would result in declining rather than increasing values for their urban investments, and they were doing (as they still are) everything possible to check the unstoppable shift.[59]

One component of their strategy is an attempt to attract new forms of industry into port cities. Advertising campaigns are constantly being created to portray large, deteriorating port cities as attractive and to present the appearance that such cities represent exciting sites for vacations and business investment. However, since east–west ports are no longer needed to participate in global business, many corporate executives have found that locating in large port cities such as New York and Philadelphia where real estate is overpriced is not an economically sound decision. Hence, corporations, their employees, and their customers are tending to leave rather than to flock to port cities, reversing the trend of previous generations.

Fuller also observed another human reaction to the phenomenon of migration from large cities. Individuals who invest in big-city real estate and realize that their investments will eventually fail, simply pretend that large-scale changes are not occurring.[60] Such behavior requires maintaining an illusion that the relationship between humans and their Planet remains static and that humanity continues operating within an east–west environment which is adequately portrayed on common World maps such as the Mercator map. Although that strategy has been successful in the past and the Dymaxion Map has never been massively disseminated to the public on a continuing basis which would provide an important global perspective, Bucky was certain that the natural comprehensive human curiosity could not be impeded forever. He felt that this is especially true of children, who are continually confronted by global issues at earlier and earlier ages, and who, like every individual human being, want to perceive and understand their domain of operation as comprehensively as possible.[61]

Because Fuller detected the large-pattern transformation of humanity from a limited east–west global alignment to an omnidirectional, predominantly north–south pattern of commerce early in his life, he was able to display that shift on the tool which he had utilized to discover the trend: the Dymaxion Map. Thus, the issue of how to visualize our emerging omnidirectional orientation was solved by Bucky before it ever became a problem. Now, each individual need only spend time examining our global environment on the Dymaxion Map to begin truly appreciating humanity's transforming relationship to the small sphere Bucky called Spaceship Earth.

In all his endeavors, Fuller's basic operating strategy was to construct artifacts (i.e., inventions) which solved specific problems and could be examined by the public. During the early years of his life, Bucky had noticed a great many people with "good ideas" who never risked action. Later, as he became more well known, Fuller was perpetually accosted by such individuals who, with the best of intentions, wanted to apprise him of their good ideas, and he became even more aware of the fact that everyone has ideas. Bucky had, however, concluded that merely talking about ideas does not support their advancement or the development of individuals and humanity.[62]

In fact, he found that the majority of people do nothing about their good ideas except engage in seemingly endless discussions. During such discussions, those with the good ideas perpetually attest to the value of their concepts and how their ideas would improve the human condition if only other people would abide by their wisdom.[63]

As he had tried out a similar strategy of talking about his good ideas for a short period, Bucky was keenly aware of that methodology and its ineffectiveness. Once he became cognizant of the abundance of good ideas being discussed, Fuller began building models, both full-size and to scale, of his concepts in order to advance them into more functional forms and to demonstrate their practicality. Through that process, he discovered that translating an idea into a practical artifact is an extremely significant step in its development.[64]

Fuller found that once an idea is manifested in a functional configuration, it can be examined by others, and only then can it be put to the crucial test: acceptance and use by society. He also learned that such acceptance does not occur overnight in most instances. Rather, an idea, and the physical manifestation of that idea, must progress through a gestation period which might last for months, years, decades, or even centuries. Still, Fuller believed that an idea has first to be manifested in a functional artifact which can be displayed and discussed.[65] He also felt that he learned much more from reducing his ideas to such practical solutions and creating artifacts than from anything else which occupied his time.[66] The Dymaxion Map and Fuller's other global viewing inventions provide excellent demonstrations of that strategy in action.

Although he had successfully invented the Dymaxion Map, Bucky was not satisfied and sought to develop other such tools. One of the most sensational such ventures was his campaign to build a sophisticated "miniature Earth" which he named the Geoscope. Because it is fundamentally a Dymaxion Map fabricated upon a geodesic sphere, the Geoscope can be considered a hybrid of the geodesic dome and the Dymaxion Map.

Prior to formally designing the Geoscope, Fuller examined cultural patterns and desires to determine if such an invention was needed and would be accepted. Through that study, he determined that at some point in the near future, large-scale global issues would be a much greater priority and that at that time his invention would be perceived as a necessity. Bucky also discovered that throughout history a great many people had worked on developing better tools for displaying and examining the whole of our Planet. In recent times, such insightful individuals and their ideas have become more evident as humanity has become more globally conscious. World-viewing artifacts invariably appear at large gatherings where historical and futuristic ideas are exhibited and discussed. Evidence of the search for better methods of displaying Earth has been seen at world's fairs and other cultural expositions, and those displays helped convince Fuller that his global viewing inventions would be appreciated and used.

During the last century, several world expositions have, in fact, featured large miniature Earth models and globes, and following Bucky's invention of the geodesic dome, expositions tended to shift from globes to geodesic domes as focal elements. Domes have been, and still are, prominent at EPCOT Center, the 1967 Montreal World's Fair, and the 1986 Vancouver World's Fair, among others. Fuller himself envisioned the Geoscope providing a natural melding of the geodesic dome and humanity's innate desire to view the whole of Earth. Accordingly, he set about designing that artifact for future generations even though the necessary technology had yet to be invented.

The first Geoscope was ingenious yet extremely simple. Bucky fabricated it with the help of a group of architectural and engineering students on the roof of the Electrical Engineering Building at Cornell University in 1951.[67]

The structure's framework was a geodesic sphere twenty feet in diameter, built of delicate wooden slats painted blue. Since the Earth's surface is mainly water, Fuller chose blue to provide the Geoscope with a sense of the Planet's liquidity. The slim blue struts produced a delicate, almost transparent quality similar to that of water. The students who chose to participate in Bucky's project were each assigned a triangular section of the Dymaxion Map to duplicate on a larger scale for a particular sector of the Geoscope. Each student would painstakingly draw his or her Map

Fig. 11–3 Students working on the Cornell Geoscope.

section on a large triangular sheet of paper cut to match a Geoscope sector, and those enlarged Map sections were used to outline the Earth's landmasses on a shiny bronze screen.[68]

The landmass forms were then cut out of the screen and attached to triangular sections of chicken wire corresponding to the triangular openings between the wooden slats.[69] That process created large triangular chicken-wire-and-bronze Map segments which were fastened to the wooden slats, thereby creating the first Geoscope.[70]

After weeks of work, Fuller and his student assistants were admiring their representation of the Earth's surface when they noticed how the sun shining on the Geoscope caused the landmasses to gleam a brilliant gold which vividly contrasted with the blue created by the sky seen through the transparent areas and the blue struts.[71] Years later, Bucky was pleased to discover that his brass screen and blue slats were the ideal materials for that presentation when, in 1969, the first photographs of the Earth taken from the Moon were displayed, and they revealed an image very similar to the appearance of that early Geoscope.[72]

The twenty-foot-diameter sphere was aligned with its axis parallel to the Earth's axis and was supported by three rigid legs atop the roof of the Electrical Engineering Building. Prior to its final installation, the Geoscope was rotated so that Ithaca, New York (its location), was positioned at the sphere's highest point.[73] In other words, the Geoscope was oriented toward the Earth in the same arrangement as would be a small lifeboat attached to the side of a large ship. In such a situation, the lifeboat is constantly aligned with the position and direction of the ship. Similarly, the Geoscope remained precisely aligned with the Earth. If the ship were to tilt in one direction, the lifeboat would react with an identical maneuver, and that phenomenon was also true of the relationship between the miniature Earth Geoscope and the Earth itself.[74]

The result of that phenomenon was accurate observation and a unique perspective. A person seated in the lifeboat would experience the same relationship between the lifeboat and its environment as was felt on the much larger ship; however, the smaller lifeboat would provide a frame of reference which was much easier for an individual to comprehend than that experienced on the gigantic ship. A person in the lifeboat could see the entire boat and its relationship to the larger environment; however, when standing on the deck of a huge ocean liner, the ship's immense size would prevent such a comprehensive perspective. Still, the orientation would be identical in both vessels.[75]

That same phenomenon occurred with the Geoscope. It provided a precise representation of Earth and its relationship to the rest of the components of the cosmos, which could be intimately examined and experienced by individuals. To create an even more vivid experience of that relationship, Fuller and his team constructed a viewing platform within the sphere. They attached a ladder to one of the tripod legs so that an individual could climb onto the interior platform. The team also built a chin rest above the platform so an observer could easily position his or her head in the exact center of the sphere and experience a view corresponding to one from the center of the Earth. That viewpoint afforded an opportunity to look out through the Earth's landmasses as well as to observe their true relationship to one another and to the visible celestial bodies.[76]

The fact that the Earth's center was be four thousand miles from the observer would seem to impede an accurate view of those relationships. However, Fuller realized that when compared to the nearest star being examined (i.e., the sun at ninety-two million miles from the Earth), the distance between the Earth's center and the Geoscope would become insignificant and that in relationship to more distant stars and galaxies, that four thousand miles would become even more inconsequential.

To clarify that phenomenon further Fuller explained that if the light radiating from the Earth were visible in the nearest galaxy, an observer at that location would not be able to distinguish the distance between the Geoscope and the Earth's center. In other words, because of its tiny size relative to other celestial bodies, anyone observing Earth from the distance at which Fuller and his team were observing other celestial bodies would find it impossible to distinguish the Geoscope from the Earth itself. Since that phenomenon would also apply to observations made in the opposite direction, the view from the Geoscope could, for all practical purposes, be considered identical to that from the center of the Earth.[77]

Asserting their natural curiosity, Cornell science and engineering students began using the Geoscope to make astronomical observations and calculations, such as the positions of stars at specific times, and compared their readings with those of local observatories. They found their Geoscope observations to be extremely accurate, once again validating Fuller's contention that the four-thousand-mile distance between the Earth's center, which is the true observation point for such calculations, and the Geoscope was insignificant.[78]

Students also relished viewing the night sky from the Geoscope's underside. They would lie with their heads below the Geoscope looking directly from the South Pole to the North Pole and would be amazed to observe the North Star directly above the Geoscope's North Pole. Some students were so eager to work with the Geoscope that they occupied the center viewing post for hours on end and were able to see the North Star apparently fixed in the same position above the North Pole while other stars seemed to move in relationship to the Geoscope's grid of slats and mesh. Those students also discovered that the closer stars appeared to the Earth's equator, the more rapidly they seemed to move past the grids of the Geoscope.[79]

Fuller explained that although their experiential observations were correct, their interpretation of the phenomenon was not. He reminded his students that the movement they had witnessed was not that of the stars but the movement of the Earth itself and primarily the Earth's rotation. Once the students grasped that simple fact, they began to realize they were, in fact, experiencing the spinning motion of our Planet.[80] Fuller himself speculated that the Geoscope may have provided human beings with their first experience of actually "feeling" the Earth's rotation.[81]

Students were so fascinated with the Geoscope that many clear evenings would find groups making observations until dawn. Bucky also loved to spend time with the Geoscope, and he found that because it was not solid, the bronze mesh used to form the landmasses caught light from the stars, providing a particularly beautiful vision. The Geoscope's combination of elements produced an amazingly true sense of stars glistening against the landmasses, and people were constantly stopping by to view the sky through it.[82]

Unfortunately, that same curious attraction also led to the Geoscope's demise. Being a sturdy geodesic sphere, the Geoscope survived New York State's weather quite well on the roof of that building. It was, however, not immune to human intervention, and one Halloween a group of mischievous students decided to try rolling the popular attraction around the campus. They reached the rooftop Geoscope only to discover that because the components had been assembled on the roof, it would not fit through the doors. Consequently, the pranksters hastily decided to transport the Geoscope by the most direct means available, pushing it over the side of the building. As it was a fragile structure designed for stationary use, that original Geoscope could not withstand abuse, much less a fall from several stories, and it was reduced to a pile of rubble on the lawn of the Cornell Electrical Engineering Building that night.[83]

Undaunted by the destruction of the Cornell artifact, Bucky continued developing other, more sophisticated Geoscopes. During 1970, he designed and supervised construction of the largest of those artifacts at

Fig. 11–4   An exterior view of the Southern Illinois University Geoscope.

Southern Illinois University in Edwardsville. At the time, Bucky was serving as Distinguished University Professor at SIU, and a consortium of religious organizations combined their resources to commission his services in designing a portion of their new, centralized religious-activities building. Bucky's contribution was a fifty-foot-diameter Geoscope which also serves as the building's auditorium.[84]

As with other Geoscopes, an individual enters into the structure to outwardly observe Universe. Fuller found that when a person arrives at the center of a Geoscope and begins to look out, he or she is immediately overcome with the sensation of being thrust outward into Universe.[85] Thus, the Geoscope models one of Universe's great paradoxes: An individual turns inward to expand outward.

Gradually, cosmic-observation Geoscopes such as those erected at Cornell and Edwardsville were displaced in Fuller's work by what he considered a far more useful configuration of the device. Following his experience at Cornell, Bucky spent three years working part time on a

Fig. 11–5   Looking up at the interior of the SIU Geoscope.

two-hundred-foot-diameter Geoscope at the University of Minnesota.[86] He was extremely dedicated to that effort because he felt that that invention would revolutionize humanity's perception of our Planet and would be of enormous benefit to human development.

Rather than an instrument for looking outward at the cosmos, Fuller's new concept of the Geoscope was as a tool for individuals to observe Earth and the relationships between human beings and our Planet. During the period from 1954 through 1956, he and a group of Minnesota students were able to complete a portion of the massive two-hundred-foot-diameter geodesic sphere which was to serve as the structure of his new miniature Earth Geoscope.[87] It was also during that period that Bucky initiated a campaign advocating the construction of a two-hundred-foot-diameter miniature Earth Geoscope in New York City directly across from the United Nations. His choice of the two-hundred-foot-dimension was not arbitrary. The new version of the Geoscope was specifically designed to display the entire Earth at a scale which would allow people to identify with details as tiny as individual houses and thus to perceive the relationship between their homes—and therefore themselves—and the entire Planet.[88]

One news event also influenced Bucky's choice of size and scale. At the time, the United States Air Force was in the process of surveying the Earth's landmasses using newly developed aerial techniques in conjunction with motion-picture cameras. The scale of the U.S. Air Force's lowest level and most detailed flight photographs was 1 to 200,000, or one inch equaling approximately three miles. At that scale, the entire Planet can be portrayed as a sphere approximately two hundred feet in diameter. Because that scale was already being used to create a mapping system which could eventually be employed to provide the Geoscope with even better accuracy, and because that scale furnished observers with a workable view of their relationship to the entire Planet, Fuller adapted the same 1-to-200,000 scale for his Geoscope.[89]

At that scale, the City of Los Angeles is 1½ feet in diameter, and a town of five thousand people emerges as a circle approximately 1 inch in diameter. An acre of land is ¹⁄₆₄ inch square (i.e., roughly the smallest detail recognizable with the naked eye), and an average house, although minute, is represented as a speck ¹⁄₁₀₀ inch in diameter. Bucky realized that people would not be able to see their homes on such a Geoscope without standing next to the exact location and using a magnifying glass. They could, however, recognize their town and perhaps their sector of a city from a short distance away, and that possibility was to provide a more intimate relationship between the observer and the Geoscope, even though it was, in reality, designed to be viewed from a distance of approximately one thousand feet.[90]

From such a vantage point, an observer's town would be visible, and an individual could realistically claim that although his or her house could not be seen, it was an actual speck which could be found with powerful binoculars. Thus, every person could have a sense of belonging on Spaceship Earth.[91]

As with so many of his inventions, Fuller's Geoscope idea of the 1950s was designed to utilize technology which he had forecast but which had yet to be invented. He envisioned the structure as being covered with either colossal rounded televisions or, more realistically, a massive number of tiny, computer-controlled lights. Those miniature lights could be instantaneously changed in color and intensity to produce the appearance of a single image, just as is now done on the stadium screens used at major athletic events.

The Geoscope was to portray patterns of historical events which occurred over such a long period that humans have never before been able to truly experience or comprehend them in a short time. For example, it could be programmed to display the Earth's changes over the last four ice ages, which were each on the order of one million years in length. When displayed using intervals of 250,000 years, that entire 5-million-year period of enormous change could be reduced to a comprehensible four-minute presentation. Such a spectacle would provide viewers with a capsule summary of how our Planet's landmasses were formed as well as some idea as to the relatively short period humans have inhabited Earth.

Even more significantly, the large Geoscope would be able to visually display continuously updated information about all aspects of Earth. Conditions such as current population, resource allocation, weather patterns, and weaponry deployment could be revealed in an instant to provide observers a firsthand experience of critical global activities which vitally affect the lives of everyone. Bucky believed that by utilizing such a Geoscope, individuals could begin to recognize formerly invisible patterns and could work together in generating unique new strategies far more comprehensive than what had been seriously considered before.[92]

He also felt that the Geoscope could be employed to examine and display strategies for success which he proposed in his World-Game Comprehensive Anticipatory-Design Science.[93] Those strategies could then be programmed into the Geoscope's computer, just as war games are now programmed into computers, and the future effects on Earth would be instantly visible on the Geoscope. With such a tool, rather than waiting for decades or centuries to learn the harmful or successful consequences of actions, humanity would be provided with immediate visual responses to proposed solutions on a scale that every individual could understand.

During the 1950s and 1960s, Fuller advanced that Geoscope closer to reality. He prepared a proposal to erect it in New York City on what was then Blackwell's Island and is now called Roosevelt Island. That island is just east of the United Nations buildings, and the Geoscope was to hang from a series of nearly invisible cables and masts so that its top would be at the same height as the top of the four-hundred-foot United Nations Secretariat Building. If it were suspended in that manner, almost any delegate looking out his or her window would be constantly confronted by an ever-changing, animated view of our entire Planet and the consequences of decisions and actions taking place within the United Nations. For instance, even the slightest armed conflict or military violence was to be instantaneously displayed with glaring red lights, so that everyone would be aware of it. Bucky believed that a continuously changing Geoscope of that magnitude would provide a perspective of our tiny Planet much different from what is normally available and would inspire people responsible for global decisions to think twice about potential actions.[94]

Fuller's U.N. Geoscope proposal in no way intimated that he had capitulated to the popular belief that politicians were useful to modern humanity. He felt as strongly as ever that politicians were becoming increasingly obsolete and that they would be totally outmoded by the 1970s. Still, he wanted to confront as many people from different sectors of the Earth as possible with a palpable display of the consequences of humanity's actions, and the United Nations seemed to be the single site at which to fulfill that desire.[95]

Although Fuller's dream Geoscope has not yet been constructed, his proposal did receive a favorable reception from a majority of the United Nations representatives. In fact, the Secretary General at the time, U Thant, was so excited by the idea that he hosted a luncheon for United Nations ambassadors from all delegate countries at which Fuller formally presented the Geoscope concept. Thant also advanced the idea on several other occasions, but the United Nations was unable to generate the $10 million that Bucky estimated would be required to complete the project.[96] Fuller wanted to see that massive Geoscope erected during his lifetime, but once again, he did not become mired in attempting to complete the project. Instead, he moved on to the next issue which he felt needed his attention, knowing he had initiated another idea that would progress through its gestation period and be available to humanity when it was vitally needed.

Since that time, the computers and mechanical technology needed to economically erect such a Geoscope have been invented. The primary obstacle which limits its practical implementation is humanity's priorities. In other words, how deeply into crisis must we plunge our species and our Planet before we feel forced to produce this sophisticated tool and benefit from the perspective it can provide?

Fuller understood that humanity has always operated within a system of semichaos which he often described as *emergence by emergency*. People invariably put off resolving issues until the ascendance to public visibility of a serious problem which focuses the general population's attention on that issue. Then, when the issue becomes popular, readily available solutions are sought and it is resolved.[97]

Fuller found that that process occurs with problems as far-reaching as eradicating major diseases and education and as insular as installing streetlights and sewer systems. He discovered that, in most instances, human nature invariably influences people to wait until an issue reaches a critically threatening level before acting upon it, and he felt that the relationship between human beings and Spaceship Earth was rapidly reaching such a level. Bucky believed that our emerging global society needs to take a giant step beyond our ancestors' survival-oriented struggle against Nature and each other. He saw that that ongoing "battle" had moved us to the brink of annihilation because of its focus on more advanced weaponry rather than on the tools for life support which he christened "livingry," and he introduced the Geoscope as an example of how technology could support rather than destroy life.[98]

CHAPTER 12

A Home at Last

Housing was always a favorite theme of Fuller's work and thoughts, and despite his reservations about industrialists, he was quick to join with them on a major project when that undertaking appeared to advance the cause of better housing. The issue of shelter and Bucky himself were both thrust into the forefront of the United States war effort during the height of World War II when the Dymaxion House project was resurrected. Since he had considered the problem in great detail well before it surfaced, and since it exemplified the "emergence-by-emergency" strategy in which he so unremittingly believed, Bucky was thoroughly prepared to tackle the problem of housing in 1944.

While working for the Board of Economic Warfare during World War II, he was given constant access to the most recent information about manufacturing in the United States, including the problems which arose as many of the young men who composed the work force were sent off to war. Housing was one of the problems Fuller observed, and during the war, it became a top-priority issue almost overnight. Housing rose to instant prominence as a solution to another issue rather than as a problem which required a solution.

Wartime aircraft production was concentrated throughout the West Coast and the Midwest at the time, and one of the major aircraft manufacturing centers was Wichita, Kansas. In a scenario typical of what had occurred at other such wartime centers, the population of Wichita had grown from 100,000 to 200,000 in approximately one week at the onset of the war as both males and females patriotically responded to a call for workers to manufacture military equipment. Small aircraft factories which had barely been able to stay in business were expanded in a matter of weeks, and housing near those plants was so scarce that beds were used on a twenty-four-hour basis, with workers sleeping in eight-hour shifts.

As the war dragged on, many of those factory workers became disillusioned with Wichita's difficult living conditions, and comfort began to overshadow patriotism and wages. The workers also began to believe that when the war ended, aircraft production would immediately drop drastically, and their jobs would be in jeopardy. Such sentiments caused a steady flow of workers to leave the aircraft industry for other fields which appeared to be more secure and less demanding. Management at aircraft plants around the country charted that problem in a single statistic: the ratio of people who quit to those who were hired. And in the spring of 1944, the number quitting was increasing dramatically while the number being hired was declining at a similar rate. The issue was so significant that it actually threatened the Allied war effort.

Both the aircraft factory executives and the War Department officials were frantically searching for a solution when someone recalled that Bucky Fuller was always championing the idea of mass-production housing built in factories and shipped intact to people's lots. As he was a government employee working for the Board of Economic Warfare, Bucky was immediately summoned to the higher echelons of the War Department and interrogated about his housing ideas.[1] Most everyone involved in interviewing Fuller was both surprised and delighted to learn that he had not abandoned the project he had initiated nearly eighteen years earlier. Instead, Bucky had continued to think about and develop his 4D (by then known as Dymaxion) House so that it would be ready when the idea had sufficiently gestated.[2]

Fuller staunchly believed that all ideas have a specific gestation period between their inception and the moment they are ready for practical application. He also felt that the gestation period is predicated upon the particular field of endeavor, and he steadfastly maintained that, for the housing industry, that period was twenty-five years. Thus, when he was approached about his 1927 idea in 1944, Fuller believed the time was right to begin working on its final stages so that it could be mass-produced for the general public at the proper moment: 1952.[3]

One of Fuller's earliest presentations of the Dymaxion House idea was to Herman Wolf, who worked in public relations for the War Production Board. Wolf then arranged a meeting between Fuller, Walter Reuther of the CIO, and Harvey Brown, president of the International Association of Machinists, which represented aircraft workers. The machinists' union was particularly excited about the possibility of supporting Fuller's project because it was their workers whose postwar fate was in jeopardy, and following that meeting, the two union leaders quickly arranged a meeting between Bucky and the management of the Wichita company they believed to have the best labor relations.[4]

Everyone involved agreed that the company, Beech Aircraft, would be the best place to establish the initial prototype project which, if successful, was to be duplicated at other plants. Everyone except Fuller felt that what they called a prefabricated house would be a viable product sought by military personnel returning home from the war.[5] Bucky, however, was adamant about his intention, and whenever the term prefabricated house would arise, he was quick to remind everyone that he was not interested in simply creating a factory-built house. In fact, when he met with Jack Gaty, a conservative man who was vice-president and general manager of Beech Aircraft in Wichita, one of the first issues Fuller addressed was the monumental scope of the project.[6]

Bucky began by reiterating that he had no interest in simply building prefabricated houses, and at the end of four hours of Fuller's exposition, Gaty was impressed enough by both Bucky and his ideas to extend Beech's commitment to the project.[7] He also requested that Fuller present his ideas to the Beech workers, and within hours an appearance had been arranged. At that meeting, all the shop stewards and workers able to fit into the International Association of Machinists auditorium listened as, once again, Bucky did not lack for words. After several hours of presentation, the prototype project was unanimously accepted by both workers and management.[8]

By the time Fuller returned to Washington, DC, a few days later, to organize his portion of the production effort word of the project had already spread throughout Wichita, and the problem of workers leaving for other jobs instantly began to subside. In fact, when Bucky returned to Wichita a few weeks later, the first thing Gaty showed him was Beech's chart of hirings and resignations, documenting the fact that the dramatic shift had begun the very week Fuller spoke to the workers.[9]

Bucky had gone to Washington to prepare for a move as well as to handle preparation for the next stage of Dymaxion House development. When he had accepted his position with the Board of Economic Warfare in 1942, Anne had remained in New York so that Allegra could complete the year at the Dalton School. Following that school year, the family was reunited in Bucky's small Washington, DC, apartment, and Allegra was enrolled as a day student in the fashionable Miss Madeira's School. When the nomadic Bucky suggested that his family once again change residences to the crowded, relatively isolated midwestern town of Wichita, which was so far from the cultural resources his wife and daughter loved, neither of them shared his enthusiasm. Thus, Allegra became a boarding student at Miss Madeira's School while Anne moved to an apartment in Forest Hills, Long Island, near her family.[10]

Meanwhile, in his continuing quest to develop the quintessential inexpensive, yet technologically advanced, home for the average family, Bucky was confronted with a necessary but unappealing task: financial organization. While in Washington, he incorporated the Dymaxion Living Machines Company, which was later changed to Fuller Houses, Inc., with R. Buckminster Fuller as chief designer and engineer. Having no desire to be involved in the daily bureaucratic operation of his corporation, especially the financial management, Bucky asked several trusted fellow government workers to serve as corporate officers.[11] Key among those individuals was Herman Wolf, who was appointed president.

To uphold his portion of the agreement with Beech, Bucky had to generate capital, and Wolf convinced him to once again begin dealing in an area in which he was extremely uncomfortable: the sale of stock. Despite his discomfort with and mistrust of the system, money was needed, and Bucky reluctantly agreed to issue 15,500 shares of stock in Fuller Houses at a par value of $10 per share. Bucky retained 5500 shares, Beech was given an option on 1000 shares, and the remaining 9000 were sold to the public. In actuality, the public investors consisted mainly of the company's officers, as well as their friends and family members. Since Herman Wolf was becoming more and more of a driving force behind the organization, many of those shares were sold to members of his family and friends in the labor unions. Although Bucky did not know what details of the project Wolf shared with others, Bucky's message to potential investors was crystal clear.[12]

He told everyone that he felt Dymaxion Houses could not be practically marketed until adequate time and resources had been allocated for development, and that the main criterion he would use to determine the exact moment of readiness was when a Dymaxion House could be installed anywhere in the World as quickly and easily as a telephone. Fuller further explained that he projected that if they worked diligently on the project, the moment of readiness would occur sometime in 1952. Accordingly, he suggested that everyone involved regard their time and money as a contribution toward solving the postwar housing problem and that investors consider their money "thrown away."[13]

Although, Bucky would straightforwardly explain that position to anyone, he had learned about corporate politics earlier and had chartered Fuller Houses, Inc. in a way which ensured him complete control. Thus, his 5500 shares carried with them veto power over all aspects of production of the House.[14]

In October of 1944, Fuller and the officers of his corporation resigned their government positions and moved to Wichita to begin production of the prototype Dymaxion House. Upon arriving, Bucky met with Jack Gaty and resolved the details of production. As he did not want the House project disrupting aircraft production, Gaty suggested that Fuller's work be set up in an abandoned building on the Beech grounds. After some discussion, Fuller agreed to use the location once it was adequately renovated, and within days, the new site of Fuller Houses was coming to life under the watchful eye of Bucky and his staff. Although small, that staff was composed of some of the finest workers in the area.[15]

Remembering another lesson from his Dymaxion Car venture, Fuller had made an initial request of only six mechanics from the Beech work force of twenty thousand. However, because he recalled his earlier problem with workers who were afraid of losing their jobs, Bucky insisted that the Beech mechanics be the most skilled workers available.[16] He felt that such workers would be irreplaceable and, consequently, they would not fear the loss of their jobs in the aircraft plant while working on the Dymaxion House.

By the time his small operation was established at Beech, word of the project had spread throughout the aircraft industry.[17] The rampant speculation was that Fuller and his team were not just developing a prototype but actually creating a gigantic new industry which would soon replace aircraft production, thereby securing the jobs of hundreds of thousands of factory workers.[18]

One of the primary problems Fuller envisioned when he first discussed the project was something which constantly plagued his work: his insistence on employing state-of-the-art materials, even if they had not even been invented when he initiated a project. Because his inventions were predominantly designed for future generations, Bucky was never hampered by the limitations of the ordinary materials which were used for a task at a particular time. Instead, he simply examined the materials

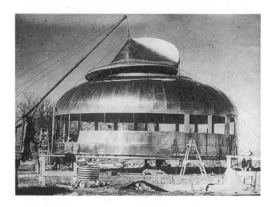

Fig. 12–1 through 12–3   Stages of erecting the Dymaxion House.

Fig. 12–4 The completed exterior of the Dymaxion House next to the tube into which all the components of the entire House were packaged for shipping.

available as well as the probable development strategy for new substances and then speculated what types of materials would be available at the time his inventions would be completed or needed by society. That was true of the high-strength metal alloys needed for the jet engines on the Dymaxion Vehicle as well as in the 1930s when he told people that his Dymaxion Bathroom would eventually be fabricated from a lightweight plastic-like material invented years later, which came to be known as Fiberglas.

Because of that stance, Fuller felt obligated to require the future use of materials which he believed would be available in 1952 and beyond and to demand that such unknown materials actually be specified for use in the Dymaxion House. His more pragmatic associates did, however, convince Bucky that only the modern materials available at that time be used, ensuring the project would not be hampered waiting for substances which had not yet been invented.[19]

Still, even those materials posed a problem. Because of the war, most industrial materials were tightly controlled, and that was particularly true of the substances required for strategic military applications such as the production of aircraft. As he was not very concerned with the politics of short-term issues such as World War II, Bucky was adamant that his Dymaxion House prototype be fabricated from the most modern, lightweight, efficient materials. By working thusly, he felt that when the House was ready for mass production in 1952, it would not have to be radically changed to accommodate newly available materials and technology.[20]

Even though they were using highly sophisticated materials such as high-strength aluminum and Plexiglas in airplane production, Beech management frowned on the idea of building a prototype that employed those materials. Since no one could guarantee that the U.S. government would allow even small amounts of any high-priority materials to be employed in nonmilitary projects at that time, Beech management was also concerned that the final product would not be practical due to lack of the necessary materials. Management was not, however, able to stop Bucky's incessant search for and insistence upon components made from restricted substances such as aluminum, steel, and Plexiglas. His requirement of exotic materials was, however, quickly resolved when the Air Corps recognized the service Fuller was providing by maintaining the morale of workers as well as aircraft production and awarded him a contract for two Dymaxion Houses.[21] The contract also furnished his project with a top-priority status equal to that of companies producing armaments, thereby permitting him access to all the materials he required—including those which had been only recently developed.

Because Bucky personally supervised the production and development of the Air Corps' two Houses and because he had been continually refining his ideas concerning the Dymaxion House since its inception in 1927, work proceeded far more rapidly than most people expected. In fact, the Air Corps Houses were completed and displayed to the public within a single year.

Although it was extremely unlike anything seen before, the prototype House shown at the October 1946 unveiling was clearly a natural evolution of Fuller's earlier work. Like his earlier house designs, the Beech version hung from a central mast; however, the newer design was completely circular rather than hexagonal. It was fabricated from the most ultramodern materials, including its primary element, the best lightweight aluminum which provided maximum strength with minimum weight. In fact, the entire structure weighed only 6,000 pounds, an amazing figure when compared to the weight of a house of that size even today: approximately 300,000 pounds.[22]

The House itself was not completely constructed on the factory assembly line; rather, all the components were fabricated and packaged at the factory into a compact unit which was shipped to a building site. Those prepackaged units with assembly instructions were similar to an enormous model kit which could be assembled on the site in a matter of days by a single person. In the Fuller tradition, each individual component was designed for ease of handling and assembly. Thus, no piece weighed more than ten pounds, so that any component could be installed by a single worker who could lift it with one hand and use the other hand to fasten the part in place.[23]

Utilizing the system he had first devised for the Dymaxion Deployment Unit, the Dymaxion House was also constructed chiefly from the top down. The curved roof sections were hung on steel cables suspended from the central mast, and then the wall sections were attached to the completed roof. Around the House's entire 118-foot circumference, residents had access to the outside environment through a sweeping series of continuous Plexiglas windows, which were slightly tinted to avert glare and were furnished with pull-down curtains for privacy.[24]

Fuller also employed his Dymaxion Deployment Unit discoveries in cooling and heating the House. Each of the panels below the windows was designed to lower eighteen inches, providing screened openings so that the occupants could establish a complete or partial ring of ventilation around the base of their home at any time. To augment that ventilation and to generate a natural airflow, Fuller designed and built a revolutionary new feature on top of his House: a massive ventilator which rotated atop the central mast.[25]

Because of its design, the ventilator looked more like the underside of a sailing ship or the tail of an airplane than the roof of a house. Having designed it to function aerodynamically, Bucky was pleased by that perception. The ventilator was streamlined and was positioned so that its closed side (which was rounded) would constantly turn into the wind like the wide end of a windsock. Opposite that side was the tapered, tail-like side with its opening, which because of its exposure to the negative pressure of passing winds literally sucked fresh air up through the house and out the ventilator.

Since they were constructing the prototype in tornado-prone Kansas

Fig. 12–5   A model of the Dymaxion House.

and Bucky wanted his house to be as safe as possible in the severest of storms, he designed the conspicuous ventilator so that it could actually lift three feet on its central mast spine and instantly relieve any great pressure differential which might arise from such a storm. During the sudden change in air pressure characteristic of powerful storms such as tornadoes, walls and windows of buildings are usually blown out, causing a great deal of the destruction. Fuller's design, however, eliminated that possibility by providing the ventilator which was akin to the emergency valve on a pressure cooker. That valve opens up to release excess steam pressure rather than allow the entire cooker to explode.[26]

The House's safety and engineering features were, however, overshadowed by the more evident benefits it provided for occupants. Inside the nearly eleven-hundred-square-foot, single-family dwelling was a spacious combination of cozy comfort and modern technology. Although the

Fig. 12–6 The interior of the Dymaxion House.

House was primarily fabricated from metal, Bucky created a warm internal ambience and captivated its occupants with touches such as floors covered by wood (over aluminum beams) and an open fireplace dominating the large living room.[27]

The House's curved ceilings also provided occupants with a uniquely varied living environment seldom seen in homes before or since. At approximately sixteen feet in height, those ceilings created a spacious feeling, and if the owner wanted to change a room's ambience, both the color and the intensity of the hidden lighting could be instantly altered.[28]

Similarly, the entire House was filled with unique environment-altering and labor-saving technology including some inventions created exclusively for the project. Among the innovations Fuller built into the House were revolving motorized closets in each of the two bedrooms, a dust-filtering system, and pressurized wall outlets for vacuum cleaner hoses.[29]

For most people, the House's most amazing aspect was its price.

After careful calculations, both Fuller and Beech agreed that the complete unit could be profitably marketed for a retail price of $6500[30] which converts to approximately $46,000 in 1989 dollars, an astounding figure at a time when the price of an average home was in excess of $12,000. Even more surprising was the fact that the $6500 figure included freight, assembly, installation, and a profit for the local dealer. The actual cost of mass-producing the House's component parts, which could be packed into less than three hundred square feet for shipment, was estimated at only $1800.[31]

Because both Bucky and Beech believed that women would be the primary beneficiaries of the House's innovations, some of the first people to tour the completed prototype were a group of twenty-eight wives of project workers.[32] Everyone was certain of their project's success, when, following their examination, twenty-six of those women commented on its spectacular beauty and all of them agreed that it could probably be thoroughly cleaned in half an hour.[33] The women unanimously agreed that they wanted to purchase a Dymaxion House.

Those women's instantaneous desire to purchase a Dymaxion House was not an unusual reaction. In fact, following the enormous publicity that the project received between October of 1945 and February of 1946 (including a major illustrated article in *Fortune* magazine), a flood of thirty-seven thousand unsolicited orders poured into both Fuller Houses and Beech. Many of those people were so excited about the Dymaxion House that they even included down-payment checks with their requests.[34]

During that period, Bucky also began another of his characteristic retreat from the "evils" of financial success and those who would attempt to profit from his ideas. His first instance of concern occurred when the trading value of shares of Fuller Houses, Inc., which had been issued at $10, jumped to $20 in a matter of days. Then, a successful Philadelphia entrepreneur named William Wasserman purchased one thousand shares of company stock in a power play which allowed him to assume the position of chairman of the board and to begin working on generating the finances required to initiate mass production of the Houses. Wasserman worked in alliance with Herman Wolf and the union people who had invested in Fuller Houses, all of whom began to envision the enormous profits they would reap as a result of commercial success.[35]

Beech engineers did their part by estimating the costs of production and distribution, including setting up machinery and a new sales organization, at $10 million.[36] Then, Wasserman and his financial people began working on a plan to raise that money. Their idea was to recapitalize Fuller Houses, Inc., by issuing 750,000 shares of stock at a par value of $1 and allowing all the original stockholders to exchange shares of the old stock for the new at a 1-to-10 ratio. The plan was to generate an additional 595,000 shares of stock which would be offered to the public at $10, producing an income of $5,950,000 as well as increasing the value of an original share of stock to at least $100. The balance of the capital needed was then to be raised through loans or the issuance of bonds. In fact, public demand for the Dymaxion House was so potent that the government even volunteered to step in with a major start-up loan.[37]

Despite their skill and savvy, the sophisticated financial planners forgot to consider one factor: Fuller himself and the self-destructive aspect of his personality which lingered from his youth and was particularly conspicuous in financial matters. The financiers had also proceeded without regarding Bucky's tendency at times to assume rigid, unchangeable positions on an issue. The gestation period of his idea was one such issue. Regardless of the fact that his 5500 shares of Fuller Houses would have skyrocketed to a value of at least $550,000 and his dream of quality mass-production homes for average people would have been realized in an extremely short time, Bucky was not about to endorse production until the exact moment he deemed everything was completely ready. He firmly believed that moment would occur no earlier than 1952.[38]

Although most people involved felt Bucky would not be able to resist the enormous financial windfall, they were quickly proved wrong. Each time William Wasserman was near the final stages of a financial arrangement, Bucky, with his ultimate control over production, would simply veto the agreement over some issue which was insignificant when compared to the massive scope of the proposed project. Yet, because Fuller's objections were valid and his contract was binding, Wasserman would simply have to make the necessary alterations and try again.[39]

Even though both sides in that confrontation contended that they were right, the truth seems to lie somewhere between the contradictory opinions. Wasserman argued that every time a deal was ready for finalization, Bucky would thwart it with new, unnecessary changes to the House. Bucky simply maintained his position of making continual improvement, was never satisfied, and felt he was simply continuing to develop the best possible prototype of the Dymaxion House, just as he had promised.[40]

Whatever his motive, the constant obstruction finally wore Wasserman down and caused him to accept a buyout offer from Fuller. That action signaled the eventual end of Fuller Houses, as other investors quickly began selling their shares and the price dropped. Even Jack Gaty sold Beech's shares as Fuller Houses fell into financial collapse.[41]

Other individuals and organizations also attempted to sway Bucky's opinion, but he quickly explained the impossibility of the task before them. He argued that the self-contained House could not be assembled and installed by skilled union laborers because every union contended that it had sole jurisdiction over the process, which was actually little more than assembling pieces and connecting external wires and pipes to available systems and foundations.[42]

He also explained that, in his opinion, the union problem was even more complicated. From his earlier experience with the Dymaxion Bathroom, Bucky was certain that every union would oppose installation of the House because, with all its wiring, plumbing, and so on installed at the factory, it would provide little initial work for local trades. Bucky

Fig. 12–7   The prototype Dymaxion House was sold to a man who did not value its mobility or the spinning vent on top. Consequently, he had the vent removed and the house permanently installed on a foundation as a lake cottage.

further rationalized his decision with his belief that the Dymaxion House was, in actuality, creating an entirely new industry, and prior to practical implementation, he felt responsible for developing all aspects of that industry, including a revolutionary new type of installation and transport truck and a system of rapid delivery and deployment.[43]

Despite the pressure from friends, stockholders, people seeking to purchase the Houses, and individuals who wanted to become distributors for it, Fuller could not be swayed from his conviction that the timing was not right. Thus, he would permit no further progress toward mass production until the 1952 date, and with no one willing to wait an additional six years, the project ground to a halt.[44]

Although Bucky was satisfied with the House's developmental progress and the fact that when the operation was closed down no one had lost any money, his stockholders were quite upset by having been deprived of the enormous profits they had envisioned in the near future. Herman Wolf had convinced many of his friends and family to invest in the company of which he was still president, and he was particularly agitated by the situation. At one point the pressure and distress became too much for Wolf, and he made the mistake of turning his anger toward Fuller's family.

As the story goes, one day following the project's closure, a distraught Wolf burst into the Fuller's New York apartment and tried to warn Anne that Bucky had been having an affair with a young woman while in Kansas. Although we will never know if that particular affair actually occurred, it made no difference to Anne, who maintained her position as Bucky's ever-faithful wife, completely surprising Wolf with her reaction.[45]

Rather than question Bucky's actions, she defended her idealized perception of reality, her husband, and his actions, telling Wolf, "You have no right to say such things about Bucky to me. And, even if it were true, it's none of your business."[46] Thus, even during a moment of great stress and in the face of possible betrayal, Anne showed a tenacious determination nearly as strong as Bucky's. The incident also illustrates how she often singlehandedly maintained the deep bond between herself and Bucky, of which he often boasted. She was, as he frequently commented, his anchor, and his passionate pursuit of the daring aspects of life on the edge was greatly facilitated by his knowledge that no matter what he might do, she would always be there for him.

Following the 1946 Dymaxion-House-project collapse, Bucky returned to the East Coast, moved into the apartment Anne had rented in Forest Hills, Long Island, and tried to determine exactly what project most required his attention. He also attempted to influence Allegra in deciding on her future and was somewhat surprised to find she had developed a mind of her own. As a result of their interaction and Allegra's determination, she was probably more influential on him than vice versa.

Having graduated from Miss Madeira's School that year, Bucky's daughter spent the summer studying ballet with George Balanchine in New York, a feat which caught her father somewhat off guard. With his biased vision and limited time spent with Allegra, Bucky had believed his daughter's primary skills and interests focused in the same areas he loved. During the brief periods he spent at home with Anne and Allegra, Bucky had witnessed Allegra's easy mastery of mathematics and mechanical skills, and he was so certain that she would follow in his footsteps that he had applied for and sought support in garnering a place for her at one of America's finest engineering schools, Massachusetts Institute of Technology. Since only six girls were enrolled at MIT at that time, Bucky himself was thrilled when Allegra was accepted by the school. He was not, however, encouraged by his wife's and daughter's reaction to the news.[47]

Upon returning to New York, it became clear to Bucky that Allegra's interest in dance was far more than a young girl's passing fancy. Only then was he, in an uncharacteristically compromising act, able to let go of his belief that he knew the best course for her future. Bucky asked his daughter to join him in a discussion of her future, and although he was somewhat surprised to learn that she was more interested in following Anne's love of the arts than his devotion to science and technology, he was willing to support his daughter's wishes.[48]

Since Bucky tended to be on the road while Anne had stayed with Allegra during her formative years, Allegra's preference seems only natural; however, it left Bucky dumbfounded. Still, he was able to swallow his ego and support the same innate wisdom of youth and abundance of choices for his daughter that he championed for all young people.

Rather than attempting to further convince Allegra that following in his footsteps was best for her, Bucky first asked if she did, in fact, want to become a dancer. When Allegra answered yes, he concealed his disappointment and demonstrated that he truly believed young people were as competent as their elders by asking what path she wanted to take in pursuing a career in dance. After further discussion, Bucky was able to at least proffer his low opinion of formal educational facilities and suggest that she might want to bypass that option and work directly at becoming a dancer. To his young daughter, who was not unlike most teenagers, such an option appeared quite glamorous, and she began to pursue a career in professional dance.[49]

Even though that interaction demonstrated how being away had left him somewhat out of touch with his family, Bucky did not change his ways. He had focused his attention on the larger context of continually supporting the most people possible at the expense of being with and supporting his own wife and child, but he felt that in working for all humanity he was, in fact, working for Anne and Allegra. Even though most of his friends and relatives asserted he should be more of a family man, Bucky was not about to sacrifice his commitment, and he continued with his mission.

He did, however, attempt to secure the love and acceptance of family and friends with gifts and trips. Specifically, when he finally attained financial wealth, Bucky was constantly giving family and friends expensive presents for no apparent reason other than that he felt they should have the items. He also flew Anne, Allegra, and later his grandchildren around the World to be with him on his excursions whenever they would accept his invitations. In that way, he attempted to make up for past deprivation without forfeiting his prominent position or the work he felt needed to be done.

In 1946, however, the only action Bucky felt to be appropriate was standing back and allowing his nineteen-year-old daughter the opportunity to pursue her career. Thus, Allegra spent several years working her way up through the ranks as a professional dancer. Although that career was far more appealing to her than classical scientific study, she soon found that her rich heritage of intellectual stimulation created a yearning for more academic endeavors. One of the methods she used to satisfy that need was to begin attending her father's local speaking engagements. Allegra had always loved listening to Bucky, and because she had grown up experiencing his intricate speech patterns, she was more able than most to comprehend his ideas.

Since she was being trained in the body movements of dance, following one such lecture Allegra decided to offer Bucky some advice predicated on her dance experience. She pointed out that he had been raised in a New England society where gentlemen frowned upon gestures as vulgar, and that as a result of that training, when he spoke it was always without much inflection, standing rigidly straight with his hands at his sides. Allegra went on to remind her father that when he spoke to family and friends in private, he was constantly using his hands to gesture, and that those movements greatly enhanced their understanding of his ideas.[50] Continuing her train of thought, his daughter explained that in studying dance, she had learned that arm and leg movements can be effective communication tools, and she suggested that Bucky might become a more effective speaker if he used his natural body movements to enhance his public presentations. She advised her father simply to let himself go and be more natural when speaking to groups.[51]

Although it was true that he had been raised in the rigid New England tradition, Bucky was always willing to try new ideas, especially when they came from someone he dearly loved. Thus, within weeks, he began implementing a less controlled mode of presentation and creating the endearing yet often frenetic style of lecturing which became his trademark. In fact, until old age impeded him somewhat, Bucky was known for spontaneously jumping up from his chair or even crawling around the stage to demonstrate and emphasize his ideas.

CHAPTER 13

Conscious Thought

 In changing his lecture style to include his daughter's dance concepts, Bucky did not alter either the context or the content of his presentations. He also continued to maintain an unmistakable distinction between his personality as evidenced in his personal life and the persona he presented to the public. Always one to divulge the complete truth as he knew it at any given moment, Bucky never concealed any of the antics of his personal life; however, he also felt that those acts were so insignificant in relationship to his mission that he seldom mentioned them in public unless they illustrated a point he was attempting to make. Also, when he spoke to audiences, Fuller was not presenting preconceived ideas, in that he was simply ''thinking out loud,'' as he called his method of sharing information. He felt he was allowing information to pass from or through himself, flowing naturally without a great deal of conscious consideration as to its content. Thus, he usually spoke of the topics which were of the most value to the largest number of people.

As always, Bucky continued to focus on spontaneous ideas rather than planned speeches, and he made a point of never using notes or thinking about what he was going to say prior to walking onstage.[1] Thus, his lengthy thinking out loud lectures usually expanded upon ideas he had spoken about or considered previously. At times, however, unique new ideas simply appeared and were injected into his lectures in a matter-of-fact manner which delighted and surprised both Fuller and his audiences.

Although his public speaking was truly spontaneous, Fuller thoroughly considered the components of the process we call *thought* so that he could better participate in that phenomenon.[2] Because he felt that all nonphysical aspects of Universe conform to the same generalized principles that govern the physical aspects of our reality, Bucky used physical geometry to clarify his discoveries concerning the thought process. He would often explain that thought is a process of consciousness sorting. During that process, the confusing jumble of our experiences are rearranged so that our minds can more easily discover relationships between the items in that unsorted patchwork of input.[3]

Fuller recognized that the Earth, the solar system, the entire cosmos, and individual human beings were continually moving in a plethora of directions. Because of that constant change, each person perceives events and objects (i.e., objects are actually fabricated from events known as energy) from a multitude of changing viewpoints which rarely coincide with the exact observations of other human beings.

Bucky was amazed by the infinite number of possible perceptions that he established were available to just one individual when he considered only the revolution and movement through space of the Earth combined with a person's daily movement. Although the continuous change and the shifting perception due to that constant motion may seem inconsequential to a person preoccupied with the activities of daily life, Fuller believed they were extremely important. He discovered that many people intellectually accept the physical motion of the cosmos and the constant change of Universe itself but that few consider their actions in relationship to the grand sweep of changing movement.

Since he was a comprehensivist concerned with all Universe, Fuller realized that the conspicuous movement of humans as well as every aspect of their environment significantly influences all action and development. Thus, he set out to become intimately aware of and involved with the vast movement and change which engulfs all humanity so that he could increase his effectiveness.[4] He truly believed, and his actions demonstrated, that by becoming aware of his constantly changing environment he could operate within it more efficiently on all levels from the most mundane tasks to the loftiest pursuits of knowledge.

Prior to doing that, Bucky found that he had to acknowledge that every individual has a unique perception of reality. When considering his actions within the context of the enormous continual motion surrounding each of us, Fuller realized that all experiences could be regarded as omnidirectionally positioned with respect to the observer.[5] In other words, from the viewpoint of any observer, all events and experiences are moving around an apparently stationary center. That center is the observer himself or herself. Thus, the framework for any observation is that the observer is at the center of Universe viewing the passing of an immense number of rapidly moving objects and events.

Since he felt that the thought process consisted primarily of successfully arranging those constantly shifting, omnidirectional individual experiences, Fuller applied his understanding of motion and the relationship between observer and observed to his exploration of thinking. He specifically set out to determine how he and all humans mentally arrange experiences during the process of thinking.[6]

Because he believed that all metaphysical phenomena—including thinking— could be modeled with physical artifacts, Fuller employed his beloved geometry to describe the sorting procedure that is thought. As in Einstein's theories, the observer is critical to Fuller's evaluation of the thought process, and Bucky established the observer (himself for purposes of experimentation) at the center of an infinite set of multidirectional experiences available to be organized.

After hypothesizing just how the thought process functions, Bucky was challenged to communicate his outlandish, yet logical, conclusions to others in a way which would stimulate rather than bore them. Consequently, he devoted many years to envisioning how thoughts and ideas are formulated and determining the best illustrative examples to communicate his insights. His interpretation is more easily understood if one uses a geometric sorting illustration he devised.

Once Fuller's illustrative model of the human thought process is grasped, an individual may begin to appreciate on a conscious level what is an unconscious process for most people. In other words, when a person understands the representative illustration of idea sorting that Fuller concocted, he or she will actually be able to think more clearly. Following such a shift, an individual may practice initiating effective conscious thinking, just as Bucky often did. That ability to distinguish and understand the human thought process not only supports an individual's thinking more efficiently, but it also aids in more effective performance within the confines of a changing environment.

Although thinking is an ongoing comprehensive procedure, Bucky divided the process into several steps for the sake of examination and to clarify his geometric illustration. The first of those steps is sorting all experiences into two categories: the huge set of experiences which are irrelevant and the very small set of experiences relevant to the issue at hand.[7] When one is examining an issue, this sorting procedure delineates the inner and outer limits of consideration so that the thinker can focus his or her attention on a comprehensible aggregation of experiences or items. A more concrete sense of those limits can be envisioned by considering all experiences which are too small and frequently occurring to be dismissed outwardly and of all those which are too large and infrequently occurring as being dismissed inwardly.[8]

Imagining a transparent sphere helps one to grasp this sorting process more fully. All experiences which are too small and frequently occurring to be considered (i.e., the smaller the experience or phenomenon, the more often it occurs) are grouped inside the sphere, and all those which are too large and infrequent are sorted outside it. The tiny set of experiences remaining are those modeled by the sphere itself, and they can be easily examined.[9] Without such a sorting process, human beings would be so overwhelmed by information that they could never complete the examination of even the most trivial issue.

For example, when a person becomes hungry and thinks about dinner, he enters into a seemingly instantaneous sorting process focused on food. The person first dismisses large experiences, such as global hunger and farming, which occur very infrequently within his experience. Next, the person excludes small experiences, such as the processing of food in the digestive tract, which occur very frequently in his life. Following that sorting, what remains is a small array of experiences relevant to the issue of his personal hunger and dinner.

If that set of thoughts is still too large for comfortable consideration, another round of dismissing begins, refining the relevant experiences to an even smaller set. In that more detailed session, the focus of the particular thought becomes even more precise. Experiences such as notable meals which have been eaten are eliminated outwardly, and experiences such as past trips to the supermarket are eliminated inwardly.

By continuous employment of this process, the thinker reaches a point of considering only a set of relevant experiences and facts, such as the last trip to the supermarket and meals which could be prepared within his time constraints. However, because the procedure occurs so rapidly, with ordinary issues "the thought" which resolves the question of that day's dinner appears as an almost instantaneous idea.[10] In the case of complex issues, thinking requires additional time and is most effective when concentrated conscious effort is applied.

If an individual possesses some understanding of the process itself, as Fuller did, he or she can more consciously work within its limitations to expedite an answer. Because thinking is a synergistic process (i.e., one in which the whole cannot be predicted by observation of its parts), by definition the thinker cannot know the result prior to its appearance.[11] He or she can, however, support the process of sorting experiences and information as a conscious effort to resolve an issue more rapidly. Through years of practice, that organizational sorting process became as natural to Fuller as walking or talking.

He defined a thought as a relevant set of experiences bounded by macroirrelevant and microirrelevant experiences which are temporarily separated inwardly and outwardly.[12] Such a dismissal does not invalidate those experiences. Rather, they are relegated to the inner and outer boundaries of consideration so that a workable number of experiences may be examined. Without such a sorting process, individuals would have to consider all their experiences in even the most trivial thought process and would never reach any conclusion.

Once an individual understands that process, he or she can begin to utilize it consciously, as Fuller did, to limit his or her focus and to resolve issues without invalidating extraneous experiences. By practicing such a discipline, any person can consciously concentrate on the experiences needed to explore a particular thought and can reach decisions much more rapidly. Thus, thinking becomes an allied tool of human development and experience rather than an obscure phenomenon which is beyond the mastery of individual human beings.

Fuller believed that even the wisest humans could not completely control the thought process.[13] Yet, whether an individual is consciously directing or participating in the thought process or not, thinking remains constant. Bucky himself practiced this conscious sorting process for years in an effort to enhance his thinking; however, he still often found himself engrossed in the spontaneous contemplation of a particular topic. When that occurred, he would generally realize that the subject of his preoccupation was one he had begun examining and had deferred many times. He might then decide to pursue that issue through to its conclusion in a very conscious fashion and, consequently, to initiate his deliberately controlled thought process.[14]

Within that operation, Fuller would seek to recognize relationships between the small number of experiences which constituted his carefully organized thought. He could then examine those experiences to determine their relationship to the macro- and microirrelevant experiences which delineated the thought's boundaries.[15] For instance, when thinking about tonight's dinner possibilities, an individual might decide to prepare roast beef. Such a decision would be based on his experience of what is available in the refrigerator as well as the time available to prepare a meal. However, when that decision is then examined in relationship to previously dismissed macroirrelevant experiences such as future meal-planning, and microirrelevant experiences such as the effects of red meat on the digestive system, the original decision may be changed.

Fuller would sometimes further extend that process to discover relationships between his thought and the essential principles which sustained it. By doing this, he was able to learn even more about the fundamental operating principles of Universe.[16] When a thought is carried to such an extreme, the thinker is provided with valuable insight into the fundamental operation of and the relationships within Universe. Such perceptions often result in a greater understanding of all aspects of life. In the instance of dinner, an individual might well find that by examining his need for food, he begins to perceive a pattern of excess eating in his life. Fuller himself so strongly believed in the importance of discovering relationships that he often defined understanding as the ability to discover relationships.[17] Accordingly, he perceived the human thought process as being intensely focused upon relationships.

He also felt that no experience or idea could be completely isolated from other experiences, ideas, or Universe in general. Rather, he believed that everything exists only in relationship to other events and beings, and without others (be they people, animals, ideas, or any phenomenon), nothing would exist.[18] To emphasize the significance of that reality, Bucky frequently recounted his axiom, "Unity is plural and at minimum two."[19]

Fuller also believed that one of humanity's fundamental missions is to find relationships between the various elements of Universe so that we humans can support the harmonious development and evolution of those components.[20] Human minds are a basic tool in that process, and Fuller was convinced that it is therefore essential that every human being understand the primary operation of that tool, thinking.

Having closed out his Fuller Houses operation as well as the possible mass production of the Dymaxion House in 1946, Bucky again sought a project which would be of maximum benefit to humanity. To support his making the best decision, he felt that he once again required a break similar to the 1927 retreat which had marked the transition between his first thirty-five years of youthful exploration and the second major period of his life, from 1927 through 1947.[21] During that second era, Bucky continued to learn, but he concentrated on discovery, contribution, and pragmatic invention. He was continually searching for and learning about the principles which govern all Universe during that time while illustrating what he had learned through his inventions.

Still, like most great humans prior to shifting their focus, Bucky felt a need to step back and examine his situation in order to efficiently move forward during what would become the fulfillment phase of his life. Since he did not feel the need to clear his mind of previous ideas, but rather to acquire new information in a concentrated manner, the 1947 break was not as severe as his 1927 retreat.[22] It did, nonetheless, signal a transition to a uniquely different period of activity for Bucky.

That third era was characterized by Fuller's single-minded focus on one invention and the two fields which contributed to the creation of that invention. Thus, the geodesic dome, along with mathematics and construction, dominated Fuller's life from 1947 through 1970.

Fuller's withdrawal from the rigors of his usual schedule was by no

Fig. 13–1 Fuller with some of the geometric models which provided him with so many insights and led to his invention of a practical geodesic dome, 1948.

means a holiday. Rather, he spent much of 1947 and 1948 studying the single field which he believed would provide the greatest benefit to his future contribution to humankind: spherical geometry. He devoted most of his time to studying and thinking about spherical trigonometry as well as improving on the conventional tenets of that subject. His theories concerning the human thought process were very useful to his studies, and he continued applying them to his beloved geometry during those two years.[23]

Once Fuller had grasped the mind's sorting process, he resolved to further expand his geometric model of human thought by discovering the minimum number of items required to establish the "insideness" and "outsideness" into which experiences are sorted. He found that a sphere, such as the clear sphere described earlier, was a workable structure for that process. However, mathematically and visually a sphere is represented as an infinite number of points, and Fuller wanted to determine what structure could provide the same sorting characteristics (i.e., insideness and outsideness) using a minimum number of points. By determining the minimum number of points required to create such a structure, he understood that he would also be discovering the minimum effective structure for the thought-sorting process.[24]

In his quest for the minimum number of items (experiences, in the case of thoughts) required to produce those two qualities, Fuller found that any two items could be joined by a line, and that any three connected items produced a triangle. Such configurations generate the characteristic of "betweenness" but do not have the three-dimensional makeup needed for the characteristics of "insideness" and "outsideness" which result in a thinkable set.[25]

Four items do, however, generate a structure with those attributes. In fact, Fuller found that four is the minimum number of items (be they experiences, points of a geometrical model, human beings, or anything else) required to delineate inner and outer boundaries. When displayed as a geometric model, those four items form a tetrahedron (i.e., a structure composed of four triangular faces). They also represent the smallest conceptual subdivision of Universe, Fuller referred to that subdivision as a "system."[26]

In Fuller's theory, a system divides all Universe into that which is outside the system and that which is inside the system, plus a small amount of Universe which embodies the system itself. Even though most people rarely consider their daily functioning on the grand scale of all Universe, almost all human activities are, in fact, predicated upon distinctions established by systems. The thought-sorting process described earlier provides an excellent example of how people regularly employ systems in daily life.

Fuller reasoned that since thoughts are systems composed of experiences and since four items constitute a minimum system, a minimum thought is composed of four experiences.[27] Those four experiences do not occur simultaneously. Rather, they are experienced by an individual at different times, and the memory of them is stored within that person's brain. Although those memories are the items which are examined for relationships during the thought process, Bucky referred to them as if they were no different from the actual experiences.[28]

He theorized that in its quest for greater understanding, the human mind continuously searches through that ever-expanding repository of experiences for those with similar qualities. When a person's mind uncovers comparable characteristics within two or three experiences, those experiences are grouped together for further study.[29] However, when a fourth experience with similar qualities is discovered and added to the others, the minimum number of items needed to create a set dividing all Universe is established, and that set has the potential of becoming a thought. In other words, with four experiences, the mind has enough information to produce a dividing structure the geometrical equivalent of which is the tetrahedron with its four points. Before any set of experiences can be upgraded to the status of thought, however, it must first arouse the mind's curious attention.

Fuller felt that the act of grouping four experiences together is usually an indication that some amount of curiosity has, in fact, been stimulated.[30] To attain the next level of thought, the human mind searches for

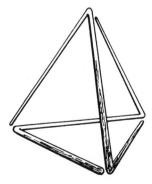

Fig. 13-2 A tetrahedron with its six sides (relationships) and four vertices (items).

the relationships between those four items. As with many of his other ideas, Fuller modeled the relationships within a thinkable set of related experiences geometrically to provide a better understanding, and he created the following analysis.[31]

One relationship exists between two items, be they experiences, dogs, rocks or anything else. Three items form a triangle, which contains three relationships or connections between the various elements. However, when the number of items is increased to four, the number of relationships (i.e., connections) escalates to six, as is modeled by the six sides of the tetrahedron.[32]

According to Fuller, those six relationships are the human mind's quarry in its pursuit of a thought.[33] Initially, individuals do not consciously recognize the final disposition of a thought. In fact, Fuller felt that if a thought is discovered prior to the completion of what a person believes is a specific thinking process focusing on a particular topic, that procedure is not "thinking." Instead, the process is simply a sorting of known experiences.[34] In thinking, the final resolution can be grasped only after the minimum six relationships are uncovered, and the thought always remains obscure until the last relationship is recognized.[35]

An individual might gain some insight into the final thought with fewer than six relationships, but the six relationships are required to generate the complete thought, just as six sides are required to construct a tetrahedron.[36] With fewer relationships, a "thought in process" might be modeled as a line or a triangle with the quality of betweenness. However, using his geometrical model, Fuller demonstrated that four experiences and six relationships are the minimum requirements which yield the characteristics of insideness and outsideness needed to define a thought.

In Fuller's theory of thinking, simply knowing about four similar experiences does not constitute a thought.[37] For example, most young children experience numerous metaphysical occurrences. They frequently have imaginary playmates, appear to communicate telepathically before learning languages, and master specific talents with little or no instruction. However, in most instances, only after children have matured and encountered metaphysical phenomena are relationships between such early childhood events seriously considered. And only then, do thoughts concerning the metaphysical aspects of life begin to surface.

Fuller felt that the combination of four items found throughout Universe is essential to all life. Since four is the minimum number of items required to divide Universe, he believed that it also provides the most efficient expression of the critical differentiation which establishes individual human life.[38] Without the differentiation of insideness and outsideness provided by a minimum of four experiences combined into a thought, human beings would be unable to perceive themselves as separate from other beings or components of Universe, and all humanity would meld into a single entity.

Fuller observed that fourness, in fact, occurs throughout Nature. Because Nature always employs only the exact amount of resources and energy necessary for any task, the majority of natural construction is predicated upon the tetrahedron and its minimal four items or points.[39] To better understand why four is so significant, Fuller considered another geometrical phenomenon, the most efficient, stable configuration in which four identical balls can be closely packed. Three balls on a flat surface come together naturally in a triangular formation with each ball touching the other two at a single point. To completely stabilize that configuration, the fourth ball nests in the indentation provided by the other three as seen in Fig. 13–3.[40]

This modest structure provides another demonstration Fuller employed to illuminate a metaphysical phenomenon using a physical artifact. By connecting the centers of this most efficient means of packing a fundamental component of Universe (i.e., the sphere), the essential regular tetrahedron is fabricated.[41]

Because of its efficiency, Fuller perceived four and its geometric manifestation, the tetrahedron, to be the most utilized structural system in Universe, and as such, it became a fundamental element of his philosophy and operating strategy.[42] Fourness is constantly employed by Nature in both physical and metaphysical situations. Once he recognized this elegant wonder of nature, Bucky regretted that most human beings do not grasp the significance of this apparently innocent, yet ingenious, phenomenon.[43]

He believed that when a significant number of individuals do begin to

Fig. 13–3   Four balls packed as closely as possible. When the centers of these balls are connected, as shown, the connecting lines form a tetrahedron.

recognize the prevalence and importance of the four-item, six-relationship configuration known in its physical form as the tetrahedron, humanity would enter into a new era of enlightenment. At that time, he felt great changes would occur in all fields of human endeavor, including education, construction, and the processes of conscious thinking.[44]

In education, the ninety-degree, right-angle geometry predicated on the square would be replaced by the Synergetic Geometry which Fuller developed and which is predicated upon the tetrahedron and on concepts which can be experienced rather than obscure principles. Unstable, inefficient right-angle buildings, which have been the basis of most human construction since the inception of permanent structures, would be replaced by the geodesic structures that were so essential to Fuller's work. Predicated on stable triangulation, which does not require the heavy buttressing of current material-intensive construction techniques, geodesic structures would provide humanity with the medium to house many more people while using far fewer resources than ordinary buildings.

In the area of conscious thinking, understanding the four-element structure would permit society to better appreciate and utilize the power of the human mind. With such an understanding, individuals could knowingly enter into an expansion of their awareness and evolution by consciously choosing to think about issues in an efficiently productive mode rather than in the haphazard manner currently employed by most people.

Another facet of thought which Fuller believed to be extremely important was imagination. The root of the word imagination is *image,* and Fuller defined imagination as the uniquely human ability to create an image. Through years of personal experience and observation, he came to believe that the human capacity to think and create order from chaos is primarily the result of the ability to employ imagination.[45]

According to his hypothesis, in thinking and establishing order, humans gather an enormous quantity of information.[46] They then utilize imagination by applying that accumulation of data to complex problems. Such an ability is especially important when new information is obtained through observation and experience. When that occurs, individuals use their minds to imagine and uncover relationships between the new information and previously accumulated knowledge, and that is imagination at work.

Fuller further felt that the process of imagination was even more miraculous in light of the fact that no human being has ever looked from outside of himself or herself. He would explain that information gathered outside a person's physical body is assembled inside that individual's brain just as news from the outside is collected within a television studio. Each person's internal "studio" is the site at which his or her "image-ination" is utilized to discover relationships between the assembled items of news.[47]

Bucky's understanding of and use of human imagination was evident in the evolution of his lectures. During almost all his speaking engagements, he employed some geodesic models to demonstrate principles and concepts, but his earlier lectures were far more cluttered with models than in later years.

With his fame in the 1930s, Bucky began receiving his earliest major requests to lecture. Having used models extensively during his 1927 retreat, he felt that they might enhance his ideas during his speaking engagements, and he soon found that models provided an excellent means of illustrating concepts as he discussed them. In fact, his lecture models became so profuse that he soon found himself towing a large trailer brimming with models to every speech.[48]

Although those models provided excellent visual aids, they continued to increase in number as his knowledge and understanding of Universe

Fig. 13–4   A portion of Fuller's extensive collection of geometric models.

heightened, and he sought innovative solutions to the problem of his growing aggregation of models. Specifically, Bucky needed a method of effectively and efficiently illustrating complex issues without having to transport truckloads of models and equipment. The development of publicly available, high-quality photography provided the solution he sought, and, always one to try the latest technology, Fuller quickly became one of the first lecturers to use slides throughout his talks. In fact, Bucky replaced nearly all his models with slides of them, thereby resolving his shipping problem.[49] Still, he was troubled by frequently confronting situations in which he was stranded without any visual aids to support his words.

For instance, at formal dinners, he would often be introduced to important individuals who would invite him to explain one of his concepts on the spur of the moment. Although Bucky could easily have fallen back on polite small talk, such instances provided him with two opportunities he was seldom able to resist, an interested individual and an invitation to communicate. After doing his best in several such situations, Fuller began to realize that he could, in fact, communicate his ideas smoothly without the aid of either models or slides. When describing ideas, he accomplished that feat by relying on his audience's sensory experiences.[50]

Bucky knew that every person has vivid memories of past experiences, and in speaking to a stranger, he would start the discussion by attempting to discover which of that person's experiences related directly to the topic at hand. He then utilized those experiences to describe abstract principles and ideas in a manner similar to the way he employed models during his lectures.[51]

The spliced-rope illustration he used to explain pattern integrities and their relationship to life provides an excellent example of that technique in action. Bucky's description was acceptable to any listener because everyone had some experience of a rope. In fact, over the years, Fuller recounted the details of that demonstration thousands of times to both individuals and groups without ever fabricating that rope or even an illustration of it. Yet, he was never questioned about not having an actual rope or even a photograph of it.[52]

By predicating his discussion upon an individual's past experiences rather than some unfamiliar abstract idea, Fuller quickly established a common ground of understanding between himself and his audience. The result of such action was that rather than alienating people with his wisdom, his mutually understood experiences communicated ideas in an easily grasped and accepted format.

Bucky's strategy was successful because human beings visualize experiences only within their minds. He knew that even his slides and models were perceived only internally, within an individual's mind. Thus, he understood that his function as a communicator was to stimulate the imagination of all members of his audience so that they, in turn, could individually construct an image of a model or an idea within their minds and visualize its demonstration of the principles he was sharing.

Fuller was certain that because his listeners were far more involved in the procedure when no visible artifact existed, communication was much more successful under such circumstances. He believed that when a person generates an image in his or her mind without visual aids, that image and idea has a far greater impact and the communication is accessed and retained at a much deeper level.[53]

Bucky came to truly appreciate the magnitude of human imagination when he was a member of an audience forced to experience an entire play without visual aids. The incident occurred in 1940, and the perpetrator was Fuller's good friend Alexander King, the art editor of *Life* magazine at the time.

King was also a playwright and a recognized authority on the theater who had written what he believed was an excellent play. The final years of the Depression, however, made Broadway production a very risky financial endeavor, and few people were interested in investing in the play. King therefore set out to bypass all the problems of finance and production. To accomplish his goal, he rented a large New York theater for one evening and invited dignitaries and critics to the "opening" of his new play. As a close friend, Bucky was also invited to the formal-dress opening night.[54]

The guests arrived on the designated night and were surprised to find an empty theater with the curtain raised and nothing onstage except two grand pianos. Bucky later discovered that the pianos were displayed only because King was required to hire stagehands when he rented the theater. Being somewhat eccentric and wanting to get the most for his money, King instructed the stagehands to position the grand pianos on either side of the stage and to move them from one side to the other several times during the performance. King himself merely used the piano benches as seats while he spoke.[55]

King's performance began when he walked to center stage and described the play's scenery to a dumbfounded audience. He then proceeded to read the entire play aloud. What impressed Bucky was the fact that many years later, he recalled the smallest details of his friend's work far more vividly than other productions he had seen in more customary stagings. He also found that phenomenon to be prevalent among others of his friends who had been members of that audience.[56]

In analyzing the situation, Bucky found that such graphic recall was a direct result of an imaging process King had stimulated in his audience. Rather than provide visual images, he had forced each person to create his or her own personal impression of the play, and since people could fashion images predicated only upon their personal experiences, those impressions were more familiar and very lucid.

For example, when King spoke of the house in which the characters resided, each person in the audience immediately generated an image of a house which conformed to his or her unique experiences. Thus, no person hesitated to accept the fact that the characters could live in that type of house. Furthermore, in creating his or her personal images, every individual introduced images far more detailed than any which could have been portrayed within the limitations of even the most sophisticated stage.[57]

Bucky also noticed that the audience was actively involved in the proceedings and consequently paid close attention to King's presentation.[58] Because his technique was similar to radio, King's strategy stimulated an interactive performance which belonged as much to his audience as it did to the author and "performer."

That, and similar experiences, helped Fuller formulate his strategy of constantly asking his audience (be it a single person or a packed auditorium) to recall common, special-case experiences from their lives. By conducting his "thinking out loud" sessions thusly, Bucky was able to establish a common ground, to stimulate his audience's individual imaginations, and to motivate them to participate and pay close attention to his message, all in a matter of minutes.

Once he had established an experience or item which was commonly recognized by everyone, Bucky could then quickly facilitate a group in collectively examining the principles upon which that experience or item was predicated. With each individual actively participating in his explorative lecture process and seeking to understand the ideas, Fuller was able to expand his audience's experience and its limited concept of reality by guiding them to unearth other special cases in which that same principle was evident. Accordingly, his audiences entered into a conscious search for additional relationships between the myriad of finite experiences stored in their brains, and their collective and individual knowledge increased in the ongoing process recognized as "thinking."

CHAPTER 14

Structures Emerge

 The conscious thinking to which Fuller applied himself during his retreat of 1947 and 1948 focused specifically on a single topic: spherical geometry.[1] He chose that area because he felt it would be most useful in further understanding the mathematics of engineering, in searching for Nature's coordinating system (i.e., the mathematical representation of the principles which govern Universe), and eventually in building the spherical structures which he found to be the most efficient means of construction.

Having observed the problems inherent in conventional construction techniques as opposed to the ease with which Nature's structures are erected as well as the indigenous strength of natural structures, Fuller felt certain that he could perfect an analogous, efficient, spherical-construction technique. He was also aware that any such method would have to be predicated upon spherical trigonometry. Accordingly, Bucky began to convert the small Long Island apartment that Anne had rented into a combination workshop and classroom, where he studied and discussed his ideas with others.[2]

As those ideas started to take shape in the models and drawings he used for sharing his insights with others, Bucky began considering names for his invention. He selected "geodesic dome" because the sections or arcs of great circles (i.e., the shortest distance between two points along a sphere) are called *geodesics,* a term derived from the Greek word meaning "Earth-dividing."[3] Bucky's initial dome models were nothing more than spheres or sections of spheres constructed from crisscrossing curved pieces of material (each of which represented an arc of a great circle) that formed triangles. Later, he expanded the concept and formed the curved pieces into even more complex structures such as tetrahedrons or octahedrons before they were joined to create a spherical structure. Still, the simple triangulation of struts remained, as did the initial name of the invention.[4]

Although Bucky's study of mathematics played a significant role in his invention of the geodesic dome, that process was also greatly influenced by his earlier extensive examination of and work within the field of construction. During his construction experience, he came to realize that the dome pattern had been employed, to some extent, by people since humans began building structures.[5] Early sailors landing upon foreign shores and requiring immediate shelter would simply upend their ships, creating an arched shelter similar to a dome.

Land-dwelling societies copied that structure by locating a small clearing surrounded by young saplings and bending those uncut trees inward to form a dome which they covered with animal skins, thatch, or other materials. Over time, that structure developed into the classic yurt which still provides viable homes for many people in and around Afghanistan and the plains of the Soviet Union.[6]

Several years after the geodesic dome's invention, Bucky would witness a surprising correlation between his dome and the yurt. In 1956, the United States government's Department of Commerce was confronted by one of those emergence-by-emergency problems Fuller had envisioned and considered years earlier. As a result of his insight, Bucky was the only person who could meet the stringent requirements set forth for the construction of an exhibition pavilion at the 1956 International Trade Fair in Kabul, Afghanistan.[7]

The Commerce Department's original plan had been simply to send a few minor displays to that trade fair, but when the United States government learned that their Cold War enemies Russia and China were sending huge exhibits, all plans were dramatically changed overnight. Commerce Department officials hastily sought a bold method of enclosing several large displays of American industry, and after they had considered and rejected several possibilities, the only viable solution they found was Buckminster Fuller and his geodesic dome. What occurred following that decision may well represent a record time for the engineering, manufacturing, and constructing of such a pavilion. Commerce Department officials signed a contract with Bucky and his company on May 23 which stipulated that the one-hundred-foot-diameter dome had to be completed by the end of June.[8]

In addition, the structure had to be light and compact enough to be flown from the United States to Afghanistan in a single DC-4 airplane. To complicate matters even further, the government stipulated that only one of Fuller's engineers could accompany the dome, and that all the labor had to be done by local Afghan workmen.[9]

Seven days after signing that contract, Bucky and his associates had completed the engineering specifications for the latest evolution of the geodesic dome, which he had invented seven years earlier during his late-1940s period of retreat. The 100-foot-diameter structure was 35 feet

Fig. 14–1 The framework of the Afghan geodesic dome being erected.

Fig. 14–2   The Afghan Trade Fair geodesic dome in place in Kabul.

high at the center and provided approximately 8000 square feet of display space. Its framework was fabricated from 480 aluminum tubes, each 3 inches in diameter and weighing a total of 9200 pounds. The actual cover was provided by a nylon skin weighing 1300 pounds which was suspended inside the frame.[10]

More significantly, that dome was designed to be erected by inexperienced personnel, including the untrained workmen who were assigned to the project and who spoke no English. Under the guidance of Fuller's engineer, those men began fastening color-coded hubs and struts together with no idea as to what the finished product would look like. However, within hours, the workers were quite comfortable and certain that they were, in fact, constructing a gigantic version of a familiar structure, the yurt.[11]

When completed, the huge dome was an amazing success. It was so strong that the workers created a problem for the United States government officials when they began sliding down the sides of the nylon skin for sport. Other Afghans were so impressed by the structure, that the dome itself was far more popular than the industrial exhibits it housed. It also attracted a much larger crowd than either the Russian or the Chinese exhibitions.[12]

Two Afghans who were particularly impressed by the United States' giant yurt were the king of Afghanistan, Zahir Shah, and his cousin the prime minister, Lt. General Sardar Mohammad Duad Khan. In fact, those men asked to purchase the dome for their country, but their request was rebuffed by shortsighted United States officials who eventually had the relatively inexpensive dome dismantled and used as an exhibition pavilion in other countries around the World. At the time, the Russian government did provide the Afghan king and his cousin with a uniquely modern gift which advanced Russian influence in that country at a pivotal moment in Afghanistan's history. Although they could not furnish something as novel as the geodesic dome, the Russians built the first single mile of paved road in Afghanistan, providing Afghan royalty with a small, but impressive, showcase for their growing fleet of automobiles as well as a clearly visible reminder of Russian "friendship." [13]

In 1948, the geodesic dome was far from the amazingly sophisticated structure it would become only a few years later. In fact, it consisted primarily of Bucky's idea and an enormous pile of calculations he had formulated. Fuller was also developing and studying the geodesic dome using small models that he built and tested in the family's Long Island apartment.

He was, however, eager to expand his understanding through the construction of larger, more practical projects. Thus, when he was invited to participate in the summer institute at the somewhat notorious Black Mountain College in the remote hills of North Carolina near Asheville, Fuller eagerly accepted. He had lectured at that radical institution the previous year and had been so popular that he was asked back for the entire summer of 1948. [14]

At the time, Black Mountain represented a uniquely innovative experiment in American education. Founded in 1933 by John Andrew Rice along with a small band of idealistic artists and scholars, the college had evolved into an educational commune where pure democracy had supplanted the formal educational structure. The school had thrived for several years, as both students and instructors built and inhabited many of the rustic structures which constituted the woodland campus.

In 1948, however, the institution was beginning to decline as a result of personality and ideological disagreements. Even so, Black Mountain's summer institutes continued to attract bold intellectuals who were not afraid to experiment with and examine audacious new ideas, and Bucky fit that mold perfectly.

Anne accompanied him to Black Mountain, and although she did not care for its rustic lifestyle, both she and Bucky did enjoy the company of many culturally stimulating fellow instructors who became their friends for life. Their comrades at the college included Willem and Elaine de Kooning, Arthur Penn, Theodore Dreiser, John Cage, and Merce Cunningham. [15]

Because he so respected the wisdom of youth and scorned the tradi-

Fig. 14–3   Bucky in costume as Baron Medusa.

tions of formal education, Fuller loved the cooperative spirit that Black
Mountain fostered between college-age students and instructors. He was
delighted by the phenomenon of faculty and students' sharing not only
ideas but the everyday experiences of meals and play. He was also happy
with the casual attire worn by both students and instructors. While it may
be the norm today, at that time, both students and instructors at Black
Mountain were often seen wearing a casual daytime uniform of overalls or
blue jeans.[16]

   Although Fuller's focus for that summer was his most significant
dome project up to that time, many of his fellow faculty members' most
vivid recollection of Bucky was not of his six-hour thinking-out-loud
lectures or the large dome he and his students constructed, but of his part
in the annual faculty play. The performance was a musical comedy en-

Fig. 14–4   Bucky (left) practicing with Merce Cunningham for *The Ruse of Medusa.*

titled *The Ruse of Medusa,* and somehow Bucky was persuaded to play the comic role of Baron Meduse. Initially, Fuller was disappointingly dull during rehearsals. Formal comedic acting was, after all, a new endeavor into which he had been coerced, and he felt uncomfortable playing a fool in front of a distinguished group of academics. He had quickly memorized his lines and could deliver them verbatim, but his traditional New England heritage and his belief that he should always think for himself impeded any emotion in those lines.[17]

Since Bucky did not feel that he could act and speak lines written by others, the play's producers, John Cage and Merce Cunningham, called in Arthur Penn to function as "play doctor" and cure the situation. Through a series of conversations with Bucky, Penn was able to devise several improvisations which allowed Fuller to include his spontaneous thoughts in the play.[18] When those scenes were included, Bucky blossomed into the joyful comic who had loved participating in impromptu shows at the Hewletts' home years earlier. As a result of that change, he was magnificent the night of the actual performance. Appearing in shoes with spats, baggy gray flannel pants, an oversized cutaway coat with polka-dot waistcoat, white socks, and a tall white top hat, Bucky personified the buffoon and, once again, demonstrated the childlike innocence which so endeared him to others.[19]

The majority of Fuller's time and attention at Black Mountain were not, however, spent in such lighthearted buffoonery. When he was not delivering lengthy thinking-out-loud lectures, Bucky's primary concentration centered on furthering an entirely new form of architecture. In his examination of traditional construction, Fuller had discovered that most buildings focused on right-angle, squared configurations.

He understood that early human beings had developed that mode of construction without much thought by simply piling stone upon stone. Such a simplistic system was acceptable for small structures, but when architects continued mindlessly utilizing that same technique for large buildings, major problems arose. The primary issue created by simply stacking materials higher and higher is that taller walls require thicker and thicker base sections to support the upper walls. Some designers did attempt to circumvent that issue by using external buttressing to keep walls from simply crumbling under the weight of upper levels, but even buttressing limited the size.[20]

Fuller found that the compression force (i.e., pushing down) which caused such failure in heavy walls was always balanced by an equal amount of tensional force (i.e., pulling, which in buildings is reflected in the natural tendency of walls to arch outward) in the structure. In fact, he discovered that if the tension and compression are not perfectly balanced in a structure, the building will collapse. He also found that builders were not making use of the tensional forces available. In conventional structures, the tension forces are not functionally employed. Those forces are, instead, relegated to the ground, where solidly built foundations hold the compressional members, be they stones or steel beams, from being thrust outward by tension. Always seeking maximum efficiency, Bucky attempted to employ tensional forces in construction. The result was geodesic structures.[21]

Because Fuller could not afford even the crude mechanical multiplier machines available during the late 1940s and he was working with nothing but an adding machine, his first major dome required two years of calculations. With the help of a young assistant, Donald Richter, Bucky was, however, able to complete those calculations. Thus, he brought most of the material needed to construct the first geodesic dome to Black Mountain in the summer of 1948.[22]

His vision was of a fifty-foot-diameter framework fabricated from lightweight aluminum, and operating on an austere budget, he had purchased a load of aluminum-alloy venetian-blind strips which he packed into the car for the trip down to the college. Over the course of that summer, Bucky was also able to procure other materials locally, but he was not completely satisfied with the dome's constituent elements, which were neither custom-designed for the project nor of the newest materials. Still, with the help of his students, the revolutionary new dome was prepared for what was supposed to be a quick assembly in early September, just as the summer session was coming to an end.[23]

The big day was dampened by a pouring rain. Nonetheless, Bucky and his team of assistants scurried around the field which had been chosen as the site of the event preparing the sections of their dome for final assembly while faculty and students stood under umbrellas watching in anticipation from a nearby hillside. When the critical moment arrived, the final bolts were fastened and tension was applied to the structure, causing it to transform from a flat pile of components into the World's first large geodesic sphere. The spectators cheered, but their cheers lasted only an instant as the fragile dome almost immediately sagged in upon itself and collapsed, ending the project.[24]

Although he must have been disappointed that day, Bucky's stoic New England character kept him from publicly acknowledging that emotion. Instead, he maintained that he had deliberately designed an extremely weak structure in order to determine the critical point at which it would collapse and that he had learned a great deal from the experiment. Certainly, the lessons learned from that episode were valuable, and his somewhat egocentric rationale was by no means a blatant lie. However, had he really been attempting to find the point of destruction, Bucky would have proceeded, as he did in later years, to add weights to the completed framework until it broke down. What had actually happened was something he and Starling Burgess had years earlier agreed no designer should ever allow to occur.

In his haste to test his calculations, Fuller had proceeded without the finances necessary to acquire the best materials.[25] Because of the use of substandard components, the dome was doomed to failure, and a demonstration of the geodesic dome's practical strength was condemned to wait another year.

During that year, Fuller's reputation as an eccentric genius grew, as did his invitations to lecture and teach about his architectural and design ideas. He was even asked to speak to the Graduate School of Design at his former alma mater, Harvard—from which he had been expelled twice. Bucky's most significant engagement that year was at the Chicago Institute of Design, where he spent a great deal of time working with students to develop his ideas. It was with the assistance of those design students that Bucky built a number of more successful dome models, each of which was more structurally sound than the previous one.[26]

Then, when he was invited to return to Black Mountain College the following summer as dean of the Summer Institute, Fuller suggested that some of his best Chicago Institute design students and their faculty accompany him, so that they could demonstrate the true potential of geodesic domes.[27]

As he was now receiving an average fee of $1,000 per lecture, Black Mountain's salary of a mere $800 for six weeks presented something of a hardship for the Fullers. However, when he assured Anne that he needed to spend the summer at Black Mountain, the financial question was quickly tabled. Having earned some substantial fees during the previous year, Bucky was also able to provide the best of materials for his second Black Mountain dome. The project was a fourteen-foot-diameter hemisphere constructed of the finest aluminum aircraft tubing and covered with a vinyl-plastic skin. Completely erected within days after his arrival, that dome remained a stable fixture of the campus throughout the summer. To further prove the efficiency of his design to somewhat skeptical fellow instructors and students, Bucky and eight of his assistants daringly hung from the structure's framework like children on a playground immediately after its completion.[28]

That summer also resulted in another breakthrough artifact for Fuller. For years, he had been searching for Nature's coordinate system, which he felt was embodied in the Energetic-Synergetic Geometry (i.e., geometry dealing with relationships rather than simply forms such as lines) he had begun to formulate during his study of spherical geometry. He had, however, not yet been able to overcome the obstacle of modeling all his mathematical theories.[29]

Fig. 14-5   Fuller (center) and his assistants hanging from the Black Mountain geodesic dome framework to demonstrate its strength.

Then, when a young sculpture student named Kenneth Snelson entered Bucky's office one day with a unique new form of sculpture, Fuller could only stop and stare. He immediately knew that he was seeing a physical representation of his idea of employing continually integrated tension in a structure. Snelson's sculpture was unique because it was built from solid struts connected by thin wires (similar to Fig. 5–2) in such a way that, as if by magic, no strut touched another.[30]

Embodied in Snelson's work, Bucky recognized many principles of Nature which he had been studying, and Snelson soon became a model builder for Fuller. Although legend has it that Snelson simply created the sculpture in a moment of inspiration, in actuality the unusual configuration had been extremely well thought out. Snelson had, in fact, listened closely to Fuller's Black Mountain discourses the year before, and those ideas had inspired his new design.[31]

Because of the unique way in which tension was totally integrated in that structure, Bucky combined the words *tension* and *integrity* to create a new word to describe it: *tensegrity*. Since they exemplify the properties of the geodesic structures with additional advantages, tensegrity structures would eventually become a primary element of Fuller's work. Because no strut touches another, any blow striking a tensegrity structure is immediately broken up and transferred throughout the entire framework, and any major damage at the point of contact is thereby inhibited.[32]

Even more amazing to most people who observe tensegrity structures is the way in which the struts, which are usually joined by thin, almost invisible wires, appear to be magically suspended in space. That phenomenon occurs because of the continuous tension which flows through the taut wires of the structure and not the compression members (i.e., struts) which provide the strength for most modern construction.[33]

Snelson's work was admired by everyone that summer. His most amazing piece was a ten-foot-long tensegrity mast which could be held vertically or horizontally without even a quiver in the structure. It could also be twisted out of shape and would miraculously snap back to its original form when released.[34]

As the summer session progressed, Bucky and his concepts became the focus for most students. They found him to be a wellspring of ideas and hope for the future as well as a playful friend. His joyous, youthful attitude is exemplified in the recollection of one student of Bucky waving good-bye to his students with one bare foot out of the back of his open convertible as he drove away from Black Mountain for the last time.[35]

One of the primary reasons Bucky was able to succeed in his production of a stable geodesic dome during 1949 was the enduring support of the one person who unquestioningly believed in him regardless of the feelings of others. Anne constantly supported Bucky, and in 1949 she again demonstrated her loyalty in a most dramatic fashion. During their many long conversations, Bucky had mentioned that he felt certain he was on the verge of perfecting the geodesic dome, but that, as was usually the case, he was shackled by a lack of finances. At the time, Bucky was earning a good income from lecturing and teaching engagements, but everything not used to sustain his family was funneled into supporting the innovative projects of the Fuller Research Institute, the name he had established to work under in 1946.[36]

Even with that concentration of funds and energy, progress in designing the dome simply was not rapid enough to satisfy him. Bucky wanted to advance development of the dome much faster, and he calculated that approximately $30,000 in financing was needed to expeditiously elevate the geodesic dome to a new level of sophistication.[37]

When Anne understood the details of her husband's plight, she responded without hesitation. She had just received another inheritance, and she suggested that she sell the stock she had acquired and "loan" the money to her husband. Thus, Anne sold $30,000 worth of stock in IBM and "reinvested" it in Bucky's ideas.[38]

Although friends and relatives considered Anne's action foolish, it paid off handsomely in both a contribution to humanity and helping to generate a sizable income for the Fuller family. With that money, Bucky was able to markedly advance his explorations into geodesics and to establish Geodesics, Inc., a company which supervised all manufacturing of and patent royalties from geodesic domes.[39] Despite a lack of confidence among the general public, Fuller's effort was rewarded financially when, within a few years of his inventing the geodesic dome, royalties from it generated most of his yearly income of over $1 million.[40] As a direct result of the independence that Anne's $30,000 investment provided Bucky, he maintained that income for much of the 1950s and continued to sustain an average annual income of at least $200,000 until his death in 1983.

Generating that income did not, however, become the focus of his life, and neither did the personal pleasures that such wealth could bring. Bucky indulged his passion for fast foreign cars, speedy sailboats, and providing friends and family with "needed" gifts for no apparent reason, but most of his income was quickly directed into developing the latest project which he felt would most benefit humanity.

The first practical large dome was built in 1950 by two of Bucky's former students working under the auspices of the Canadian Division of Fuller Research. It consisted of a fifty-foot framework constructed in Montreal because the needed aluminum alloys were still under wartime rationing in the United States.[41] The next major dome project was a much smaller 1951 model that was specifically designed to endure cold environments, was built for the Arctic Institute, and was used extensively at Baffin Island, Labrador.[42]

During the early 1950s, Bucky also became aware of the commercial viability of his designs and felt that major corporations would soon want to enter the geodesic dome business. Realizing that without substantial protection his invention was vulnerable to being stolen by almost any large, well-financed corporation, Bucky decided to seek patents on the geodesic dome.[43]

His primary motive was, however, unique in that it was not financial. Rather, in his continuing effort to use himself in demonstrating the significance of individual human beings, Fuller wanted to be certain that his accomplishments as an individual were documented. Obtaining a patent was one reliable method of ensuring that the magnitude of his achievement as an individual would not be swept aside in a wave of large-scale dome-building.[44]

Within years, Bucky's intuition proved to be correct. Once he had succeeded in practically demonstrating the effectiveness of the geodesic dome, several major corporations attempted to circumvent the patent that he applied for in 1951 and was granted in 1954. Bucky had, however, obtained the services of the brilliant patent attorney who had helped him obtain a patent on the Dymaxion Map years earlier. That man, Donald Robertson, thoroughly understood Fuller's predicament and produced such a superior patent that years later, when major corporations wanted to build domes, they had no choice but to pay a fee to and to acknowledge Fuller. Appreciating the importance of Robertson's patent work in documenting the geodesic dome as his invention, Bucky would recount, "If I had not taken out patents, you would probably never have heard of me."[45]

With his impregnable patent, Fuller became the sole source of licenses for anyone seeking to manufacture geodesic domes. In later years, some people would criticize his use of the patent process as contradicting his commitment to constant work on behalf of all humanity. When directly questioned about that issue, Bucky would explain that he employed patents to document his demonstration of the power of individual human beings as opposed to that of organizations, but that, more importantly, he constantly backed up his words with action.[46]

Although he required large corporations to pay for the use of his geodesic dome patent, which most of them first attempted to circumvent, he never charged individuals who had no capital and were attempting to create new dome companies or simply to explore new dome ideas. Whenever he was approached by an individual with no money and dreams of building geodesic structures, Bucky invariably supported that person's initiative and creativity by, at minimum, waving the license fee that anyone building domes was required by law to pay him. Generally, those dome builders were youthful idealists with a vision of a new type of society, and they felt, as Fuller did, that the geodesic dome was a key element of that dream. Thus, their work was extremely important to Fuller, and he sometimes went so far as to provide them with some financial aid for their endeavors.[47]

Even before the protection of his geodesic dome patent was granted, Fuller and his invention were elevated to international prominence when the first conspicuous commercial geodesic dome was produced. That structure was erected in 1953 as an answer to a Ford Motor Company problem believed to be insoluble. During 1952, Ford was in the process of preparing for its fiftieth anniversary celebration the following year, and Henry Ford II, grandson of Henry Ford and head of the company, decided he wanted to fulfill one of his grandfather's dearest wishes as a tribute to the company's founder. The senior Ford had always loved the round corporate headquarters building known as the Rotunda but had wanted its interior courtyard covered so that the space could be used during inclement Detroit weather.

Unfortunately—but fortunately for Bucky—the building was fairly weak. It had originally been constructed to house the Ford exhibition at the Chicago World's Fair of 1933, but Henry Ford had so loved the building that he had had it disassembled and shipped in pieces to Dearborn, where it was reconstructed. Having been designed as a temporary structure, the fragile Rotunda building could not possibly support the 160-ton weight that Ford's engineers calculated a conventional steel-frame dome would require.[48] Under such pressure, the building's thin walls would have immediately collapsed.

Still, Henry Ford II was a determined person, and he wanted the courtyard covered. Consequently, Ford management and engineers continued searching for an answer until someone suggested calling Buckminster Fuller. By that time, Fuller's work was drawing international attention, and although his geodesic dome had yet to be proven effective in an industrial project, desperate Ford officials decided they should at least solicit Bucky's opinion.[49] When he arrived at the Detroit airport, Fuller was greeted by a Ford executive in a large limousine, who treated him like royalty and quickly escorted him to the Rotunda building for an inspection. After a short examination of the ninety-three-foot opening requiring a dome, Ford management asked the critical question: Could Fuller build a dome to cover the courtyard? With no hesitation, Bucky answered that he certainly could, and the first commercial dome began to take shape.[50]

The Ford executives next began to question the specifications of Fuller's plan. When they asked about weight, he made some calculations and answered that his dome would weigh approximately 8½ tons, a far cry from their 160-ton estimate. Ford management also requested a cost estimate and advised Fuller that because of the upcoming anniversary celebration, the dome had to be completed within the relatively short period of a few months. When Fuller's price was well below Ford's budget, and he agreed to construct the dome within the required time frame, he was awarded a contract.[51]

The agreement was signed in January of 1953, and Bucky immediately began working to meet the April deadline.[52] The somewhat discredited Ford engineers who had failed to develop a practical solution were, however, not at all certain of the obscure inventor's fantastic claims and begin working on a contingency plan which would prevent further embarrassment. To protect their reputations further, those engineers secretly contracted another construction firm to hastily haul away any evi-

Fig. 14–6 The Ford dome under construction above the courtyard of the Rotunda Building.

dence of Fuller's work when he failed.[53] The Ford engineers were once again proven wrong when the dome was successfully completed in April, two days ahead of schedule.[54]

Actual construction of the dome was a marvel to behold. Reporters from around the World gathered to witness and recount the architectural effort as well as Ford's anniversary celebration. Because the courtyard below the dome was to be used for an anniversary television special and business at Ford had to proceed normally, Fuller's crew was provided with a small working area and instructed to keep disruptions to a minimum.[55]

Ford management was also concerned with the safety of both the dome workers and the people who might wander beneath the construction. They anticipated that problems would arise when Ford employees, television crews, reporters, and spectators gathered below to observe the construction workers climbing high overhead on the treacherous scaffolding, but once again Bucky surprised everyone. Instead of traditional scaffolding, he employed a strategy similar to the one he had developed years earlier for the quick assembly of the Dymaxion Deployment Units.[56]

Because the sections of the dome were preconstructed and then suspended from a central mast, no dangerous scaffolding was required. The construction team worked from a bridge erected across the top of the Rotunda courtyard. Like the Dymaxion Deployment Units, the Ford dome was then built from the top down while being hoisted higher and rotated each time a section was completed. The dome was assembled from nearly twenty thousand aluminum struts, each about three feet long and weighing only five ounces.[57] Those sections were preassembled into octet-truss, equilateral-triangular sections approximately fifteen feet on a side. Since each small segment weighed only about four pounds and could be raised by a single person, no cranes or heavy machinery was required to hoist them to the bridge assembly area.[58]

Once on the working bridge, the identical sections were riveted into place on the outwardly growing framework until it covered the entire courtyard. Upon completion, the entire 8½-ton structure remained suspended on its mast, hovering slightly above the building itself until the mooring points were prepared. Then, it was gently lowered down onto the Rotunda building structure with no problem.[59]

To complete the project, clear Fiberglas "windows" were installed in the small triangular panels of the dome. Because Fuller had not yet developed or determined the best means of fastening those panels, they would eventually be a primary cause of the destruction of the dome and the building itself. Since it was the first large functional geodesic dome,

Fig. 14–7 The construction bridge used by workers to assemble the Ford dome. Working on the Ford dome as it rests suspended from its central mast high above the courtyard. The bridge on which workers assembled the sections to create the dome from the top down is seen in the foreground.

Fig. 14–8 Looking up at the Ford dome as it nears completion.

many aspects of the Ford dome were experimental. They had been tested in models, but how the dome and the materials utilized would withstand the forces of Michigan winters could be determined only by the test of time. The Rotunda building dome did perform successfully for several years before the elements began taking their toll and leaks between the Fiberglas and the aluminum began to occur. Still, with regular maintenance, that problem was not serious, and convening corporate events under the dome became a tradition. One of those events was the annual Ford Christmas gathering.[60]

In 1962, numerous leaks in the dome were noticed as the Christmas season approached, and a maintenance crew was dispatched one cold late-autumn day to repair the problem. The temperature was, however, too cold to permit proper heating of the tar they used for the repairs, and in a common practice, the workers added gasoline to thin the tar. They were warming the tar with a blowtorch when that potent mixture ignited during the repair process, and the building quickly caught fire. As the building had never been planned as a permanent structure, it was not long before the entire Rotunda was engulfed in flames which destroyed the first commercial geodesic dome, the singular structure that, more than any other, had catapulted Fuller to public fame.[61]

Following the Ford dome success, Bucky was flooded with offers. Chief among those invitations were requests to speak and teach at institutions where students and professors alike were hungry to learn about the amazing geodesic dome and the man who invented it.[62] Although he attempted to satisfy as many of those requests as possible, Bucky's primary focus remained on further developing and learning about geodesics. Despite the fact that most of the projects which followed the Ford dome were for paying customers, they usually served to allow him to further his research while also solving his clients' problems. During those times, Fuller experimented with materials such as plastics and plywood and even developed a practical dome made of paperboard.[63]

Although those first paperboard domes tended to wilt in wet environments, Bucky felt that in the near future such structures could be coated with a layer of plastic which would support them in providing extremely efficient, inexpensive, yet comfortable housing. He so believed in the material that he designed and produced a series of paperboard domes which became even more efficient when he began printing assembly instructions directly on the sections as they rolled through the papermaking machinery.[64]

Fig. 14–9   A paperboard geodesic dome being erected.

Another unique dome designed by Fuller was the 1953 pneumatic model for use in extremely cold climates. That dome was constructed of two layers of strong vinyl skin quilted together to form a number of distinct sections which were filled with air to erect the dome. That innovation not only provided an insulating layer between the occupants and the outdoors, but because it eliminated a rigid framework, it also made the dome extremely portable and durable. The pneumatic dome was then successfully employed in scientific expeditions where people and equipment were airlifted to and lived upon the ice floes near the North Pole for weeks at a time.[65]

Fuller's next major dome projects were the result of a rather strange alliance. Although he had supported the United States' actions in both world wars, Bucky was a man of peace who truly believed that war would become obsolete in the near future when adequate resources to support all

Fig. 14-10   The interior of the paperboard geodesic dome.

humanity were available. Still, he responded to almost anything presented to him which could be used to support the success of humanity. Hence, when the United States Marine Corps approached him for assistance, he viewed their request as simply another opportunity to test the practical application of principles and accepted the invitation. That association was extremely beneficial to both partners, providing the marines with unique solutions to difficult problems of shelter and Bucky with testing opportunities and income. Thus, the United States Marine Corps quickly became one of Bucky's largest dome clients as well as his ally, providing resources and support for several new types of domes.[66]

The Defense Department was also generous in asserting its enormous purchasing power in support of Bucky's protecting his patent against the large corporations which were attempting to circumvent his rights as an inventor. Whenever such a challenge occurred, Defense Department officials simply informed the problem corporation that they would seriously consider stopping purchase of that company's products if Fuller was not completely acknowledged and financially compensated for his dome patent. That tactic, in conjunction with the extremely strong patent, proved successful in every instance and helped Bucky to remain involved in nearly every significant dome erected during the 1950s and 1960s.[67]

Fuller's initial association with the marines was the consequence of yet another apparently unsolvable problem. For several years, the Marine Corps had sought a rapidly assemblable, lightweight structure which could be delivered by air and used for sheltering first-strike troops and their equipment at remote sites. The corps had invested a great deal of time and money in attempting to develop such a structure but had discovered no practical solutions until a Marine Corps officer read about the Ford dome and contacted Fuller. Within weeks, a working relationship was established, and in 1954, that association resulted in the fulfillment of one of Bucky's fondest dreams.[68]

In January of that year, Bucky was invited to North Carolina to witness a Marine Corps helicopter easily airlift and transport a thirty-foot geodesic dome at a speed of nearly sixty knots. On January 29, that historic event was documented on the front page of *The New York Times*. Fuller's 1927 vision of preassembled, air-deliverable homes and buildings had been proven practical, and although the idea had yet to be marketed commercially, the visionary Fuller realized that he had laid the groundwork for that potentiality.[69]

The United States Defense Department was also the client whose need resulted in the single most prolific large geodesic dome. Once again, that dome was developed from Fuller's solving a problem believed to be unsolvable.

*

During the early 1950s, the Cold War politics between the United States and Russia elevated long-range missiles into a primary topic of concern for the American people. To protect the country from missiles routed over the North Polar regions, the United States Air Force proposed constructing an early-warning system across the far reaches of northern Canada. Approved by Congress, that system became known as the Distant Early Warning or DEW Line. It was to stretch for forty-five hundred miles along a route through the harsh regions of northern Canada just above the Arctic Circle.

In those days, when the total devastation of nuclear weapons was not

Fig. 14-11 A marine helicopter airlifting a geodesic dome.

understood, the air force plan was welcomed by concerned citizens, who felt it would provide a useful fifteen or twenty minutes' warning of a surprise nuclear attack. Once the plan was accepted, the primary impediment was not financial, but one of discovering a method for protecting delicate radar installations from the fierce Arctic environment without impairing their signals, as was the problem with metal structures. The protective structure also had to be lightweight, air-deliverable, and quickly erected in order to provide efficient protection in the hostile environment.

The air force had engaged a group of prominent MIT-based scientists and engineers to work on the problem, and their initial solution was a pneumatic, igloo-like vinyl structure which maintained its shape through compressed air. That idea did not, however, fare well in practical tests when hurricane-force Arctic winds quickly sucked the pressurized air out of the structure, causing its collapse.[70]

After considering other unsuccessful alternatives, the scientists began contemplating the merits of a possibility which had confronted many of them on a regular basis for years. Fuller had been lecturing and teaching at MIT since 1948, and with the assistance of students, he was continually erecting geodesic dome models around campus. In fact, he sometimes put up his structures in hallways and stairwells just to entice skeptical faculty and students into examining and experiencing the potential of geodesics and tensegrity.[71]

Yet, despite his enormous commercial success, Fuller and his ideas remained on the fringe of academic acceptability, and professors, who simply did not understand his newly discovered methods of construction, were not about to embrace or acknowledge the potential inherent in Bucky's work. Fuller himself was not surprised by that reaction because it was similar to what he had experienced thirty years earlier when he attempted to implement the Stockade System of construction.

Then, as in 1954, he had found that engineers and others who worked within the limits of a prescribed set of rules and definitions were generally unwilling to stretch their imaginations in considering possibilities which they had not been taught during their formal training. With such a mindset, those individuals would certainly not consider the possibility that something as peculiar as the geodesic dome or someone as eccentric as Buckminster Fuller would have a potential solution to a major defense problem.[72]

Still, the MIT scientists and engineers were confronted with a deadline and no practical solutions. Consequently, they solicited Fuller's opinion and were impressed enough to request his submission of a formal proposal. As usual, Bucky's response was predicated upon the most modern ideas and materials. He proposed an almost spherical geodesic dome with both struts and skin fabricated from Fiberglas which later became known as a *radome* because it was designed to house radar.[73]

The radome was unique in that its skin and struts were fabricated in one-piece, triangular panels which were then easily assembled into the finished product. Although that structure became commonplace at airports, military installations, and other locations where radar is used, it was initially rejected by the scientists and engineers who felt it could not possibly withstand the harsh Arctic environment.[74]

Confident of his work, Bucky simply suggested testing a prototype, and that same year, a thirty-foot radome was built for experimentation. The skeptical scientists decided to first determine just how much pressure the dome could withstand before it crumbled, and because Mount Washington in northern New Hampshire was known to have some of the strongest constant winds in North America, they selected it as the test site.[75] Since other test structures erected on that mountain had customarily been destroyed within a matter of minutes, the scientists were ready with stopwatches to record the exact number of seconds the dome survived. They

Fig. 14–12   A radome being tested on Mount Washington.

were, however, shocked to discover the radome standing several hours later, undamaged by nearly two-hundred-mile-per-hour winds.[76]

In fact, that dome remained on Mount Washington, oblivious of the strong winds, ice, snow, and cold, for nearly two years, until it was finally removed by the scientists. During the initial phases of their testing,

Fig. 14–13   A standard radome protecting a radar installation.

the scientists also learned that the geodesic dome's shape was even more beneficial than expected, naturally shedding both ice and snow.[77]

Still, they were determined to discover exactly how much pressure a radome could withstand before it collapsed, and they devised a more rigorous test in which the center of a radome was connected by sturdy cables to a massive steel hook embedded in a gigantic cement piling buried deep in the Earth. A gauge to measure strain was installed along the cables, and increasing tension was applied in an effort to pull the dome down. A great deal of pressure was applied, but the steel hook was yanked out of the cement block before the dome collapsed. Even when more durable equipment was installed, the testing apparatus always gave out before the dome could be destroyed. Consequently, the radome's ultimate strength was never measured.[78]

Following those tests, it was obvious that a radome could withstand any of the natural Arctic forces while permitting radar signals to pass through. Hence, production began in earnest, and since he received royalties from each of the thousands of radomes installed over the next few years, that single project greatly contributed to Fuller's income, which skyrocketed to an annual sum of over $1 million during the early 1950s.[79]

Although the majority of that income was simply recycled to finance Bucky's newest experimental projects, some did go to support his love of the best that society had to offer, including fine hotel rooms when he traveled, dining at the best restaurants, and the needs of his family and friends. Since the Fuller's had only one living child, Allegra's welfare was a constant focus for both Bucky and Anne.

Thus, when she decided to seek additional formal education at Bennington College after working with George Balanchine's dance company, her parents supported her decision. Bucky was particularly pleased with Allegra's choice of Bennington because of its somewhat controversial program, which encouraged students to spend a portion of their college career outside the formalized setting and to learn from practical experience and the "real world."[80]

Allegra decided to devote some of her "out period" to a job at the International Film Foundation, where she could augment her love of dance theater with an understanding of a medium that could document dance. It was while working in the film industry that Allegra demonstrated the famous Fuller rebelliousness when, in an act similar to her mother's decades earlier, she fell in love with and decided to marry a young director named Robert Snyder.[81]

Just as the impoverished Bucky had initially experienced rejection from Anne's older aristocratic relatives, Allegra's marriage to the Jewish "Hollywood movie type" was not greeted with great jubilation by her family.[82] This was especially true of the sophisticated Anne, who, although claiming not to be anti-Semitic, had been raised in what had been the extremely anti-Semitic community of Lawrence, Long Island.[83]

The actual reason for the family's initial rebuff of Snyder was, most likely, not his religion but his profession. To the traditional New England Fullers and Hewletts, Bob Snyder was simply one of those fast-lane Hollywood people who did not fit into their cultured existence and could certainly not maintain the lifelong, family-oriented marriage that members of their society were expected to perpetuate. Despite such objections, Allegra, like Anne and Bucky, made her own decision and married Snyder on June 30, 1951, with her father's approval and her mother's rather reluctant consent.[84] Years later, the family members were all pleased to find themselves wrong about Snyder, as he and Allegra remained married and raised two children.

The first of those two children and Bucky's first grandchild was Alexandra Fuller Snyder, born November 1, 1953. Two years later, on April 28, 1955, his only grandson, Jaime Lawrence Snyder, was born. Coincidentally, two years after that birth, the Fuller artifact which may well be the most influential on young children's lives began commercial production, just in time for use by Bucky's own grandchildren. In 1957, small geodesic playdomes began making their way onto playgrounds around the World, providing children with opportunities to experience and enjoy the natural structure of a geodesic framework at an early age.

CHAPTER 15

Expanding Globally

Following his successes at Ford and with the radome, Fuller's status as a designer, architect, and intellectual was firmly established. Yet he remained a radical thorn in the side of the Old Guard, especially those whose primary interests were financial and those who dominated the construction industry. Bucky's feelings toward most conventional architecture and the people who championed it is epitomized in the following whimsical song, which he wrote for a group of his students who loved to sing while they worked on dome projects.

Roam Home to a Dome
(Sung to the tune of "Home on the Range")

There once was a square with a romantic flare,
Pure Beaux Arts, McKim, Meade and White;
In the mood that ensued, he went factory-nude
Mies, Gropy, Corbussy, and Wright.

Roam home to a dome
Where Georgian and Gothic once stood;
Now chemical bonds alone guard our blondes.
And even the plumbing looks good.

Let architects sing of aesthetics that bring
Rich clients in hordes to their knees,
Just give me a home in a great circle dome
Where the stresses and strains are at ease.

Roam home to a dome
On the crest of a neighboring hill
Where the chores are all done, before they're begun
And eclectic nonsense is nil.

Let modern folks dream of glass boxes with steam
Out along super-burbia way;
Split-levels, split-loans, split breadwinner homes
No down money, lifetime to pay.

Roam home to a dome,
No banker would back with a dime,
No mortgage to show, no payments to go,
Where you dwell, dream, and spend only time.[1]

Although usually presented as a parody, the ditty provides a glimpse of Fuller's true feelings toward the construction establishment. He genuinely believed that the ideal life of "dwell, dream and spend only time" should be available to every person, and he perceived the geodesic dome as supporting that vision. He was also well aware of the cold reception that the dome and his other radical inventions had received from traditional leaders and organizations who firmly believed in maintaining the status quo.[2] Hence, even his million-dollar-per-year income of the 1950s did not hinder Bucky's incessant commentary on the importance of moving beyond accepted societal opinions and sharing the abundance of Earth's resources among all humanity.

In Fuller's view, the establishment's negative reaction to geodesic domes was typical of the belief systems which constantly confronted him. However, with the mid-1950s creation of the DEW Line of radomes, geodesic domes began to proliferate globally and to be employed for many diverse functions. Because of his controlling patent, Fuller himself was intimately involved with most of those domes.[3]

Some of the most visible modern domes are those used as arenas and sports facilities, and following his success at Ford, Bucky became highly involved in the first such structure. At the time, Walter O'Malley, owner of the Brooklyn Dodgers baseball team, was so inspired by the Ford project that he decided to look into the possibility of installing a dome above his team's stadium. The innovative O'Malley was one of the first people to believe that such a project was possible and to realize the practical significance of weatherproofing a stadium in that fashion.[4]

Always interested in new and unique projects, Bucky accepted

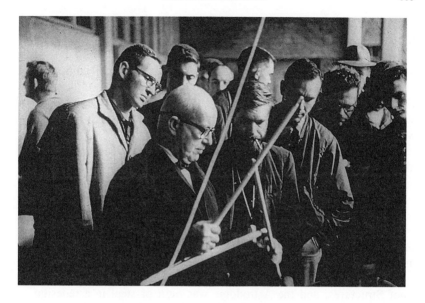

Fig. 15–1   Fuller working with students, 1959.

O'Malley's invitation to work on a dome plan for the Brooklyn stadium, and in an act typifying his respect for young people and their ideas, he solicited the assistance of the Princeton University School of Architecture. At Princeton, a team of students and faculty worked closely with Bucky over the next two years developing several possibilities. Their work was, however, abruptly terminated in 1957 when O'Malley negotiated a deal to move his team to Los Angeles where inclement weather was not an issue.[5]

Bucky's work was nonetheless useful. The possibility of erecting a domed stadium had received an enormous amount of attention and had captured the American people's imagination. As a result of that publicity and a recommendation from O'Malley, when Judge Roy Hofheinz sought national recognition for his attempt to acquire a major-league baseball franchise in Houston, Texas, he contacted Bucky to discuss the possibility of a Houston geodesic dome stadium.[6]

In 1959, Fuller and Hofheinz signed an agreement, and Bucky began work on what would eventually become the Astrodome. Along with numerous other projects, Fuller worked on the Houston dome for several years, and in 1962, Hofheinz was on the verge of signing a formal contract to utilize one of Bucky's geodesic dome designs when he abruptly

backed out of the deal and signed a contract with another company, Roof
Structures.[7]

As Fuller would later recount, Roof Structures had been attempting
to circumvent his geodesic dome patent for years. The small company had
been able to pursue that effort with the support and urging of the gigantic
U.S. Steel Corporation which wanted geodesic domes to be constructed of
steel rather than the lighter, more durable aluminum which Fuller knew
was much more efficient for large structures.[8]

In a scenario typical of those which continually haunted Fuller and
his radical ideas, he was again confronted by the short-term power of
profit-oriented big business. The management of U.S. Steel were keenly
aware that they would lose a great deal of prestige as well as future
contracts if the first stadium dome was constructed of aluminum, and
when they learned that Bucky was about to receive a contract for the first
covered sports arena, they resolved to devote even more money to thwart-
ing that effort. The Roof Structures–U.S. Steel alliance used its financial
and political power to cajole Hofheinz into signing an agreement with
Roof Structures, and the Astrodome was built of steel in a redesign of
Bucky's original plans. Because of the changes in his design, Fuller's
argument for the efficiency of aluminum as well as his geodesic patent
rights were excluded from the process, as was any acknowledgment of his
input into the project.[9]

Such a ruse could not, however, remain hidden forever, and Bucky
was eventually acknowledged as the Astrodome's primary designer. At
the stadium's opening in 1965, Judge Hofheinz's son was responsible for
publicity, and he constantly emphasized that the structure was actually a
modified geodesic dome initially designed by Buckminster Fuller.[10]

Bucky had always been adamant about using the most effective,
modern materials available. He employed lightweight, strong aluminum
in his large structures, and those who worked with and learned from him
tended to pursue a similar strategy. It was therefore no surprise that Don
Richter, one of Fuller's closest assistants at the Chicago Institute of Design and on other projects, took a job as a designer for Kaiser Aluminum
Company.[11]

In the mid-1950s, that company was one of many directly managed
by Henry J. Kaiser himself, and when Kaiser happened to spot a small
model of a geodesic dome on Richter's desk, he assumed it was one of his
company's new products and questioned the young designer. Richter ex-

plained that the object was a geodesic dome model and was soon expounding the merits and possibilities inherent in geodesic structures. The enthusiastic engineer's fervor and ideas provided a vision of possible new ventures to Kaiser, and at his behest, Kaiser Aluminum purchased the right to manufacture geodesic domes from Fuller.[12]

Kaiser's first venture was designed by Don Richter and served as a prototype for a series of aluminum domes which were assembled from preconstructed sections efficiently manufactured at Kaiser factories. The initial dome was to be an auditorium for Kaiser's Hawaiian Village Hotel in Honolulu.[13]

As it was the first such project, the erection time for the 145-foot diameter dome could only be estimated. Engineers working on the structure projected that workers would require several days to assemble the sections which were to be shipped from Kaiser's Oakland plant, and

Fig. 15-2 The Kaiser Hawaiian geodesic dome.

Fig. 15–3 The concert being performed inside the Kaiser dome less than twenty-four hours after construction began.

Henry Kaiser himself relied on that estimate in determining when to arrive for the grand opening of his newest project.[14]

However, when the dome sections reached Hawaii and assembly began, it became unmistakably apparent that the timing estimates were well off the mark and that the dome could be completed within hours rather than days. Envisioning a potential media event, Kaiser's public relations people decided to hire extra crews and to continue the erection process throughout the night. They also contacted the conductor and manager of the Hawaiian Symphony Orchestra to ask if a concert in the dome the following day would be possible. Perceiving the media value of such an occasion, the conductor summoned the symphony, and the announcement of the free concert was made on local radio stations as the auditorium dome took shape.[15]

That structure was completed in twenty-two hours, and one hour later, a capacity crowd of 1832 people enjoyed a concert by the orchestra.[16] One of the people who was not in attendance was the dome's owner. Henry Kaiser simply could not reach Hawaii on such short notice, and by the time he arrived on his preplanned schedule, the festivities had long since ended.[17]

Another Kaiser miscalculation occurred when the engineers considered the sound qualities of the dome. They had assumed that since it was constructed of massive aluminum panels, the sound quality inside would be somewhat metallic, and they contracted the services of a noted acoustical engineer prior to the dome's construction. When he arrived in Hawaii following the dome's grand opening, the acoustical engineer agreed with the conductor of the symphony orchestra, who commented that the acoustics were the best he had ever experienced inside a building.[18]

Kaiser engineers could have saved themselves a great deal of time and effort had they simply discussed their concern with Fuller. As they later discovered, the acoustical quality inside a spherical structure, which conforms to Nature's patterns of most efficient construction, is exceptional. In the Kaiser dome, sound radiated throughout the building and was actually focused toward the center, far exceeding the sound quality of conventional buildings with corners that trap sound and cause unnatural

Fig. 15-4   The Union Tank Car geodesic dome.

Fig. 15–5   The interior of the Union Tank Car geodesic dome.

reverberations. The acoustics of the Kaiser design were also enhanced by the three-section, triangular panels which acted as acoustical mirrors, reflecting sound uniformly back inward toward the audience.

Another major dome project of the 1950s designed and constructed by Fuller's company was the Union Tank Car dome in Baton Rouge, Louisiana. Union Tank Car's operation required a great deal of space in which to maneuver railroad cars while they were being reconditioned and repaired. Ordinarily, that process was l.ampered by conventional rectangular structures with congested linear tracks running through them. The company's management felt that the dome's circular structure might provide a more workable environment, and when consulted, Bucky agreed. He was then awarded a contract and began designing what would be the largest clear-span (i.e., without any columns) structure built by human beings until the 1982 construction of the larger geodesic dome housing Howard Hughes's *Spruce Goose* airplane in Long Beach, California, which is shown under construction in Figs. 15–6 through 15–10.[19]

When completed, the Union Tank Car dome fulfilled everyone's expectations. With a diameter of 384 feet and a central height of 128 feet, it provided ample space for a revolving circular track segment which radiated from the central hub of the dome like a spoke.[20] Railroad cars could be moved onto that track spur and shuttled to an outer stationionary spur for work. Because of the massive open space provided by the dome, new cars coming in could be positioned without disturbing cars being worked on.

Fuller was particularly proud of that dome because it also conformed to his belief in the importance of recycling materials, as it was constructed mainly of scrapped steel salvaged from the same railroad cars which the facility was designed to service.[21] Union Tank Car went on to build other such domes, using an even more unique system which employed a gigantic pneumatic "balloon" that was gradually inflated as construction advanced. In that system, the balloon lifted the dome to higher and higher levels as it was assembled from the top down.[22]

The apolitical nature of Fuller's work and contributions to humanity were conspicuously evident in another 1950s geodesic project. Kaiser Aluminum had received a contract from the United States government for a flamboyant dome to house the United States pavilion at the 1959 American Exchange Exhibition in Moscow, and like so many other domes, the structure itself generated far more interest than its contents. One particular incident which occurred among the displays during that exhibition also generated a great deal of attention in the media. While escorting Soviet Premier Khrushchev through the exhibition, then Vice President Richard Nixon became embroiled in an argument which was later dubbed the "kitchen debate" because it had occurred in a modern kitchen display.[23]

Following his examination of the American exhibition, Khrushchev was particularly impressed with the golden dome Kaiser had built and commented that he would like to have Fuller come to Russia to teach Soviet engineers. Soviet engineers also requested and were granted the right to examine and measure the geodesic dome in great detail, and as a gesture of goodwill, that dome was sold to the Russian government which installed it as a permanent exhibit in Moscow's Sokolniki Park.[24]

Khrushchev, like most people, was apparently not aware of Fuller's impact on Soviet engineering when he made his comment. Bucky had already traveled to the USSR and had learned firsthand just how well known he was to Soviet engineers. A year earlier, he had been invited to serve as the engineering representative in a United States delegation to the USSR led by the famous editor and writer Norman Cousins. During that

Fig. 15–6 through 15–10   The steps in the on-site fabrication of the geodesic dome housing the *Spruce Goose* in Long Beach, California. Even though it remains the largest dome in the World, it was still erected in the Fuller tradition from the top down.

Fig. 15–6   The framework for the top is assembled around a central mast.

Fig. 15–7   Upper framework is covered by panels, and lower level and panels are prepared to be fastened. Note, no panels are positioned on the lower area of the photo where the *Spruce Goose* airplane is to be moved into the dome.

Fig. 15–8   After the completed upper section is hoisted on the mast, many of the lower sections are put into place.

Fig. 15–9 The World's largest airplane is moved inside the World's largest dome. The ship beside the dome is the *Queen Mary*.

Fig. 15–10 The completed dome beside the *Queen Mary*.

trip, Fuller was surprised to be the only member of the group invited to a dinner sponsored by a group of Soviet engineers, and upon arriving, he learned that the reason he was the sole American at the dinner was that it was being given in his honor.[25]

During the event, Bucky was told that the Soviets had, in fact, been studying his work with great interest for nearly thirty years and that they felt his engineering ideas could make an enormous difference within their society by providing inexpensive shelter and housing as well as other innovations.[26] They also admitted that Fuller's ideas were so radical that having them accepted by the old-line politicians who controlled most Soviet activities and policies would probably be even more difficult than their gaining acceptance in Western society.[27]

When Fuller was asked to share his thoughts on innovation and how to implement it, he agreed that geodesic structures could most certainly contribute to the welfare of any country, especially ones such as the Soviet Union in which harsh weather and limited resources were so prevalent. He also recounted the importance of his ability to operate as an individual without the restraints of any bureaucracy or government. Specifically, Bucky asked the Soviet engineers what would happen in their country if they acted as he did, pursuing roundabout methods of obtaining scarce, newly developed materials controlled by the government. The question received a hesitant laugh from his audience, as they agreed that any Soviet engineer attempting to employ the same individual initiative that Fuller frequently demonstrated would not be around long to report the story.[28]

Despite that reaction, Fuller continued to chide the Soviet engineers in the same way he advised students and others interested in making contributions to the future. He suggested that the most successful operating strategy any person living on Earth could employ was to work as an individual in cooperation with other individuals, but not under the domination of archaic bureaucracies.[29]

During the same year as that initial Soviet Union journey, Bucky also made the first of what would eventually be over forty trips around the World in response to invitations. One of his stops on that tour was to give three lectures in New Delhi, India, where he could not help but notice a striking woman wearing a beautiful sari who was given a seat of honor in the front row for all three lectures. What intrigued Fuller more than anything was how intent that woman seemed to be on understanding his every word.[30]

Following the final lecture, she was introduced as Indira Gandhi, daughter of Prime Minister Nehru. When Gandhi inquired about a small model Bucky used in his lecture, he was pleased not only to explain it in great detail but to present it to her as a gift. When Gandhi then asked if Bucky could possibly visit her home to meet with and explain his ideas to her father, he was thrilled and graciously accepted.[31]

Later, Fuller arrived at the huge residence, which had been constructed by the British during their rule and was used to house Indian politicians. A servant led him past the public rooms to a much smaller room which served as the reception area of Nehru's private suite, and Bucky was left alone to wait for further instructions. However, instead of someone's telling him what was expected, Nehru simply opened another door and quietly slipped in. Standing near that door in his usual white tunic, the prime minister bowed to Bucky with hands folded in the traditional Hindu greeting. Following that lead, Bucky made a similar bow of greeting and, without a word from Nehru, began thinking out loud. As Nehru stood silently, Bucky continued speaking for the allotted 1½ hours, at which time he stopped. Nehru then bowed with folded hands and silently left the room.[32]

Following that monologue, Bucky discussed the meeting with Indira Gandhi, and she explained that whenever her father was listening to something or someone he felt was extremely important, he would stand silently as he had done with Fuller. She further clarified that Nehru felt that when he acted thusly, his mind would not drift and he could receive maximum benefit from the speaker.[33]

Gandhi also found Bucky to be enormously interesting as well as congenial, and she invited him to visit whenever he was in India, which he did on several occasions. Both she and her father particularly enjoyed listening to Fuller's integration of spiritual and scientific disciplines because they found it quite similar to their own beliefs.[34]

Fuller was provided with a unique view of Indira Gandhi during another of his visits. He was in India in 1964 when Prime Minister Nehru suffered a stroke, and upon hearing the news, he immediately called Gandhi, who was, by then, a close friend. Although she was distraught, Gandhi asked him to come over right away, and Bucky was pleased to be of use to her. When he arrived at her home, Gandhi's devastation was obvious, and Bucky did what he could to comfort her. During their conversation, he also asked if she would consider continuing her father's political work if he did not recover.[35]

Gandhi emphatically replied that she had no interest or talent in that area, and Bucky felt her sentiment was genuine. When Nehru died later that year, she was able to maintain her apolitical position despite a public outcry for her to become politically active. Yet, within two years, in an action similar to Bucky's strategy of doing what was needed for the welfare of the most people, Gandhi applied herself to the political arena, ran for prime minister, and won. Over the years, she and Bucky maintained their close relationship, and because he felt that she was not acting from the self-centered motives he generally found sustaining other politicians, she was one of his few true politician friends.[36]

Throughout all of his projects and travels, Bucky continued accumulating a growing archive of information documenting his life as well as the global changes he had witnessed. That amalgamation continued to expand with Fuller's popularity, and in 1959, at an age when most people retire, he finally began establishing a formal base for his operation and the data he had collected. The motivation for that formalization was twofold. First, he truly wanted the material and ideas he had accumulated to be easily accessible to future generations, and second, the administration of Southern Illinois University at Carbondale provided him with an opportunity to more effectively pursue his work in association with their institution.[37]

SIU offered Bucky a position as research professor. That post included staff assistance as well as an office and the facilities to conduct his research on a continuing basis. The university also allocated enough space to house his archive, which he tentatively bequeathed to SIU.[38]

In return, Bucky agreed to lecture and to conduct seminars for students whenever his busy schedule permitted. What the university really acquired was the prestige of having Buckminster Fuller on its faculty, and his stature helped transform SIU. During the thirteen years of his tenure, SIU student enrollment increased from five thousand to over thirty thousand, and a growing number of prominent academicians and gatherings were presented at the university simply because of Fuller's presence.[39]

Still, Fuller's schedule did not provide much time for normal academic life.[40] He rarely spent a total of more than six weeks per year in Carbondale, and since he was traveling around the World the majority of the time, it was appropriate that the small office which served as his first true global headquarters was situated above a travel agency.[41]

It was also fitting that prior to moving to Carbondale, the Fullers had one of the earliest geodesic dome homes built for them. The house was a small, thirty-nine-foot-diameter white plywood dome with two bedrooms, and it finally allowed Bucky to fulfill his dream of being able to "roam home to a dome" on the rare occasions he found himself in Carbondale.[42]

With the dawn of the 1960s, Fuller's focus began a gradual transformation from physical inventions to metaphysical ideas. He had always believed that the difference between the physical and the metaphysical was simply that metaphysical phenomena could not be directly experienced by human beings, but in the 1960s he began an ever expanding public display of his prowess as a philosophical thinker as well as an engineer, a designer, and an architect. As society radically shifted during that decade, Fuller's perspective proved correct, and both he and his ideas were increasingly accepted by people from all walks of life.

Bucky saw society rapidly entering into a new era which would eventually be termed the *information age,* and he realized that as humanity advanced into that period, nonphysical phenomena (especially knowledge) would increasingly become more important than physical phenomena (especially muscle or military force).[43] In other words, ideas and not the inventions they spawned would be recognized as the critical components of innovation. Accordingly, his thinking-out-loud lectures and writing proliferated in response to public demand.

Bucky's poetic ability was also publicly acknowledged by the school which had formerly chastised him for his wild lifestyle and lack of interest in traditional education when, in 1962, he was awarded the Charles Eliot Norton Chair of Poetry at Harvard.[44] Harvard's chapter of Phi Beta Kappa also honored the formerly twice-expelled student by presenting him with its key.

By 1963, the demand for Bucky's ideas had reached a feverish pitch, and he was able to publish his first books since the 1938 *Nine Chains to the Moon.* With typical Fuller vigor and excess, he did not simply release one or two books but published four during that year alone. Those works were a book which included a great deal of poetry, *No More Secondhand God;* two books of prose, *Ideas and Integrities* and *Education Automation.*

In conjunction with his speaking engagements, which had mushroomed to well over one hundred per year at sites around the World, Bucky's books and articles advanced his ideas to the forefront of contemporary thought. Nowhere were those ideas more well received and studied than among youthful audiences who did not perceive Fuller as an aged inventor-philosopher. Rather, they saw him as he truly was, a radical progressive interested in the welfare of all humanity who presented pragmatic possibilities for human achievement and fulfillment.

His largest audiences always convened on college campuses, where students proved to be far more receptive than staff or faculty who were generally steeped in traditional thinking and concepts. And it was to youthful students that Fuller's message of the hope which was their birthright was geared.[45]

Bucky had personally observed his own potential being systematically molded into a traditional cast by society, and even though he had evaded being shaped in the image others felt would be best for him, he realized that most young people were not so fortunate. Because of that experience and his ongoing observations, Fuller constantly shared his view of a perpetually changing environment which created an entirely new set of circumstances for each generation of children.[46]

Bucky had specifically devoted a great deal of his life to surveying the effects on individual human beings of dramatic shifts in technology. By examining the ease with which young children accepted phenomena such as airplanes, computers, Moon landings, and other unanticipated changes, Fuller came to understand that each of those innovations had marked a dramatic change in what was accepted as the norm by the general public. He also realized that because children are not entrenched in a belief system and can more readily appreciate the wisdom exhibited by a changing environment, they are usually the first to accept a new phenomenon or artifact as natural and normal.[47]

Although Fuller's hypothesis of what is natural asserted that only individuals born following the discovery of a new innovation could completely accept that change, he believed that, in general, the younger a person was, the more flexible and adaptable to change he or she would be.[48] To illustrate that point, Bucky often recounted stories from his own youth.

During that period, the norm was devoid of modern technological conveniences such as automobiles, telephones, electric lights, refrigeration, and indoor plumbing. Despite a lifetime of being deeply involved with the leading edge of technology, Fuller, like all human beings, could not help but retain some connection to that simpler era of his boyhood. In fact, he was one of the family members who most vehemently argued that the family retreat on Bear Island should be kept in the rustic state he remembered from his youth.

Other family members also supported his desire to maintain Bear Island without modern technology and creature comforts. Thus, innovations such as electricity, indoor plumbing, and the modern high-speed communication have yet to reach the Fullers' island domain, an ironic twist for the retreat of a man who championed the mass production and easy accessibility of modern technology. That rustic island, with its undisturbed, turn-of-the-century environment, continued to provide Bucky with a protected retreat from the grind of fast-paced modern civilization throughout his life. No matter how far he traveled or how many projects he became involved with, nearly every summer several weeks were set aside for rejuvenation at the Bear Island complex.[49]

There, he could reestablish his perspective, remind himself of his earlier insights into Nature, and determine just how much society had changed. His comparative observation was aided by the Bear Island retreat, the one setting which, more than any other, reminded him of his youth and sustained his awareness of Nature in its purest sense.[50] As he would write, "My teleologic stimulation first grew out of boyhood experiences on a small island eleven miles off the mainland, in Penobscot Bay of the state of Maine."[51]

In later life, Bucky realized that during his early years people had been inherently remote from one another. He learned that customs and traditions had been developed over thousands of years during which people had lived in isolated tribes and communities.[52] Within those small communities, individuals had been in almost constant contact with one another.

That daily proximity to a small group of people and remoteness from most others spawned behavior patterns which had been codified over the centuries into laws, rules, and traditions. Although many of those laws and traditions were, and still are, no longer germane to the prevailing conditions, most older people continue complying with them.

Fuller came to understand that such antiquated belief systems are powerfully ingrained within a society and tend to endure into future generations regardless of their relevance. For example, because he was born before humans could fly, flying was always somewhat of a remarkable phenomenon to Bucky. Even though, during the last decades of his life, he circumnavigated the Earth several times each year via airplane and was a pilot, Fuller's conditioning and youthful memories had established flight as far more alien to him than to his descendants.[53]

He clearly remembered the dangers involved in early flight and the admiration bestowed upon the brave men and women who flew airplanes and returned safely. During that period many airplanes crashed, and the unmistakable risks generated a widespread fear of flight. Even in the later years of his life, when flying was documented as the safest means of mass transportation, Fuller found that many older people, like himself, still held onto some of their subconsciously conditioned belief that flying was inherently dangerous.[54]

When confronted by people reluctant even to consider his ideas about generational influences on human beliefs and the human ability to accept innovation, Fuller would ask them to recall their initial reaction to the Moon landing of 1969. Following that monumental feat, he found that most people began behaving as if it were a routine event. At the time, Bucky also observed people's attitudes as they quickly shifted from astonishment to asserting that landing on the Moon was a simple accomplishment whose "time had come."[55]

In further examining human reactions, Fuller confirmed his feeling that the Moon landing, like other momentous achievements, could be considered natural and logical only by individuals born following its occurrence. Because they study such events and their development as history rather than regard them as major transitional moments in their lives, individuals born following an event's occurrence are endowed with a unique new perspective. Fuller found that no matter how people attempt to alter their perspective, those born prior to the first Moon landing always viewed that feat—and any other such events which had occurred following their birth—as an unnatural surprise which they pretended was natural. He also believed that adults tend to pretend events are natural out of vanity. They simply do not want to appear ignorant or, worse, wrong.[56]

In later years, Fuller himself was often confronted by the specific issue of young children's perception of the 1969 Moon landing. Children born after that date frequently wrote to him asking the question, "If humans can send people to the moon, why can't we create a World that works for everyone?" Such queries only served to support Bucky's observation that children have a unique modern sense of what is normal and natural.[57] He also felt that the seemingly naive question illustrated that as a result of their perspective, children generally cannot understand why the problems adults believe are complicated cannot be easily solved.[58]

Fuller believed that because modern children generally experience the positive, life-supporting aspects of human technological achievements, they feel that increasing such successful action is natural and normal. Although a child's view appears naive to many adults, within that juvenile attitude Fuller discerned a tenacious hope for the future of all life on Earth.[59]

His studies and observations led Bucky to theorize that humanity had reached a moment in its evolution when individual human beings could consciously choose the path of future development, and that a more child-like route would serve rather than hinder progress.[60] The path of action Bucky himself selected to endorse and employ included returning to the natural, youthful perspective of endless curiosity and potentialities that everyone has experienced at one time or another.

Applying that youthful perspective of cooperation and questioning for most of his life, Fuller was able to produce a remarkable number of diverse inventions, books, articles, and lectures, as well as to explain the underlying thoughts and principles of his work. Still, even his childhood curiosity and insight had not always been nurtured in the manner he would later recommend as essential for all children. Like most young adults, Bucky had been expected to grow up and take on responsibilities.

However, during the period when adults had pressured him to conform to society's rules, Bucky was able to continue observing and learning from everything he encountered. That constant attention to his environment and the rapidly occurring changes within it would eventually support his developing insights that were far beyond those of most other people. Thus, when he consciously resolved to return to a curious youthful perspective predicated upon personal experience rather than to accept the dogma and tradition of the past, Fuller was somewhat prepared for the opposition he encountered.

Another example of generation-based conditioning which Bucky observed was his attitude toward automobiles. When he had entered Harvard in 1913, only three of the seven hundred freshmen owned automobiles, and one of those was the property of the son of Ray Stanley, inventor and manufacturer of the Stanley Steamer.[61]

Bucky himself did not purchase an automobile until the early 1920s, and that car's tires usually blew out after only about one hundred miles, forcing him to make frequent stops. Removing and repairing those tires was a formidable task, requiring at least one hour per tire. Even the simple act of starting his car was generally a major undertaking, with Fuller often having often to remove the spark plugs and prime them with gasoline just to start the engine.[62] Although those and other automobile problems monopolized a great deal of his time, they also established an intimacy between owner and machine which is rare today.

The cars of that era provided Fuller and his peers with direct, personal experiences of the automobile's unreliability and caused them to drive with utmost caution. Because of their important function and the fact that they tended to wear out quickly, brakes also constituted a major problem in early automobiles, and Fuller observed his own resistance to change as better brakes were developed. Although he drove for most of his life, Bucky persisted in maintaining a somewhat tentative driving style, not completely trusting even the most modern brakes. Constantly more-or-less unsure of even a modern vehicle's roadworthiness, Fuller allowed a larger-than-usual distance when following other cars, especially in his later years, when the average speeds of cars increased greatly. That habit inevitably invited younger drivers who were accustomed to reliable brakes and faster speeds to pass him with great confidence.[63]

In his incessant observation of human nature, Fuller noted that those younger drivers had been conditioned to expect mechanical success from their vehicles, just as he had been conditioned to expect mechanical failure. He perceived that variation in driving technique as another demonstration of irreversible generational differences which have to be respected by individuals of every age if we are to achieve universal human cooperation.[64]

Fuller believed that his experiences with the automobile and the airplane were palpable examples of the manner in which different circumstances affect the behavior and perception of people born during different eras.[65] They also demonstrated the natural resistance of even the most daring older people to new technology. He felt that an apparently safe, logical action or innovation is, in fact, uniquely appropriate only within the context of a specific time frame, and that an innovation is thus thoroughly comfortable only for the generation born following its appearance.

Fuller also observed that universal phenomenon within his own family. His daughter Allegra was born in 1927, the year Lindbergh flew across the Atlantic Ocean. At the time, flying was still an uncommon experience limited to an elite minority, and people rarely saw an airplane in flight. In his later lectures, Bucky would speak of his thoughts during one such unusual encounter with flight. He was wheeling Allegra through Chicago's Lincoln Park in her carriage on a bright fall day in 1927 when a small biplane appeared overhead. Although an airplane overhead is a common experience in modern Chicago, it was an strange event at that time, and it sparked Bucky's thoughts. He began considering his young daughter and her relationship to the airplane, and he was soon engrossed in thinking about another large pattern of human development and realizing that his daughter would grow up in an era when airplanes in the sky would be considered normal.[66]

Years later, Bucky would again observe the changing norm in his own family when Allegra's daughter, his granddaughter Alexandra, was born in 1954 near New York City. Alexandra's first home was Riverdale, just north of Manhattan Island. The apartment was on the top floor of an aged wooden house with a glass porch, and most westbound flights leaving La Guardia Airport took off directly over that house.[67]

Every few minutes Alexandra would hear a very loud noise overhead and someone would say "airplane" to her. Thus, it was not surprising to Bucky when her first spoken word was not mom or dad, but *air*. Family and friends would carry Alexandra onto the glass porch to show her the airplanes flying overhead, and she saw thousands of airplanes before she ever noticed a bird. Thus, to her, an airplane in the sky was far more normal and natural than a bird. Looking down from that house, Alexandra could see streets jammed with automobiles, and an environment filled with automobiles also became a familiar component of Alexandra's reality.[68]

By observing the conditions in which his granddaughter was being raised, Fuller again witnessed the enormous differences between the environmental experiences of modern generations and what he had encountered just fifty years earlier. During his youth, the appearance of even a single automobile had produced sizable crowds of spectators within a few minutes.[69]

Bucky was also very cognizant of Alexandra's early education and, in particular, the children's picture books she was being shown by adults. Those books usually contained illustrations of common farm animals. Adults would show young Alexandra the animal pictures, announcing that such animals were outside, but she rarely saw live farm animals as a young child. Bucky realized that to Alexandra, animals were simply images in books, and that in her young mind, a picture of a pig was about the same as a picture of a polio virus. She did, however, notice that adults enjoyed showing her the books, so, in Bucky's opinion, she laughed along with them even though the images contained no real meaning for her.[70]

Alexandra's unique concept of what was natural provided Bucky with another indication of the rapid transformation occurring within the generational perspective. During Alexandra's childhood, a few people were still born and raised in farming environments where chickens and pigs were a daily experience, but the predominant environmental experience in the United States had shifted from rural to urban. The fundamental catalyst for that shift had been the development of labor-saving machinery following World War I, especially farm equipment.[71]

New machines permitted farmwork to be accomplished more efficiently than was possible using only animal and human muscle. When mechanized farm equipment was combined with other new developments, such as refrigeration and rapid transportation, individuals were provided with a new freedom to live long distances from agricultural production areas and still be provided with food.

The rural lifestyle had been dominant since primitive human beings had shifted from being hunter-gatherers to being cultivators of crops, and its predominance over the centuries has remained an extremely powerful influence on what is now regarded as normal and natural. Fuller found that only within his lifetime had basic agricultural production become mechanized, causing the majority of the American population to relocate into urban centers. Living within those urban areas, humans found their perception of what was natural quickly becoming unrelated to and insulated from rural influences such as vegetation, earth, and animals. Because of that shift, the norm for Alexandra's generation was to first experience animals through books rather than in the flesh.[72] Today, that experience has again changed, as many modern children first experience animals and the outdoors through the technology of television.

Fuller believed that such transitions in common experience were neither good nor bad, but that they had to be acknowledged and accepted if humans were to live together in harmony on a Planet which is becoming more and more densely populated. He realized that change had been, and would continue to be, constant, but that humanity's newly discovered technology had increased the speed of change to such a rapid velocity that individuals had to be extremely conscious and observant just to notice even the most conspicuous shifts.[73]

Operating in harmony with such rapid transformations requires an enormous amount of attention to detail. Bucky felt that people had to observe their environment ceaselessly if they were to anticipate the changes which would affect their future, and he successfully demonstrated that technique throughout his life.

Still, he found that most of the population persisted in following old precepts rather than examining ideas, traditions and laws within the context of a rapidly changing environment, as he did.[74] Nearly all of those current laws were conceived by and for individuals who lived in small rural communities and had frequent contact with their neighbors but little contact with the outside world. Within that environment, wars were limited to local disputes between neighboring nations or tribes, and news of a war traveled so slowly that it often arrived at a neighboring village after battles had ended. Currently, however, any warlike action attracts an immediate response from an informed populace which appreciates the threat of total planetary annihilation.

Fuller saw that current laws, procedures, and bureaucracies were—as they still are—becoming increasingly extraneous, inappropriate, and ineffective as modern communication and transportation brought people closer to their fellow inhabitants on our small Planet. He was one of the first to express the opinion that individual citizens of Earth must demand innovative new operating systems which will accommodate the unprecedented changes we are experiencing—something he felt only young people could easily appreciate.[75]

In his constant work with young people, Fuller found that the supposedly recent phenomenon of the "generation gap" was not predicated on the hostility of a younger generation toward the older generation, as was generally believed. He felt that a generation gap has always existed as a result of the younger generation being born into a new natural order which is unnatural to older people and which will itself eventually change.[76]

Fuller felt that the question humanity now faces is whether all people can live together in harmony by predicating action on what is natural at a given moment and an anticipation of what will be natural in the future. He demonstrated that one individual could shift his or her perception to do just that. A substantial portion of his legacy is the example of a single individual willing and able to adapt in order to contribute to the well-being of all of humanity. Within that legacy of constant adaptation to change is a model of the potential every individual has to transform himself or herself as humanity continues its increasingly rapid development.

In recounting the trials and tribulations of his life, Bucky invariably began by asserting that he was simply an ordinary individual who responded to the circumstances he encountered. He would continue by contending that his audience had the potential to accomplish as much as, if not more than, he had achieved. He would then proceed to explain precisely how he had realized such success, using his own life to explain the principles he found to be operative throughout Universe.[77] One theme which invariably received an enthusiastic response from young people was that of individuals thinking for themselves.

Fuller argued that organizations, be they educational, religious, political, industrial, or any other form, and the individuals who control them must sustain a belief system of scarcity because an environment of total abundance would relegate such institutions to extinction.[78] Within traditional systems, individuals are generally encouraged to follow the dictates of former generations rather than to think for themselves and to predicate their actions upon experiences of the moment.

Fuller felt that because they were deeply entrenched in the financial constraints of that belief system, most adults devoted a great deal of time to the issues of job security and obtaining the necessities of survival for themselves and their families rather than to thinking for themselves and determining what needed to be done in a rapidly changing society. Within our system, most people consider the wishes of their "boss" of paramount in importance, further relegating independent thought to a less significant status.

Bucky realized that in the relatively static, slowly changing societies of the past, such a focus had not been nearly as important as it was becoming in the rapidly transforming society which technology had advanced. He also felt that the single most significant hope for the future rested with the children who had not yet been indoctrinated into the system or impeded by bosses and the belief that they had to earn a living.[79]

Fuller frequently discussed his perception of humanity's drastically transforming environment and relationship of people to their environment in terms of how the process of news dissemination had shifted. His theory of that change began with the contention that men, like all male mammals, have generally been inclined to roam over much larger areas than women or the young.

Out of necessity, mothers tended to remain near the home, taking care of the young, while fathers usually operated autonomously, hunting and gathering food. In the case of humans, the man would return home with items, and the woman, with her less vociferous authority and her wisdom, would determine what to do with them. Fuller felt that the most important nonmaterial item men brought home was the news. For most of human existence, roving males have gathered and disseminated most information concerning the outside world for women and children. Families learned from those traveling men, but their information was presented with a masculine interpretation.[80]

The primary authority recognized by children raised in such an atmosphere were parents who educated children based on the information that their father brought home. Children were also generally taught to respect the authority and wisdom of older men such as a king or a grandfather. Thus, the information obtained by children was usually tainted by the prejudices of older men and the traditions of past generations. That tradition continued in the majority of societies throughout most of recorded human history and began to change radically worldwide only during this century. Because he was born at the turn of the century, Fuller was able to witness that shift and, consequently, to be aware of its significance.[81]

As a young man he noticed that if adults regularly used certain phrases or spoke with a dialect, those practices were passed on to children as "the" proper manner of speaking. Because of that practice, languages were very locally oriented. In the early-twentieth-century United States, with its almost total immigrant population, that orientation resulted in the dialects of small neighborhoods' adhering to the customs and accents of the country from which the residents had emigrated, despite those neighborhoods often being right next to one another.[82]

Still, even in that rapidly changing culture, the father continued to bring home the news from beyond the immediate neighborhood and to supply his family with information biased in favor of his belief system and the traditions he had been taught. Then, Bucky witnessed that condition almost instantly shift one May afternoon in 1927 when many fathers returned home from work to hear their children recount that the man on the radio had said that a man had just flown nonstop across the Atlantic Ocean.[83]

Fuller felt that the phenomenon of women and children listening to the radio at home and learning about Lindbergh's exploits before men at work had heard the news or had seen it printed in the newspapers was, ultimately, far more significant than the journey itself. He also realized that those events were similar in that neither of them had been anticipated by the general public, both had been the result of modern technology, and both had altered the course of human development dramatically.[84]

He believed that the radio and other modern forms of communication had provided most people with a freedom unmatched by other inventions.[85] In our information-oriented society, communication and information represent power, and the radio shifted a great deal of power away from men, who had always dominated humanity with muscle and military authority, and toward women. More importantly, children were no longer as prejudiced by their parents' traditional beliefs.

Bucky also theorized that the radio announcer's swift advance to a position of immense authority in society had created another significant change which had influenced youth more than any other segment of society. That change focused on our language and, ultimately, acted as a catalyst for more rapid human integration. Radio announcers, and later television personalities, attained their positions because their speech patterns were devoid of regional vernacular and, therefore, could be understood by the greatest number of people. Consequently, as children grew older and perceived authority, wisdom, and insight to be coming over the airwaves rather than from their parents, they imitated radio personalities rather than mom or dad.[86] The result of that shift was a more universal language and another step toward a single global human species in which individuals do not view themselves as different from other human beings.

During this century, children also began to learn more about distant places formerly relegated to obscure references in textbooks, and they started to develop a compassion for other members of the human species, regardless of where those people happened to live. Bucky appreciated and participated as that empathy and concern for the entire population of Earth was graphically demonstrated by one particular generation.

During the 1960s, students began to speak out publicly, especially in the United States. The generation which had grown up with the knowledge provided by early television had entered college, and campuses were becoming a forum for public discussions of global issues as an increasing number of students voiced concern for fellow human beings. Nowhere was that expression of questioning and protest more visible than at the University of California at Berkeley, where the "free speech movement" was born. And there, just as on so many other campuses, the radical ideas of Buckminster Fuller provided valuable insights into possible solutions to the growing global problems being examined. During the free speech era, Fuller, who was nearing seventy, presented many of his lengthiest and most inspired thinking-out-loud sessions.[87] At that time he also became such an endearing figure that he was nicknamed "Grandfather of the Universe."

The students who listened to and read Bucky's words usually accepted his statement that he was an ordinary individual, but they also tended to be inspired by his demonstration of just how much any individual, including themselves, could accomplish if they set their minds to it as Fuller had done. Since he had challenged authority for most of his life, Fuller empathized with the plight of 1960s students. He also advised them that lasting social change could best be accomplished through controlled technological innovation focusing on the welfare of individuals rather than on the destruction of warfare.[88]

Many who listened to and read Fuller's ideas took his thoughts to heart and concentrated on the technological innovation he had championed since the early 1950s. As a result of such interest, the geodesic dome entered the world of the hippie.

Bucky had always been somewhat of a Bohemian artist, and he was not opposed to his invention's taking hold among young people searching for a unique, inexpensive structure which could serve as home, a greenhouse, or in a multitude of other functions. Hundreds of the students who had worked with Fuller or had attended his lectures during the 1950s had followed his suggestion and had begun to work at the leading edge of design technology, but the students of the 1960s applied Bucky's design-science ideas in a much more direct manner. They simply began building and experimenting with rudimentary domes, and the geodesic dome soon came to be associated with the openly liberal lifestyle of the 1960s generation.

Geodesic domes began springing up on remote hillsides and in communes around the United States. Some of the most well-known such domes were erected on the outskirts of Trinidad, Colorado, at a site which became known as Drop City because of the tendency of people to simply drop in for a period of time. Drop City was originally established by a group of students who heard Fuller lecture in Boulder, Colorado, and were so inspired by his philosophy of independent thinking and action that they decided to establish a self-contained community using geodesic domes as homes.[89]

Following the Fuller strategy of recycling resources, those young people built most of their unique structures with whatever materials they could acquire inexpensively or could scavenge. Some were constructed of plywood covered with tarpaper fastened to chickenwire by bottlecaps and used glass from old cars for windows. However, the domes which gained the most attention were those fabricated from the tops of scrapped cars.[90]

Drop City thrived for several years, not so much as a stable community but as a place where people could come and go as they pleased. It also became known as the stopover for young people traveling from the East Coast to the San Francisco area during the hippie period of Haight Ashbury.[91]

Because he tended to traverse many disparate elements of society, Bucky was also involved with more traditional events of the 1960s at the same time that he was lecturing and encouraging radical students. One of the most conspicuous such ventures was the United States Pavilion at the Montreal World's Fair, Expo '67. In 1965, the United States government had requested Bucky's help with the project, and he had proposed construction of the ultimate Geoscope, which he had begun designing years earlier.

His proposition consisted of a Geoscope housed in a gigantic geodesic dome. The Geoscope was to unfold and transform into a huge, computer-controlled Dymaxion Map which would display global events and resources updated to a given moment. Not possessing the comprehensive global vision of a Buckminster Fuller, government officials decided that the World's Fair display would be more acceptable and less controversial if it presented the positive attributes of the United States rather than any global scenarios which might tarnish the image of the United States as a benevolent global giant.[92]

The officials did, however, resolve to use Fuller's geodesic dome idea to house their display, and Bucky was commissioned to design that structure, which, when completed, represented the culmination of geodesic dome designing for him. To assist him in that effort, Bucky selected Shoji Sadao, a former student with whom he had formed Fuller and Sadao, Inc., an architectural design firm.[93]

That dome was unique in many ways. A three-quarters sphere, it was as tall as a twenty-story building; yet, rather than being constructed from conventional plans and blueprints, it was fabricated almost entirely from a complex set of mathematical tables. Those figures provided workers with the exact specifications needed to stamp out precise stainless-steel-alloy components which were assembled to create the dome's frame.[94]

Fuller himself wanted to construct the dome using the same system of lightweight aluminum struts bolted together that he had employed in some of his most successful earlier domes, but once again, his opinion was overridden by the dictates of financial and corporate powers. Those individuals insisted that steel components be permanently welded in place. During construction as well as years later, those bureaucrats would come to regret not following Fuller's visionary as well as pragmatic suggestions. Had they allowed him to proceed without restraint, the dome would have been assembled far more rapidly and would have been finished ahead of schedule rather than still being worked on as the Fair opened.[95]

When Expo '67 closed, the United States government immediately received several offers to purchase the geodesic dome which had housed its exhibit, but because of the welded joints, it could only be left in place and sold to the city of Montreal for one dollar. Had that dome been built of aluminum sections bolted together, as Fuller recommended, it could have been easily disassembled and moved to another location by the purchaser. However, with heavy steel components welded together, the only options were leaving it where it had been erected or demolishing it. Thus, the city of Montreal received a bargain dome, even though it was of great interest to several other parties and could have been sold at a profit.[96]

Eventually, the construction methodology resulted in the Montreal dome's destruction. Rather than simply replacing sections which failed or

Fig. 15–11   Fuller discusses the Expo dome with President Johnson.

adjusting bolts when problems arose and leaks occurred, as would have been the case with bolted aluminum sections, workers were forced to reweld problem areas, many of which were several stories up on the dome's surface. During one such repair session, sparks from the equipment ignited a panel of the Plexiglas windows which served as the skin, and within minutes, the fire had spread so rapidly that firefighters could only watch as all but the dome's steel skeleton was destroyed.[97]

Still, during its glory, the huge "skybreak bubble," as it came to be known, was an engineering marvel to behold and proved far more popular than any of the Fair's other exhibits. Inside the dome, the United States displayed various exhibits under the title "Creative America," but the crowning achievement of creativity was the dome itself.[98]

In designing the enormous structure, Fuller innovatively solved many unique problems. Primary among them was comprehensive environmental control. To maintain temperature control with such a huge amount of clear window space, Fuller devised a system of light sensors which raised and lowered shades on many of the clear panels in response to the amount and direction of the incoming sunlight. He also installed a system using over 250 small electric motors connected to a central computer which could open or close the individual triangular panels in response to weather conditions.[99]

Inside the dome, the structure's enormous size created a unique environment in which the separation between outside and inside was almost invisible. That near invisibility was, in fact, a phenomenon which Bucky had been considering for many years. He felt that if gargantuan domes could be installed over sections of cities, the amount of energy saved would be monumental while simultaneously providing the residents with much more comfortable living conditions.[100]

For example, Fuller calculated that a single one-mile-diameter dome enclosing the mid-Manhattan area between Twenty-second and Sixty-second Streets and between the Hudson and East Rivers would have a surface area of only $\frac{1}{84}$ of all the buildings it would enclose. That dome would reduce the heating and cooling requirements of the buildings it covered to $\frac{1}{84}$ of the current requirements. In addition to regulating the temperature, such a dome would also control the precipitation and the air quality within the enclosed environment.[101]

When Bucky proposed such radical solutions, many people balked at the possibility of constantly living inside any structure. Once the Montreal dome was completed, he would refute that argument by explaining that

Fig. 15–12 and 15–13 The Montreal Expo '67 geodesic dome under construction.

Fig. 15–14   Completed exterior of the Montreal Expo '67 geodesic dome.

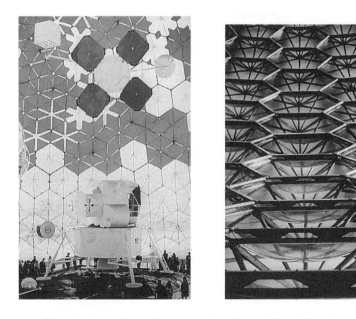

Fig. 15–15 and 15–16   The interior of the Montreal Expo '67 geodesic dome.

even at the relatively small scale of the Montreal dome, the structure itself was beginning to become invisible to the human eye. He would then explain that at the mammoth scale of a one-mile-diameter dome, only the lowest sections of the dome would be visible to people situated near the dome's outer edges. Most sections of such a massive structure would be so distant from any point on the ground that they would be invisible to the naked eye, and from the inner area, even the lowest sections of the dome would be so distant that they would be imperceptible.[102]

The idea of a "skybreak dome" was by no means new to Fuller, and he often explained that it might be applicable at sizes as large as several miles in diameter and as small as individual dwellings. In fact, that concept was so important to Bucky that he had initiated work on it in 1949, right after he had invented the geodesic dome.[103]

At that time, he had worked with a group of students to develop a geodesic dome which would allow people to literally "live in the garden." The skybreak dwelling which they designed was to cover the complete grounds of a residence rather than simply to enclose a number of rooms as is normally the case. That larger dome was to have transparent sections which could be covered to provide shade or privacy; however, the dome would also provide constant environmental control and protection from rain, snow, wind, and insects as well as temperature. Within that dome, people could build rooms for privacy if they desired, but the entire interior was designed to be filled with trees and plants so the inhabitants would have the advantages of nature year-round. Thus, the concept of living in the garden.[104]

Although that idea has yet to be brought to fruition, Bucky did demonstrate the feasibility of a skybreak dome at the Montreal World's Fair. When it opened in 1967, people were stunned by the beauty and practicability of the structure.[105]

Because 1967 was the Fuller's fiftieth wedding anniversary, Bucky dedicated the Montreal dome to his wife, Anne. On May 28 he brought her to tour the fair and to view what he described to her in the following statement: "I have inadvertently brought about the production and installation of our own Taj Mahal as pure fallout of my love for you."[106]

CHAPTER 16

World-Game Abundance

 Although Fuller had devoted a sizable portion of his life to the examination of global resources including energy input and output, during the 1960s that issue became much more personal. Having always maintained a vigorous lifestyle, he suddenly noticed his level of energy decreasing and decided to see if anything could be done about that change.[1]

It was then that Bucky seriously considered his own body. The fairly affluent financial status he had attained since inventing the geodesic dome had contributed to his reaching a weight of nearly two hundred pounds, far too heavy for someone five feet six inches tall. Logically examining the situation, Bucky decided that either old age or his weight was slowing him down, and that if the problem was weight, he could still do something about it.[2] With typical Fuller scientific exactness, he examined the issue of food and energy as it applied directly to himself and his weight and determined the most efficient method of losing weight while still meeting his energy needs. Bucky resolved that since the majority of the energy utilized on Earth originated from the Sun, he should attempt to acquire the most highly concentrated solar energy available to him as food.[3]

He understood that solar energy is transformed into vegetation through photosynthesis, and that animals, including humans, consume that vegetation to acquire the energy. Always seeking the most effective means of operation, Bucky decided that he would follow the process one step further, resolving that he would be most effective if he ate more beef because cattle had eaten the vegetation and transformed solar energy into a more protein-concentrated food.[4] Hence, he began adhering to a regular diet of beef even though that diet often caused consternation among many of the holistic, health-oriented individuals to whom he regularly spoke during the balance of his life.[5]

To those who would question his regimen of approximately one pound of steak at breakfast, lunch, and dinner, he would explain his rationale that lean beef provided maximum energy with a minimum of calories. The personal results Bucky attained by following that diet were difficult to dispute. When he began eating primarily thin strips of London broil augmented with salad and fruit three times a day, his weight dropped back down to 140 pounds, identical to when he was in his early twenties. He also found that his energy level increased dramatically.[6]

Contrary to one popular legend, Bucky's diet was not restricted to beef and jello three times per day. Even though he tended to eat beef regularly from the mid-1960s on, he retained his spirit of experimentation and was willing to test almost any diet recommended to him.[7] Still, he continued using whatever worked best, and the beef concentration proved most successful at maintaining his maximum energy and healthy weight.

Although he did apply his insights into resource management and distribution to his personal eating habits, Bucky's primary use of such information was global. He had studied history in great depth, searching for the reason most humans operated from a context of scarcity rather than the true abundance which he perceived humanity was about to reach during the end of the 1960s.

In his examination, Fuller looked back as far as sufficient written records were available and found a pattern emerging from that period eight thousand years ago in Egypt. At that time, the average life span was twenty-one years, and leaders came into power during what we now call adolescence. One factor which contributed to the short life was the limited amount of life-supporting resources such as food, medicine, and even shelter. Average people were taught, and quickly believed, that life on Earth was simply a qualifying test for the *great afterlife,* where resources were abundant and the suffering they were enduring on Earth did not exist.[8]

One person who did not suffer in ancient Egypt was the pharaoh. As most people know, pharaohs were worshiped as gods or, at least, individuals who had direct access to God. They were therefore provided with a lifestyle far superior to that of ordinary citizens. Pharaohs were endowed with the finest goods available in the hope that they would use their godly influence to provide better lives for the ordinary citizens over whom they ruled.[9]

Among the common people's greatest hopes was that their beloved pharaoh could reach "the other side" following death so that he could take other citizens of Egypt with him to the land of plenty following their deaths. Fuller found that the people of that era were told that in order to make the journey and organize things for them on the other side, their pharaoh needed many tools and resources. Consequently, the people built enormously fortified stone pyramids to protect the pharaoh and his required resources from vandals after death.[10]

Because such monuments required massive amounts of resources and took years to complete, the pharaoh or his father (because the job often required an entire lifetime) would select the kingdom's most inventive individual to design and oversee the project. Such a man might have become the Leonardo da Vinci of the time, but he frequently went unnoticed in history because all his talents were concentrated upon the single structure.[11]

Fuller found that those designer-planners had uncovered generalized principles such as leverage, which they applied to create greater efficiency throughout the project. They also tended to be aware of the poverty and the squalid living conditions which were devastating their workers and hampering the project.[12]

Fuller speculated that one such designer might have noticed that the banks of the Nile were lush with crops while only a short distance inland, his workers were dying of starvation. Again employing principles that would support the construction, that designer may have instructed some of his laborers to build irrigation ditches which soon resulted in a larger food supply, better fed workers, and hence, more efficient construction.[13] What most people of the era may not have realized was that such innovation was the result of increased knowledge and was actually creating a better life on Earth for an increasing number of people prior to their crossing over to the afterlife.

If everything went according to plan, the pyramid was completed prior to the pharaoh's death, and when that event occurred, the "lucky" designer-planner was executed and buried in that pyramid as a reward for doing an outstanding job. Supposedly, that burial was a great honor which quickly allowed him to reach the other side with the pharaoh. The problem with the Egyptian system was that each successive pharaoh required a lavish burial site, and a new, innovative designer-planner had to be found in every generation. Even though the previous designer had been buried, because the Egyptians had developed written records, a new designer could utilize his predecessors' ideas. The effect of such constant innovation and the accumulation of knowledge was a new order in which resources were plentiful enough to permit the similar grand burial of high-ranking nobles, ensuring their reaching the afterlife. Bucky felt that that innovation marked the beginning of the Second Dynasty in Egypt as well as a significant shift in the development of humanity. In the ensuing years, the increasingly effective use of resources provided even more individuals with the splendid regal burials that ensured their entering the glorious afterlife.[14]

In nearly every society, Fuller found a similar belief that the suffering of this life was a test for a magnificent afterlife and its abundance. He also discovered that because of increasing knowledge, each successive generation of humans was usually able to do more with fewer resources. In addition to more people's being supported at higher standards of living, more individuals were also able to be fittingly prepared for the advantages of that afterlife.[15]

Bucky felt that the next great societal metamorphose in his theoretical scenario occurred among the ancient Greeks and Romans who discovered that they had enough resources and knowledge to prepare their ruler and nobles as well as all the rich middle-class citizens for the afterlife. During that period, huge numbers of slaves were used to construct large mausoleums for the wealthy middle class and the nobles.[16]

As knowledge compounded exponentially from one generation to the next, religious organizations, which had always been closely associated with government but had sought to maintain the greatest domination over individuals, split from government. They did, however, retain the power to determine who would be allowed into the glorious afterlife. During that period, prophets such as Buddha, Muhammad, and Christ began preaching a new concept: the ability of all people to reach the glorious afterlife if they followed a particular religion's tenets. That idea was known as *redemption,* but in Fuller's view it was nothing new or unusual. Rather, redemption was simply another demonstration that humanity's knowledge was increasing and was providing the opportunity to do more with fewer resources. Proponents of redemption asserted that humanity had attained a level of development which provided the opportunity for everyone to reach the afterlife. Life on Earth remained a struggle for most people, but they could look forward to the afterlife if they followed the dictates of the church and its leaders.[17]

Following that transformation, organized religions began erecting enormous buildings for worship, in which average people could prepare themselves for the afterlife. They also coerced worshipers to bring resources to those houses of worship in support of the formalized church. The primary reason people were able to provide the church with resources was not some divine miracle but humanity's continuing increase in knowledge, which allowed people to do more with less. Specifically, farmers had learned how to use simple tools and domesticated animals to expand the acreage they could cultivate, while craftspeople were inventing other tools which supported them in turning out more products.[18]

The concept of redemption signaled that the increase of production through expanding knowledge had reached a level at which the afterlife of everyone was ensured, and humanity's focus shifted toward daily life on Earth. The plight of many people had improved over the centuries, but the majority continued to subsist on meager rations. In fact, even kings and nobility sometimes had to go without because of a poor harvest or a raid by a neighboring kingdom.[19]

Still, with the renewed focus on earthly existence, the first to be guaranteed life-supporting resources were monarchs. That privilege came to be known as the *divine right of kings*. Later, during the period when the Magna Carta was written, human efficiency in providing life support had increased so that sovereigns and nobility could be guaranteed sustenance. And, during the Victorian period, knowledge and production had compounded to such an extent that the physical needs of even a small but growing middle class could be ensured.[20]

The Victorian era brought Fuller's brief explanation of historical perspective up to the time of his birth, and from that point on he relied on his personal experience as well as on what he learned from books to understand the changes of history. Bucky had seen that the innovations of modern technology had radically transformed his surroundings very quickly. He read and appreciated the ideas of industrial leaders like Henry Ford, who believed that mass-production technology would provide adequate means for supporting everyone on Earth.[21] As a youth Bucky had witnessed the change that such technology had generated. He could recall neighborhood artisans painstakingly manufacturing goods such as furniture and clothing one piece at a time.[22]

He also remembered that because of the limited production, those goods were expensive and generally available only to the rich. There simply were not enough craftspeople to satisfy the needs of a rapidly growing New England population, even though that population was far less dense than that of the rest of the World. Bucky witnessed mass production dramatically alter that situation. The transition did not, however, eliminate the need for craftspeople. In addition to maintaining the small-scale production of a limited number of products, many of them embraced the emerging new field of industrial technology. They continued to support the tradition of expanding upon the knowledge of the past by becoming the designers and fabricators of the machine tools which were the mainstay of modern technological innovation.[23]

While studying traditional economic theories, Fuller also stumbled upon the philosophy of the sole person who he felt was used more than any other to justify the conditions of starvation and poverty from which major segments of society suffered while others lived in relative abundance. That man was the British economist Thomas Malthus, who lived during the late eighteenth and early nineteenth centuries, the height of the British Empire's dominance of the sea and much of the Earth's landmass.[24]

Malthus was an economist and a professor at the East India Company College just outside London, the one institution charged with educating all the civil servants managing the British Empire's global holdings. The East India Company College also became the sole repository for global information gathered by British citizens and civil servants around the World. His access to such a multitude of data made Malthus the first person to study a comprehensively complete inventory of the World's vital and economic statistics.[25] Using that data, in 1798 he wrote the classic work *An Essay on the Principle of Population,* which he revised in 1803. Those elaborate dissertations documented the fact that the human population was increasing far more rapidly than the relatively scarce production of life-supporting goods and services. From that data Malthus concluded that the scarcity of goods that people experienced would increase unless something was done to curb the rising population.

Fuller found that, over the years, Malthus's conclusions had become widely accepted and had been combined with the theory of evolution proposed by the famous British scientist Charles Darwin during the mid-nineteenth century. Darwin's hypothesis was popularly reduced to "survival of the fittest," and when coupled with Malthus's documentation of the increasing scarcity of resources and the expanding of population, it supported escalating competition for resources among rival sectors of the human population. The most evident such competition was among economic systems, principally socialism versus capitalism.[26]

Malthus had not, however, considered the synergistic consequences of combining new knowledge with resources. Still, because of his unique access to information, in the ensuing years most educated people continued to utilize Malthus's pessimistic perspective and ideas as a foundation for future predictions.[27]

Fuller recognized the flaw in that logic and argued that Malthus's basic premise was incorrect. He believed, and later documented, that the ever-increasing technology which Malthus could not even imagine, much less utilize in his predictions, was creating an entirely new environment capable of supporting all humanity.[28] Like Malthus's ideas, Fuller's theory was also predicated on global resource information. However, Bucky's information was far more comprehensive and clearly indicated that humanity's constant addition of new knowledge to physical resources was producing a set of circumstances never before seen in recorded history. Such new knowledge would permit every individual and industry to do much more with the same resources and would consequently enable more people to be supported.[29]

Bucky clearly documented that the rate at which technology was increasing life-supporting production was much greater than the rate of population increase. Using the global resource inventory he had accumulated, Fuller also predicted that in 1970 humanity would reach a level at which every human being could be adequately supported while using only the resources available on Earth, if humanity shifted to concentrating on the production of life support rather than destructive weaponry.[30]

Fuller's accumulation of data began in 1917 during his naval career. While reexamining his life in 1927, he christened his collection of information the "Inventory of World Resources, Human Trends and Needs." Bucky's mid-1930s work also helped support his insight into global resources. Working for Phelps Dodge Corporation from 1936 through 1938, he was required to create a comprehensive study of global resources and industrial trends in order to identify how Phelps Dodge's primary product, copper, fitted into those trends.[31]

His position at Phelps Dodge provided Bucky with access to the most complete private and public economic records ever available. Realizing the significance of his economic insight and research, *Fortune* magazine requested Fuller's assistance in compiling its tenth-anniversary issue for publication in February of 1940. Bucky accepted that invitation, and with the help of a staff specially hired for him by *Fortune*, he greatly increased his resource data and helped produce an issue which focused on the subject he suggested: a comprehensive inventory and comparative data study of World resources versus United States resources.[32]

That issue was one of the most successful in the history of *Fortune,* requiring three separate printings. In fact, Henry Luce, then chairman of the board of *Fortune's* parent company TIME, Inc., later announced that the resource issue had produced a profit for the magazine and had moved it out of its habitual deficit situation.[33]

Following publication of that issue, Fuller accepted a position with the Board of Economic Warfare and later worked on production of the Dymaxion House. It was while working on the Dymaxion House in 1945 that he was again contacted by *Fortune's* management concerning the data he and his staff had compiled earlier. In 1945, *Fortune* was transferring its operation to a new building and wanted to reduce its archive. Consequently, the magazine's management offered Fuller all the resource files he and his staff had accumulated for the 1940 issue. With that acquisition, Bucky's personal resource inventory became one of the most comprehensive in the World.[34]

Because he continued to update his information, Fuller was able to maintain an accurate accounting of global resources long before most people were even interested in issues such as starvation in Africa, the pollution of the environment, or the housing of the homeless. He was also able to utilize that data in determining past and future economic trends.

Fuller's theories concerning economics were as sophisticated as his ideas on other topics, and they adhere to his belief that individuals rather than organizations possess the ultimate power to influence the development of humanity. He felt that, like all human beings, the people involved in the business community would be far more successful if they concentrated on cooperation rather than competition and worked on projects which would be of long-term benefit to the most people rather than focusing upon short-term financial gains for a small number of shareholders.[35]

As a result of his accumulated data and experience, in the late 1940s Bucky began discussing a fictitious corporation which illustrated his fundamental feelings about most business ventures. The hypothetical company, Obnoxico, was first publicly mentioned in a 1947 response to a comment that he was not trying to earn a living because he was incapable of doing so. Always willing to respond to a challenge, Fuller replied that he was clever enough to make vast amounts of money in only a year with his mock company Obnoxico. He explained that Obnoxico would have only one function: to make money by exploiting the most sentimental weaknesses of humanity.[36]

To emphasize the direction he perceived the majority of businesses moving in in their quest for increasing profits, at a time long before items such as pet rocks and lava lamps became popular fads, Bucky mockingly proposed his own fad item. The primary product of his fictitious Obnoxico company required that on the last day a baby wore a diaper, parents carefully remove it from the child, repin it empty, and stuff it full of tissue paper to match the baby's shape. The molded diaper was then to be packed and sent to Obnoxico, where it would be gold- or silverplated before being returned to the parents, who would fill the shiny, rigid diaper with ferns and hang it from the rear window of their car.[37]

Once Bucky had announced his ludicrous product, many of his friends joined in and began suggesting other apparently ridiculous items. They also began sending him advertisements for new products whose primary or sole reason for existence seemed to be making money. Among the items he received notice of were plastic pebbles for the garden and the newly emerging plastic flowers.[38]

Bucky carried his scenario one step further and, only then, began to appreciate just how easily money could be made if one was willing to exploit the emotions and greed of fellow human beings. When he suggested that Obnoxico was considering opening stores to sell such items and that in exchange for contributing products, he would provide individuals with shares of stock in the company, the entire joke quickly got out of hand. Once people sensed the possibility of making money from such an endeavor, greed and competition set in. He was besieged by new ideas, and it was clear that a majority of people who had learned of the impending next phase seriously wanted to participate in the corporate venture. At that point, he could easily have issued stock and made a fortune with what had begun as a simple joke.[39]

Fuller, however, had more important things to do with his time, and he continued to live by his assertion that an individual can "make money or make sense, the two are mutually exclusive."[40] That statement has often been misinterpreted. Fuller did not feel that people should not make money. Rather, he believed that individuals should concentrate on projects which made sense, and that both the people and the projects would then be supported financially. Thus, he continued to select projects which made sense to him, and in the process, he and his family were supported.

Bucky did, nonetheless, witness the rise of the Obnoxico stores he had proposed in the late 1940s. "Gift shops" filled with items that have little or no inherent value to society but a great deal of appeal to the emotions of a susceptible public have proliferated in shopping malls throughout the United States and are now spreading worldwide. Their primary purpose, like that of their merchandise, is to make money. Bucky was not opposed to the enjoyment of gifts and greeting cards. However, he did feel that an economy which wastes a sizable percentage of its production capacity on items such as toilet seats with dollar bills embedded in plastic or a multitude of stuffed animals while thousands of humans beings starve to death daily or live without adequate shelter does not yet possess the comprehensive perspective required for the success and survival of humanity as a global species and is not supporting the true function of humans on Earth.

Obnoxico was an extreme example of the waste which can occur when profits become the primary motive for action. Fuller was not opposed to the institutions of organized commerce. He simply felt that the primary function of business and industry should be supporting the welfare of humanity and all life.[41]

He also believed that the large multinational corporations which emerged during the 1970s represented a significant hope for the expanded success of humanity. Because of his comprehensive perception, Fuller was cognizant that major companies which had established themselves in specific countries were quickly surmounting national boundaries to do business wherever they chose. He observed that those large corporations instantly communicated between locations in all nations and were not confined by obstacles such as visas, passports, or borders, as is true for individual human beings.[42]

Because of the global freedom they enjoyed, Fuller concluded that large corporations might well be unknowingly laying the groundwork for the future success of humanity. He expected a form of corporate global networks to come to the forefront of public attention and need at some time in the future when only such a comprehensive system, which was already in place, could solve seemingly out-of-control worldwide problems such as pollution, the efficient allocation of resources, or energy requirements.[43]

In such an "emergence-by-emergency" scenario, large corporations might rapidly provide the efficient global solutions which would be possible only within the context of a comprehensive worldwide awareness and operating strategy. However, before that could occur, the primary motivation of such corporations would have to shift from profits and toward the more comprehensive considerations which continually stirred Fuller to action.

Although the comprehensivist Fuller was always working on a multitude of projects at any one time, he tended to focus his work in specific areas for particular periods of time. The era of intensive geodesic dome exploration and invention which began in 1947 concluded in 1967 with the Montreal Expo dome.

At that time, much of Fuller's attention shifted to creating a similar summary resolution for the enormous amount of work, insight, and information he had accumulated concerning global resources and energy. He was particularly interested in creating a conspicuous public presentation of global resources prior to 1970, the year he felt humanity would be able to do so much more with fewer resources that we could adequately support everyone living on Earth. He named that presentation "World Game" in contradistinction to the "war games" which military leaders had been playing out for centuries. Bucky himself had personally experienced the strategies of war games in great depth while serving in the U.S. Navy during World War I as well as while working on the Board of Economic Warfare during World War II.[44]

He had witnessed the massive amount of time and resources that military and government leaders devoted to destroying other human beings and their property, and he felt that, in the approaching era of abundance, a similar concentration should be directed toward life support or, as he called it, "livingry." Livingry is the antithesis of weaponry.[45]

Since he traveled around the World on a regular basis each year, Bucky was aware of the global consciousness that more and more individuals were displaying, and he wanted to formalize his resource inventory, making it more accessible to humankind. He was particularly concerned with providing unique new opportunities and ideas to young people and students, many of whom were involved in the campus rebellions of the 1960s.

With the consent of its administration, Fuller began utilizing the facilities of Southern Illinois University, where his headquarters was located, to establish World Game as a participatory conference-workshop. The first such conference occurred early in 1969 at the New York Studio School. The event was a prototype test in which Bucky supported twenty-six students in "playing the game" of attempting to determine the most effective means of employing the Earth's resources to benefit the most humans.[46]

From that initial endeavor, Fuller and his assistants learned just how much effort producing effective World Game conferences required, and they regrouped for another presentation. Following his natural tendency of trying to provide more than most people could feasibly assimilate with little attention to the concerns of bureaucracies, Bucky reverted to his early 1960s concept of how the World Game should be presented. At that time, he had proposed a form of World Game to the United States government officials who had solicited his ideas for the Montreal Expo.

Within the Expo '67 geodesic dome, he had planned to construct his optimum Geoscope. It was to be viewed by visitors entering the dome as simply a one-hundred-foot-diameter miniature Earth globe, but it would slowly change into the triangles of the icosahedron so that observers could appreciate the fact that no distortion occurred during that transformation. Those triangles were to unfold and flatten into a football-field-sized Dymaxion Map of the Earth covered with millions of tiny computer-controlled lights. That Map was to be employed in displaying all the World's resources and events as well as in playing out possible scenarios for utilizing resources.[47]

Although the idea was rejected by the government officials, Bucky continued to insist that such a device was essential to the welfare of humanity. Accordingly, he initiated a plan to create that Geoscope at his SIU headquarters.[48] After careful calculations, Bucky announced that the installation would cost approximately $30 million, and the fact that no practical financing plan existed did not cause him to hesitate for an instant. Confident that if he was doing what needed to be done, Universe would support his projects, Fuller organized and conducted the first bona fide World Game seminar at SIU in July of 1969 with the assistance of a young Ph.D. candidate, Edwin Schlossberg. Although that program was conducted without the aid of the magnificent Geoscope Bucky envisioned for the future, the participants did much more than simply listen to Fuller or examine his extensive archive of resource data during the two-week seminar.[49]

They were divided into research and design teams which focused on specific aspects of the increasingly efficient resource allocation Fuller defined as *ephemeralization*. Bucky often discussed the practical aspects of that phenomenon during his discourses on doing ever more with ever less weight, time, and energy while maintaining a specific level of performance.[50] Thus, he inspired as well as educated team members and helped each of them to achieve a new level of enthusiasm and understanding. Those research and design teams were unique in that they worked in cooperative competition with one another. Although they were each competing to uncover the most efficient answers to the global problems posed, their solutions were all predicated on universal cooperation, which would eventually support the most effective solutions for all human beings.[51]

Energy was of primary concern to those first World Game "players," and it continued to be a critical component of all subsequent World Game sessions. Luckily, from its inception, World Game participants could rely on Bucky's accumulated data and his insights into energy issues. One of the primary insights he shared during sessions was just how much available energy was wasted by human beings and how simply that inefficiency could be resolved. As always, Bucky's focus was the process he described as "comprehensive, anticipatory design science."[52]

In the case of energy, he suggested that comprehensively examining usage would provide unique solutions. Specifically, Fuller found that scientists were developing more efficient forms of converting fuel into necessary services, but that those efficient methods were not reaching the general public rapidly because of the profit motive guiding most companies.[53]

Fuller often used engines to illustrate what he saw occurring. The internal combustion engine, which powers the majority of our automobiles, is only 15 percent efficient. In other words, only 15 percent of the fuel pumped into a car's gas tank is actually used to move the vehicle. More recently developed engines, however, are far more efficient and would be of great benefit to both consumer and, more importantly, to the environment. However, the massive commercial application of such engines would also require large amounts of investment capital, and the short-term profit for corporations would be reduced. For example, turbine engines are 30 percent efficient, jet engines are 65 percent efficient, and modern fuel cells are 80 percent efficient, wasting only one fifth of their fuel.[54]

Fuller, who had become aware of that issue in the 1930s, had always been willing to share his findings with concerned individuals. But in an era when fossil fuel was believed to be almost unlimited, few people were interested in conservation measures or devising more efficient engines.[55] Most people also tended to be initially unwilling to accept his notion of just how inefficiently we human beings, who believe ourselves to be so intelligent, are operating. After all, when Bucky proved that in 1980 all energy use in United States was only 5 percent efficient, that figure was also difficult for most people to accept.[56] However, after an examination of the data that he and the World-Game participants had compiled, the truth of Bucky's message became obvious, and the hope that his design science provided became much more attractive.

Fuller and the World Game provided thousands of individuals with a new understanding of energy possibilities and the continuing energy wastefulness of human beings. Few people considered occurrences as common as cars idling at stoplights and in traffic jams. Yet, when Bucky said that each such car, which has at least one hundred horsepower, represented the energy of a minimum of one hundred horses jumping up and down in place, human technological wastefulness became much more undeniable. As Bucky loved to suggest, it was as if the American population was constantly paying for a national stable of hundreds of millions of invisible horses performing a useless, invisible tap dance under the shining hoods of their cars.[57]

The importance of energy became even more vivid when Fuller discussed the concept he christened *energy slaves*. That concept had greatly assisted him in uncovering and illustrating just how rapidly humanity was approaching a level of prosperity at which everyone could be provided with the necessities of life. Fuller created the concept of energy slaves in order to produce a standard for precisely measuring the rate at which the technological advantage of humanity was increasing. As a pragmatic scientist and engineer, he felt that such accurate measurement was essential to examining and planning the actions of humanity so that those endeavors would be most effective.[58]

Fuller utilized the standard created by the United States, the Swiss, and the German armies to make calculations which were predicated on the following logic. Those armies had determined that an average man could do approximately 150,000 foot pounds of work in an eight-hour day. That figure provided Bucky with the mathematical amount of "advantage" that a single person could generate during a specific period of time using only muscle.[59]

Prior to the invention of machinery or the harnessing of animals, the amount of work advantage available to one human being was precisely what he or she could produce with muscle, approximately 150,000 foot pounds in eight hours. In 1810, the United States was inhabited by approximately one million families. One million actual slaves were also confined within the country for use as workers. Although most of those slaves were held by a minority of families, statistically the United States had an average of one human slave per family.[60] During the eighteenth and nineteenth centuries, the industrial revolution allowed individual human beings to begin realizing the advantage of mechanical devices, but the truly rich continued to employ human slaves or servants to support their seemingly lavish lifestyle.

In 1950, Fuller combined the concept of slaves working to enhance the lifestyle of human beings with the specific figure of 150,000 foot pounds of effort per day to create the standard amount of work that one machine would have to produce to be considered a mechanical or energy slave. He then calculated the total planetary energy being produced by the two main industrial sources (fossil fuel and waterpower) during one year and divided it by 150,000 foot pounds to determine that humanity had generated and used 85.5 billion energy slaves during 1950.[61] Fuller also found that the distribution of those vitally important energy slaves was very geographically lopsided, dividing the "have" localities from the "have-nots."[62]

For example, the average inhabitant of North America had at his or her disposal 347 energy slaves, while residents of Asian countries averaged only 2 energy slaves per person and in Central America the figure was so small that he reported it as zero. When Fuller examined the relationship between standards of living and the number of energy slaves citizens had at their disposal, the correlation was obvious.[63] Clearly, the greater the number of energy slaves, the better the quality of life. Furthermore, Fuller concluded that the number of energy slaves available to individuals was not related to politics or national boundaries. The industrial revolution had advanced the quality of life similarly, though at different rates, in Germany, the United States, and the Soviet Union while not affecting most of the African nations until recently.[64]

Fuller believed that technology could not be dominated or controlled by any form of government, and that accordingly, it tends to affect geographical areas uniformly. He also felt that technology can be communicated and used only by those who have studied and understood scientific disciplines, and he found that those people are rarely politicians.[65]

Fuller felt that politicians were of little use in the radically new environment he saw emerging as early as the 1950s because humanity's problems were not political. Rather, he perceived the key question facing humankind to be how to get three times as many energy slaves working, and to get them working for all the families of the World. That radical perspective was also reflected in his ideas concerning ownership and resources. When he spoke to World Game participants as well as other people, Bucky would share his notion that ownership was illogical in our

Fig. 16–1   Fuller working with World-Game participants.

emerging new society, and after examining the facts, many people began to agree with him.[66]

In Bucky's view, what was needed to facilitate the creation of an environment of abundance and sufficiency for everyone was not the hoarding of resources which was so common, but a shift from an ownership-oriented economic system to a service-oriented system.[67] A person need only observe one of the marinas from which Bucky often sailed during the last years of his life to understand the onerousness of ownership he perceived in nearly every aspect of modern society. At such marinas, hundreds of boats sit idle for days or weeks at a time and are generally used by their owners only on weekends or holidays. Those boats represent valuable squandered resources which could be housing thousands of people on a daily basis, but they are simply maintained as hoarded possessions of the rich.

When he addressed the first SIU World-Game seminar, Bucky used his experiences with automobiles to illustrate his feelings about ownership. He explained that he had owned fifty-five automobiles during his sixty years of driving, but that he no longer held title to even a single car. That dramatic shift in operating strategy was a direct result of his flying from city to city and requiring a car at each destination. In each location, Bucky, like so many travelers, would simply rent a car, and that experience helped him to realize that, although the majority of those cars had been used by others prior to him, each rental car was, in fact, so well maintained that he almost felt he had been given a new car in every city.[68]

Bucky also discovered that he was delighted to be relieved of the burden of automobile maintenance as well as to participate in a system in which other individuals could use that same "transportation resource" when he did not need it.[69] Such a notion may appear absurd to those of us who, like Bucky, were taught that acquiring goods is necessary to prove oneself a valuable and worthwhile citizen; however, within a cosmic context of abundance, sharing resources so that they are the most efficiently employed is logical, natural, and necessary.

Fuller's audiences would sometimes argue that they were willing to share transportation with others, but that they would not be willing to share more intimate possessions such as homes or beds with fellow human beings. In answering those comments, Bucky would simply share his global travel experiences. He explained that he was constantly checking into hotels, where he would be provided with a room that had been used by many other fellow human beings over the years, yet appeared almost new to him. He observed that no one ever complained about sharing beds or rooms in hotels, which sought to maximize their efficiency by filling as many rooms as possible with whoever needed them at a given moment. Such examples illustrate Bucky's concept that one significant method of increasing the life support available to humanity is to maximize the use of the available resources. They also provide insight into his hypothesis that such a behavioral and perspective shift will transform society from a product to a service orientation. Rather than buying items like cars and houses, people will lease a transportation or housing resource only when and where that resource is needed, just as telephone service is purchased only as it is needed.[70]

Although many people now own their telephones, they do not own the cables which run from one location to another. Instead, they rent time on those cables when they need it. In a society predicated upon the absolute abundance that Fuller proved to exist, people would carry that process even further. When individuals operate from the knowledge that there are adequate resources to support everyone on Earth, payment for the use of resources will simply become obsolete.[71]

On a global scale, Fuller's strategy of contributing to the welfare of all humanity through a more effective, cooperative use of the available resources is clearly illustrated in the worldwide energy network he presented at the 1969 World-Game seminar. Having studied the history of energy production in great detail, Fuller understood that from the days of the first waterwheel, transmission of the energy produced had been a major problem. Even in factories constructed on riverbanks to exploit the power of rushing water, as the distance between machinery and power sources increased, the amount of energy lost in transmission multiplied greatly. That factor became much more evident following the practical application of electricity. Originally, transmission efficiency was so poor that power plants could serve sites located only within miles; however, in the early twentieth century, scientists discovered methods which allowed effective transmission of electricity as far as 350 miles. That innovation permitted local electrical generating plants not only to provide power to more distant sites, but also to connect with one another and share power.[72]

Engineers soon found that sharing power and electrical production capacities greatly increased efficiency because individual generating plants were usually designed to meet peak production needs, which occurred for only a few hours each day. Consequently, during most of a twenty-four-hour day, a plant's capacity was not fully utilized, or in the case of facilities such as water-powered plants which could not be stopped, electricity simply went unused.[73]

By applying 350-mile partnership links, distant electrical plants could send their excess production to other localities during times of less demand. A transmitting area would later be compensated with power from the receiving facility when the demand pattern was reversed. That phenomenon was particularly effective if localities spanned time zones creating greater differences in demand patterns.[74] For example, the afternoon peak demand caused by industrial production, weather conditions, and the increased activity of people during the daylight hours differs by three hours between cities in the eastern and western time zones. Accordingly, when a 3 P.M. peak demand occurs on the West Coast, it is 6 P.M. in the East, and demand is decreasing. That same type of extreme demand variation was also exploited when areas displaying extremely different weather patterns were connected. Thus, using new technology, engineers constructively employed available, yet previously underutilized, resources and capacity.

Bucky became extremely interested in that potential during the 1950s, when more-with-less technological design increased the practical delivery capacity of electrical systems from 350 to 1500 miles. Always searching for the most effective means of supporting the most human beings, Fuller quickly realized that the new transmission technology presented a potential for humankind far beyond anything previously considered.[75] With such expanded electrical transmission, distant countries and entire continents could share excess electrical production with far removed communities. More significantly, the discrepancy between day and night power requirements could be exploited.

Specifically, Fuller hypothesized that if the Earth's entire population could be connected through a gigantic cooperative electrical network, the nighttime power-production capacity, which is generally underutilized, could be transmitted to the daytime side of our Planet where it was needed.[76] By creating such a communal exchange, vast amounts of resources would be conserved as power plants would be fully utilized on a twenty-four-hour basis. With such increased efficiency, the cost of power would fall, as would the need to exploit more resources in building additional power plants.

Upon hearing Fuller discuss the merits of that possibility, most audiences questioned how such a cooperative network could be established. Always seeking solutions to future problems, Bucky was prepared for the challenge. His examination of World geography and subsequent invention of the Dymaxion Map had provided him with the needed global knowledge as well as a format to display planetary solutions.[77] He would exhibit

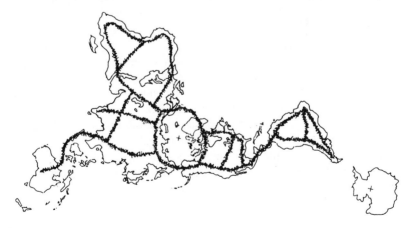

Fig. 16–2   Fuller's route for the global energy grid which would cooperatively connect most of the World's electrical production facilities, thereby dramatically enhancing the efficiency of global energy production.

a Dymaxion Map on which all the locally existing energy networks were displayed, along with another such map on which the most logical ultra-high-frequency transmission lines were shown. In his capacity as global design scientist, Fuller had drawn those lines without regard to national boundaries as seen in Fig. 16–2.[78]

He did, however, appreciate the issues which such a cooperative effort would generate and was willing to discuss the feasibility of his revolutionary idea with the very people he felt were becoming increasingly outmoded: politicians. One such discussion occurred during the mid-1970s between Bucky and his friend Canadian Prime Minister Trudeau. After examining the plan in detail, Trudeau wholeheartedly endorsed it. In fact, he was so enthusiastic about the proposal that he took a copy of it to General Secretary Brezhnev of the Soviet Union. Later, Trudeau reported back to Fuller that after having their scientists and engineers examine the plan, the Soviets also endorsed it, although only informally.[79]

Formality, however, is another issue. One of the key elements of the electrical grid network would be a direct connection between the United States and the Soviet Union via the Bering Strait. Such a hookup was, and continues to be, one of the most politically sensitive components of the plan. Even though pragmatic scientists, engineers, and economists on both sides agree that the connection would benefit everyone involved, political implications continue to overshadow what could be one of the most productive projects in the history of humankind.[80]

This deadlock once again demonstrates Bucky's conviction that politicians and all political systems impede rather than support the development and success of humanity. It is also demonstrating the power of individual initiative. In nations around the World, concerned individuals are beginning to appreciate and support this endeavor, and a grass-roots movement to create such a cooperative network is gradually growing in size and stature. Engineers and scientists, as well as ordinary individuals who have examined the facts, are coming forth to argue that those in positions of power should consider and support the global energy network, and that, should it be installed, that network would synergistically nearly double the energy available to humanity the moment it began operation.[81]

Energy was not, however, the only focus of the World Game. Bucky's inventory divided all the socially abundant resources into eight primary categories so that the participants could more easily work with them. As usual, Fuller's comprehensive perspective is reflected in the somewhat cryptic phraseology of his categories:

1. Reliably operative and subconsciously sustaining, effectively available twenty-four hours a day, anywhere in Universe.—Gravity and love.
2. Available only within ten miles of the surface of the Earth in sufficient quantity to conduct sound.—The complex of atmospheric gasses whose Sun-induced expansion on the sunny side and shadow-induced contraction on the shaded side of Earth together produce winds and waves.
3. Available in sufficient quantity to sustain human life only within two miles above the Earth's spherical surface.—Oxygen.
4. Available aboard Earth only during the day.—Sunlight.
5. Not everywhere or at all times available.—Water, food, clothing, shelter, vision, initiative, friendliness.
6. Only partially available for individual human consumption and also required for industrial production.—Water.
7. Not publicly available because used entirely by industry.—Helium.
8. Not available to industry because used entirely by scientific laboratories.—Moon rocks.[82]

Although these unique groupings were pragmatically utilized by World Game participants to determine the most effective planning scenarios, their primary purpose was not just for use in World Game seminars. Fuller created the categories to help pinpoint scarce resources and maintain a precise inventory of their location for the benefit of all humanity.

He felt that exacting computer inventories would provide the only means of accurately establishing which resources were rare, their relative degree of scarcity, and how much of a resource was needed for what purpose and at what location. He also realized that a single resource might be scarce at one location and abundant at another, necessitating policies unique to specific sites.[83]

For instance, oxygen is abundant at sea level, yet above eleven thousand feet it is quite scarce and has to be obtained from compressed tanks for breathing. Dealing with other resources in a similar manner would however, be unwise and impractical according to his calculations.[84]

Helium is a gas which is scarce everywhere on Earth but possesses properties which make it uniquely valuable in specific industrial applications. Although compressing the World's entire helium stockpile into billions of small bottles and distributing one to each inhabitant of Earth would disperse it equally, only a tiny minority of people would have any idea how to utilize their helium, and the resource would be wasted.[85]

Fuller's World Game was predicated upon constantly maintaining an accurate inventory of all scarce resources and providing the exact amount of a resource needed for a life-supporting task to the site where it was required. He envisioned that job being accomplished through the use of computers, and he realized that humankind was rapidly approaching a level of technological effectiveness which would make such an inventory possible. He also appreciated the fact that such a program would require a global cooperation among governments never before witnessed in the recorded history of humanity.[86]

Bucky did not, however, believe that the generally self-centered bureaucratic governments or the people who controlled them would be willing to subjugate their perceived requirements to those of the entire population of Earth simply because everyone would prosper in the long run. Rather, he felt that the global problems of humanity would continue to increase until they became so gigantic that they could be handled only by massive emergency measures which would be demanded by grass-roots movements of individuals.[87] In the midst of such an emergency, the true significance of the World Game would emerge, and government leaders would be forced to submit to the solutions it could provide.

The World Game, or some form of global resource management similar to it, could then be established to support the best interests of all humanity within a cosmic context of abundance. Clearly, a program as potent as the World Game would not divert resources to something as destructive and adverse to life support as weaponry or the military. With livingry rather than weaponry as the underlying focus of planning, other serious issues relevant to modern society would become more obvious and of higher priority.

Having developed a tremendous interest and expertise in the area of construction, Fuller devoted a great deal of the World Game to shelter. During the 1960s and 1970s, long before homelessness became a public issue, he once again addressed the problem. This time, rather than invent another unique house, Bucky spoke of resource allocation in a truly abundant society. At World Game sessions and in other forums, he related his cosmic perspective of the dramatic shift of 1970, when, according to his calculations, because of technological innovation, humanity reached a moment of being able to support the entire population of the Planet. Fuller often expounded on the senselessness of people's having to work just to earn a living or, more appropriately, to work to earn ''the right to live'' in the period of abundance he felt humanity had entered. Instead, he argued that individuals should simply do whatever they wanted to do. Having observed human nature for decades, Bucky was certain that given such freedom, individuals would ultimately opt for doing what needed to be done out of a desire to contribute to society as well as to fulfill their individual missions here on Earth.[88]

Fuller's analysis suggested that a tremendous amount of resources would become available for housing following the 1970 transformation to abundance and global management. In his scenario of abundance, when people were no longer forced to spend many hours working at jobs which did not produce any life support, all the beautiful new office buildings with their modern plumbing and heating could be converted to housing. For years Bucky had been appalled that such modern structures were left empty during the night hours while homeless people roamed the streets without a place to sleep or were sheltered in substandard slums, and he believed that resource reallocation within a context of abundance would greatly support resolution of that issue forever.[89]

The World Game was created to facilitate the transition of humanity from the context of scarcity, which has always plagued us, to a new level of prosperity in which everyone can be supported sufficiently. Bucky felt that it required only the proper moment of implementation.

Perhaps the World Game will never be formally introduced to the general public on a large scale. It will, however, have served its purpose as the spark of an idea which required a gestation period prior to becoming formally manifest. In any event, Bucky's World Game is far from over. Issues such as global starvation and homelessness have come to the forefront of public attention, and Bucky's World Game lives on in the budding programs developed to eradicate such problems.

CHAPTER 17

Legacy to the Future

 The years of 1969 and 1970 marked a monumental transition for humanity. It was also the time in which Fuller, who was then seventy-five years old, experienced the final great transition of his life. Although he did not cease work on any of the projects he felt required his attention, Fuller shifted his focus to completion rather than to the development of new undertakings.

He also devoted a significant amount of time to a transitional retreat similar to those which marked the other major shifts in his life during 1927 and 1949. As the 1949 break had been less time-consuming and intense than the 1927 interlude, so, too, the 1969 break was less intense than the one in 1949. In fact, to most people, his retreat appeared to be nothing more than enjoying time on the new sailing sloop he had purchased and christened *Intuition*. That time did, however, mark a decided shift in Fuller's concentration.

After completing the 1969 World-Game seminars, he felt his inventory of global resources had been exposed to the public in such a way that others were excited enough to take it over, and he began allowing others to manage and operate that aspect of his work. Although continuing under Fuller's guidance for the next decade, the World Game began functioning as a separate entity, whose managers frequently called upon Fuller for insight and wisdom. Characteristically, Bucky might actually have continued with his World Game studies and planning had it not been for a series of events which convinced him that his revolutionary ideas simply could not be carried out rapidly enough for him.

Although it appeared to be a coincidence, the fact that the initial SIU

World Game seminar occurred at the same time that the first humans were landing on the moon was perceived by Fuller as an obvious indication that his ideas and the plan of the Greater Intellect were synchronized. For the first time in the history of humanity, human beings could actually stand on another of the celestial bodies and observe the Earth in its entirety, just as Bucky had always demanded that it be regarded within the World Game and his other endeavors.

He also ardently believed that every person living on Earth should have an opportunity to comprehensively view our home Planet as well as the distribution of its resources and population. Thus, Bucky felt that a $30-million World Game installation at SIU was an essential element of humanity's development.[1] In the spirit of compromise, Bucky did agree to undertake the project in gradual increments, and the Illinois state legislature allocated $3 million in funding for the project, provided Fuller and his associates could raise an additional $9 million. That project was just getting under way when the controversial nature of Bucky and his revolutionary ideas again surfaced and shifted the course of his life. Specifically, in 1970 the wave of campus unrest which had begun during the mid-1960s on the West Coast reached the Midwest and the Carbondale Campus of SIU, creating an air of tension among faculty and students alike.[2]

Although the brunt of the student protests were against the United States' involvement in Vietnam and Southeast Asia, with his comprehensive perspective Fuller felt that the problem was far more sweeping. He believed that the Vietnam crisis was merely one component of the much larger Cold War between communist and capitalist alliances which had been compounding since the end of World War II.[3]

As Fuller perceived the situation, the countries involved in that dispute were devoting billions of research dollars to the discovery of more effective methods of destroying "the enemy." That increased weaponry production had grown to such an extent that some countries would be able to destroy the entire Planet within minutes, and much of the research for such devastation was occurring where research had traditionally occurred, at university laboratories. Bucky believed that as a part of the Cold War, not only were both sides allocating huge sums of money to the formulation of new weapons, but they were also devoting a great deal of money to the destruction of each other's primary source of new scientific discoveries, the universities. He felt that such foreign intervention was a major factor contributing to the unrest on American campuses.[4]

Whether such interference was occurring or not, college students during the late 1960s and early 1970s were adamantly opposed to military research on campus, and they often vented their opposition at an institution's president. At SIU's Carbondale campus that man was Fuller's good friend, president Delyte Morris. Morris had been instrumental in establishing Bucky's position at SIU, and he had continued to lend enthusiastic support to Fuller's projects as well as to those of other liberal professors with vision. Students, however, perceived Morris as a figurehead of the military–industrial complex which was responsible for the United States' aggressive military posture, and they called for Morris's resignation.[5]

Peaceful protest soon escalated into more revolutionary behavior as rumors about Morris were spread, and his home was vandalized. Within a short time, the pressure of such action, along with legislators' calling for force to be used against the students, resulted in Morris's resigning his position.[6]

Without Morris's support, Fuller's projects and those of other progressive professors were doomed. The Illinois legislature withdrew its offer of $3 million in World Game funding, and, rather than being an asset at SIU, Bucky found himself regarded as a liability. In the midst of that stressful situation, Martin Meyerson, president of the University of Pennsylvania, came forth with a most intriguing proposition. He had known Bucky since 1948, when Fuller taught at the Chicago Institute of Design and Meyerson was head of planning for the City of Chicago. An architect, Meyerson had served as dean of the School of Environmental Design at the Berkeley campus of the University of California, and because of his compassion for and deep interest in students, he had been appointed chancellor of that institution during the 1960s. When he accepted the position as president of the University of Pennsylvania in 1971, Meyerson immediately set out to achieve something he had wanted for years: having Buckminster Fuller associated with an institution he headed.[7]

Because Bucky thoroughly trusted and respected Meyerson, he was willing to seriously consider the possibility of moving his center of operation, even though he was seventy-five years old. Bucky reasoned that since he would not stop his nearly continuous global lecturing and consulting or spending at least several weeks each year isolated with family and friends on Bear Island, the location of his office was not really very significant. Consequently, he entered into negotiations with Meyerson. At his advanced age, Bucky was quite certain about his wants and needs. He sought the best possible support for sharing his message as effectively and rapidly as possible with the public and for making certain that the research he had instituted continued in earnest. He also wanted to be associated with an institution which would support proper housing of his archive so that it would be available to the public.[8]

One of their initial negotiating sessions quickly concluded when Meyerson asked just how much money Fuller would require to continue his research projects and, in his characteristic disregard for finances, Bucky casually replied about $300,000 or $400,000 per year. Having only recently attained his position at the university, Meyerson did not have access to such funding, and he began searching for a different strategy.[9] He found his solution in the newly formed University Science Center, which was being built as one element of a Philadelphia redevelopment project. The University of Pennsylvania was only one participant in that project, and Meyerson enlisted the support of three other member institutions to join the University of Pennsylvania in sponsoring Fuller's move to Philadelphia. The other three members of the consortium which eventually convinced Bucky to move were Bryn Mawr, Haverford, and Swarthmore Colleges.[10]

The combined resources and prestige of those four institutions persuaded Bucky that his work would progress much more rapidly in Philadelphia. Thus, he accepted the position of World Fellow in Residence at the University City Science Center and shocked the administration at SIU by announcing that he and the vast resources he had accumulated over the years were leaving their university.[11]

When the move was completed in 1972, Fuller was given a magnificent reception. Mayor Frank Rizzo designated Bucky as an honorary citizen of Philadelphia, an honor never before accorded to anyone other than foreign heads of state. The mayor also presented him with a gold case containing an insignia of citizenship, which he jokingly mentioned would get Fuller out of any trouble he might get into with the police. Although he felt he was past the age when such an honor would be of practical use, Bucky was reminded of the New York City honorary police captain badge he had received fifty years earlier and the number of times it had extricated him from traffic problems with police officers.[12]

As with most of Fuller's actions, the move to Philadelphia was fraught with controversy. Although no formal contract had been signed, he had informally promised his extensive Chronofile archive to SIU, and its relocation to the University Science Center marked an end to that agreement. Still, Bucky felt that he had to respond to the changes of the moment, and he believed that the Philadelphia office would provide a much more efficient base for his work.[13]

The move itself was difficult for both Bucky and Anne. Although he appeared to be in better health than his frail wife, both had reached a stage of life when physical demands were becoming more and more of a strain. Additionally, Bucky felt obligated to maintain much of the support for several of the research projects he had initiated as well as for his headquar-

Fig. 17–1 Fuller at the 1973 Chicago Museum of Science and Industry exhibition of his work. Above him is a replica of the 1964 *Time* magazine cover on which he was featured and next to him is a model of Triton City; also shown in Fig. 17–2.

ters office.[14] The primary financial sustenance for those endeavors resulted from his ongoing worldwide lectures which continued to number well over one hundred each year.

That breakneck pace of travel and speaking finally caught up with Bucky in November of 1972, when he simply collapsed. Believing that he was suffering a heart attack, Bucky was rushed to a hospital, where doctors determined that he was simply exhausted. His fervor in doing all that he felt needed to be done for as many people as possible had finally taken its toll.[15]

One of the endeavors which may have contributed to the decline of Fuller's health and energy level was the process of compiling and completing his ideas about mathematics. He was adamant that the geometric math he had christened "Synergetic-Energetic Geometry" (later renamed Synergetic Geometry) as well as his thoughts on Nature's fundamental coordinate system should be properly recorded prior to his death. Because of that desire, an informal compilation project had been initiated during his 1927 period of silence, but formalizing that endeavor into a book had to wait until the end of his life when Fuller felt his ideas would be more complete.[16]

Realizing old age was creeping up on him and aware of the massive nature of the task, in 1961 Fuller had signed a contract to complete such a book for the small advance of $2500. The publisher, Macmillan Publishing Company, had patiently waited for several years without seeing any activity from Fuller on what was to be a seventy-thousand-word book based on his most recent lectures. With every intention of producing the book, Bucky had begun maintaining bits and pieces of his best ideas and writings. Those tentative chapters were kept in an imposing black leather briefcase stamped with the initials "B.F." which was protected by Bucky and his staff with the same caution afforded to secret documents by military organizations. That briefcase was always kept near Fuller and was never mailed or checked with baggage when he was traveling.[17] When its contents were later examined by Ed Applewhite (Fuller's eventual collaborator on the book which became *Synergetics* and explained his mathematical concepts) he was amazed to discover that the outward precautions were not a true indication of its contents. Inside, he found notes, ideas, and writing which were simply stuffed into the satchel for future analysis.[18]

Applewhite had first met Bucky and his family during the mid-1930s at the age of fifteen. Ten years later, Bucky asked him to join Fuller Houses as director of personnel working on the Dymaxion House in Wichita, and Applewhite accepted.[19] Following the demise of that project, Applewhite went to work for the government and later the CIA.

In 1969, when he visited Bucky at his Carbondale office, Applewhite had just retired from his post at the CIA and was looking for work. During that trip, Fuller offered Applewhite the opportunity to manage all his affairs.[20] Having experienced Fuller's frenetic energy and tendency toward independence and complete control of his projects over the years, Applewhite was aware of the pitfalls such a position might include and declined the offer. He did, however, agree to collaborate with Fuller on gathering together all the pieces of Synergetic-Energetic Geometry so that it could be published in one volume. Several other people had failed at that task, but Applewhite felt confident that he could apply the organizational skills he had acquired working with codes in both the navy and the CIA to master the formidable assignment.[21] Still, Bucky's philosophy and ideas would additionally require interpretation, translation, and explanation before they could be even partially understood or appreciated by average readers—a task which will most likely continue for decades.

Having initiated the World-Game seminars in 1969, Bucky felt that he had completed the global-resources aspect of his life and was ready to move on to a culmination of his mathematical ideas. He was also aware of his advancing age and deteriorating health. Hearing aids had become as necessary as his thick glasses in helping him to fully experience his physical environment and to maintain his breakneck pace of lectures, writing, and projects. Most of his books were, however, not uniquely composed as printed material. Rather, they tended to be edited transcriptions of his thinking out loud lecture sessions, which he felt should be preserved in written form.[22]

Although somewhat successful, those books were primarily purchased on the strength of Fuller's personality. After hearing one of his lectures or learning about him through the media, people wanted to examine more of his ideas, and books appeared to be the best means of doing so at a slower pace than the one used by Bucky in personal presentations. Even though Fuller's ideas, like Nature itself, were elegant in their simplicity, his books tended to be as convoluted as his lectures. And although *Synergetics* was created in a much different manner, Fuller's requirement of expressing precisely what he felt was true at a given moment with a lesser regard for the reaction of his audience resulted in that work's retaining a similar quality.

Many aspects of that book's production are as significant as the work itself because they represent a microcosm of the Fuller personality while providing a glimpse of the difficulties he constantly confronted as he struggled to benefit the most people possible within the few short years he knew were left to him. The process clearly portrays a man who at the age of seventy-six was constantly aware of what he felt needed to be done in the best interests of humanity and who also realized he had to produce a minimum income of $200,000 per year to support his staff and the projects in development.[23] Under those pressures, Applewhite and Fuller agreed to only four guidelines for *Synergetics:* All the geometry was to be expressed in written words. Applewhite was not to be listed as a coauthor. All the text was to be in Fuller's phraseology. And, the book was not to make any attempt to conform to technical textbook or academic standards.[24]

Following those guidelines, Applewhite began a thorough search through Bucky's accumulated Chronofile material, including both written and recorded works and other collected memorabilia. He was particularly astonished after he opened the revered briefcase containing the most recent version of the *Synergetics* manuscript, only to find an amalgamation of diverse writings including the most recent first chapter of the book written as fifty pages of blank verse entitled "Brain and Mind." Applewhite successfully argued that poetry was not suited to such a work, and after several lengthy discussions, "Brain and Mind" became a section of *Intuition,* one of Fuller's books of poetry.[25]

Applewhite then began the heroic task of assembling and organizing all of Fuller's written and spoken thoughts pertaining to geometry and soon found that separating a single topic from the message of a comprehensivist was not only difficult but impossible. Thus, *Synergetics,* although outwardly focused upon the mathematics Bucky had been developing for decades, also includes a great deal of his philosophy, understanding of Universe, and tendency to explain phenomena in much greater detail than most people would ever care to study.

Applewhite's strategy for creating the book was quite simple, and it resulted in a monumental resource which was not only used in the production of *Synergetics* but will also be of immense value to future generations interested in Fuller's work. He instituted his work by typing relevant quotations on 5″ × 8″ index cards and sorting them into topics.[26] That accumulation of wisdom eventually became the *Synergetics Dictionary,* which was published separately as four massive volumes in 1986.

Since Fuller remained in Philadelphia only a few days each month, Applewhite was forced to catch Bucky whenever and wherever he could. Thus, collaborative "writing" sessions were held in hotels and restaurants around the World, as well as in both Applewhite's home and Fuller's office and home. A major portion of the work on *Synergetics* was also done via long-distance telephone conversations, with Fuller calling Applewhite at almost any hour of the day from sites as diverse as La Jolla, California; Tokyo, Japan; New Delhi, India; and Windsor Castle in England.[27]

Prior to a meeting, Applewhite would prepare and organize triple-spaced sheets of Fuller's quotes in the order he felt would be most logical.[28] Then, in typical Fuller fashion, Bucky would begin amending the pages to include his most recent thoughts on the topic. Bucky's work was always done longhand on single sheets.[29] He somehow managed to commit all his notes about what was typed on a page to that page, even if it meant a nearly illegibly tiny handwriting. Noticing that quirk, Applewhite soon initiated a policy of allotting massive margins so that Fuller would write in script large enough to be deciphered.[30]

The changes were then retyped and the new sheets submitted to Fuller, at which time he would again amend them.[31] Although Bucky maintained that he required only seven changes to a draft, that process might have continued indefinitely had not a new agreement been reached with Macmillan Publishing. Under the new contract, specific deadlines were created and the advance for *Synergetics* was increased to three payments of $25,000.[32] Because Macmillan had already been waiting for nearly a decade without seeing a single page of manuscript, the new contract also permitted the publisher to withdraw from the agreement if, after reading portions of the work, the editors determined it did not suit their requirements. Fuller and Applewhite were able to honor the first deadline of the new contract, and Macmillan accepted the book after its editors examined the first third.[33]

Even though Applewhite himself had limited access to Fuller over the next years, each succeeding section was submitted to the publisher on time, and the completed manuscript was delivered in March of 1974, just over four years after Applewhite had agreed to work with Bucky. Believing he had completed the monumental task, which had occupied almost all his waking hours during that time, Applewhite then boarded a Polish freighter for a much needed, isolated eleven-day trip to Germany.[34] However, upon returning to the United States, he was surprised to discover that his work was far from completed.

From Bucky's perception, no task was ever completely finished. Rather, it was accurate for a particular moment in time, but when that moment passed, his sense of integrity required him to revise and rewrite any elements he felt were no longer thoroughly accurate. Thus, when Macmillan began returning galley proofs to be checked for simple errors, to the consternation of everyone, Bucky acted as if he were dealing with manuscript pages and began making massive alterations. Such action should not have surprised anyone who had worked with Fuller and had witnessed his almost total disregard of the formalities of bureaucratic processes. Still, people involved in the project, especially Applewhite, quickly reminded Bucky that any large changes would be charged to him.[35]

Displaying the nearly fanatical need to disregard such financial formalities which had often plagued and limited his effectiveness, Fuller continued with the alterations he felt were "essential," and they eventually resulted in his paying for over $3500 worth of additional changes.[36] Those massive modifications also created production delays while adding to the somewhat agitated atmosphere surrounding the project at Macmillan.

That animosity was compounded by internal changes at Macmillan which resulted in *Synergetics* being passed from editor to editor. When the project's completion was finally in sight, the publisher's executives scheduled *Synergetics* as its featured book for their 1974 fall list of new books even though the only people who had actually read the final work in its entirety were Applewhite and a few copy editors.[37]

Prior to the final printing, however, another aspect of the drama which so often surrounded Bucky's projects surfaced. Cutting finished galley proofs into pages and binding them into what appear to be finished books is a standard practice in the industry. Those galley "books" do not usually include illustrations or photographs, and *Synergetics* was no exception. However, without its essential artwork, the book, which was inherently difficult to understand, appeared even more convoluted.[38]

To make matters worse, numerous copies of the galley books were circulated throughout Macmillan so that they could be inspected by various departments, including the sales, marketing, editorial, and legal departments. When the managers of those areas attempted to read and assess the book, they realized the enormity of their problem.[39] Some people, including Ed Applewhite, were beginning to speculate that *Synergetics* might have been more appropriately issued through a technical or specialty publisher. In fact, after reviewing the book, many of the Macmillan managers were seriously considering printing it as a college textbook rather than a popular or "trade" publication.[40]

Such a move would not have been imprudent. With its 876 pages, 150 illustrations, and dozens of tables, charts and diagrams, *Synergetics* was meant to be studied rather than read. Furthermore, its structure is that of a textbook, with numbered sections rather than chapters. For example, Section 400.00, "Systems," is divided into subsections such as 410.00, "Closest Packing of Spheres," and 420.00, "Isotropic Vector Matrix." To further "clarify" topics, subsections such as 410.00 are divided into subsubsections such as 410.01, "Nature's Coordination," and 410.10, "Omnitriangulation of Sixty Degrees."[41] Below those subsubsections are subsubsubsections, such as 410.011 and 410.02, which are not titled and present the actual text paragraphs. Each and every paragraph of that massive volume is thus numbered and focuses on a particular topic.

While the entire work does contain a massive amount of knowledge (much of which has been interpreted and presented throughout this book), in *Synergetics* that information is presented in the form of a mathematical dictionary rather than a popular work. And, although many people need and use a dictionary, few are moved to read one from cover to cover. To make matters worse, *Synergetics* has no index, and therefore a person attempting to reference a particular topic is forced to search through pages and pages of material.

This is not to say that the book is not precisely as Fuller felt it should be. Always the comprehensivist, Bucky had succeeded in his attempt to combine science, poetry, and philosophy in a single work. Such an astonishing accomplishment would be praised were it not for the fact that few mortals are as comprehensively oriented as was Fuller, and thus, most people were reluctant to respond positively to, much less to acquire and read, such an epic volume.

What *Synergetics* needed was another volume or at least a few chapters of extensive translation and explanation. What it received from the management at Macmillan during the early summer of 1974 was cancellation. Unable to decide about the work, Macmillan managers simply removed it from their production schedule and continued an internal debate for the next several months. Finally, during the late fall, work on the book again began to progress. That decision was, however, so controversial that it could be made only by the company's chairman of the board.[42] He resolved to attempt reducing the company's losses on the work by releasing it as a general trade book on the strength of Fuller's name and reputation. He also decided to cut costs by eliminating the color printing from the previously planned thirty-two pages of color drawings (which would have added a tremendous amount of clarity to the work) as well as by increasing the retail price from the planned $12.50 to $25.00, a very high price for 1975.[43]

As a result of those decisions, Bucky's attempt to complete the mathematical coordinate facet of his life and to publish his findings culminated in an initial printing of fifteen thousand copies of *Synergetics* in April of 1975.[44] Subsequent printings followed, and although it was considered successful by everyone who had worked on the project, the book remains an enigma to most people, including the critics who reviewed the first edition. Rather than seriously commenting on the work, most of them were unable to understand it and simply relegated the book to the category of an interesting "intellectual toy."

Engineers, architects, and scholars tend to prominently display the hefty volume on their bookshelves. However, on examining the book, one might find the binding barely broken. It is a marvelous collection of what Bucky believed to be all he had learned in his search for Nature's coordinate system through 1974, and it has been somewhat of an inspiration for many of the insights provided throughout this book. It is also a massive resource for those willing to consider, among other things, the possibility that most of the mathematics taught through the centuries is erroneous, and that Fuller had, in fact, discovered a cosmic mathematics predicated upon Nature.

Synergetics is not, however, easily accessible to the general public. Within months of its completion, Fuller had recognized that problem and had entered into a process of rectifying it. In characteristic fashion, he had actually begun rewriting the book during the production process, and that rewriting continued even after the book had been published. When the possibility of altering the finished work was rejected, he simply started work on the second volume.[45]

Ed Applewhite was also involved in that work. Thus, *Synergetics 2* follows the same convoluted style of *Synergetics*. Fortunately, Fuller had discovered the flaw in the earlier work, and in a continuing attempt to clarify and make his ideas as accessible as possible, he divided many of Applewhite's numbered sections, subsections, subsubsections, and subsubsubsections into narrative scenario sequences which are easier to follow. Scenarios such as "Child as Explorer" and "Complex Humans" provide readers with a much more accessible entrance into Fuller's ideas and help somewhat to overcome the "mathematical dictionary" presentation of the two volumes.

Both books were also greatly enhanced by the inclusion in *Synergetics 2* of a complete index of the two volumes. Still, expectations were already set from *Synergetics,* and readers unable to cope with that volume were usually not willing to attempt reading what appeared to be a similar work. Unwilling to have those books relegated to the back shelves of libraries, following the 1979 publication of *Synergetics 2,* Bucky mounted an ongoing campaign of advising his audiences to read the second volume before taking on the first book.[46]

In both *Synergetics* books, readers will find a detailed recounting of many of the theories and hypothesises presented throughout this book. Fuller believed that both physical and metaphysical phenomena as well as the general principles which governed all Universe could be modeled by means of physical illustrations and could be explained particularly well with geometry (i.e., his own synergetic geometry, which is predicated on the triangle rather than right-angle squares and rectangles). Thus, a reader will find those volumes filled with esoteric, almost poetic, interpretations of reality as well as complex mathematical descriptions of the means through which Nature demonstrates and configures everything to conform to generalized principles.

With the 1975 publication of *Synergetics,* Fuller essentially completed his work with mathematics, geometry, and the search for Nature's coordinate system. He had already culminated his concentration on geodesic structures with the Montreal Expo dome. Not that the comprehensively curious Fuller totally abandoned those, or any other, topics. Rather, he shifted his primary focus to the next project he believed to be most important: the completion of publicly chronicling his philosophy and general observations concerning the development of humanity. His primary method of relating those thoughts was, as usual, the written and spoken word.

Specifically, Bucky began working on one final book which would embody his philosophical convictions as well as his perception of the development of human civilization: where we had come from and what we were moving toward. Fuller also sought to include in that book his strategies for solving the most significant problems of the future.

The book he created was *Critical Path,* and like *Synergetics,* it was written with the assistance of a collaborator, Kiyoshi Kuromiya. *Critical*

Path, like most of Bucky's other books, tends to drift back and forth through topics, just as Fuller did in his lectures.

Within a verbal format, Bucky was able to "weave" the threads of a lecture tapestry, the whole of which was often evident to his audience only following the lecture's completion, if they perceived it at all. Thinking in front of people, he could rely on his sense of the audience's understanding, his personality, and his ability to share a broad context filled with details that generally went unnoticed.[47] However, in the written format, those ephemeral qualities of communication were not available, and his books generally tend to fall short of delivering the unadulterated message that was presented in speaking.

Even with the support of Fuller's other staff members, producing *Critical Path* was a formidable task for Kuromiya and the aging Fuller. It was, however, published in 1980 and marked the culmination of the final stage of Fuller's lengthy process of completing and recording his work for posterity as well as his documenting critical solutions to major issues. *Critical Path,* more succinctly than ever, documents his philosophy and proposals for most efficiently achieving human success in the shortest period of time, many aspects of which have been recounted throughout this book.

When, in 1980, both Bucky and Anne reached their mid-eighties and Anne's health was deteriorating, the family decided that another move would be advisable. Thus, the couple moved to a house in Pacific Palisades, California, within two blocks of Allegra and Bob Snyder's home. Their grandson, Jaime, also moved into that house to help care for the couple. While Anne and Jaime generally stayed home, Bucky continued his frenetic pace of travel and speaking engagements as well as spent a great deal of time in Philadelphia where his headquarters remained.

Bucky was no longer an obscure, eccentric character. His years of considering solutions to humanity's future problems had brought him worldwide esteem. Thousands of people attended each of the one-hundred-plus thinking out loud lectures he continued presenting every year, and despite his increasing age, those sessions were consistently hours in length, exhausting audiences long before Fuller was through.

The talks were, however, reaching a new and different audience. Rather than being limited to scientific and technologically oriented individuals, Fuller's message was increasingly welcomed by people with interests in humanistic and psychological issues as well as those concerned about the future. Although many of those audiences may have had difficulty with the technical and mathematical nature of his writing, they could empathize with his message that future success and prosperity for all humanity was a genuine possibility. They also appreciated Bucky's life as a demonstration of what one individual could accomplish if he was determined to operate comprehensively and unselfishly on behalf of all humanity.

And although with his advancing years a certain amount of rigid egotism surfaced, because most people clearly observed and admired his commitment he was seldom confronted on the issue. For his part, Bucky was simply more interested in presenting the issues and facts that he felt were most important (as he perceived them) as efficiently as possible. Thus, a question about some personal aspect of his life would most likely result in an hour or longer monologue on humanity's evolution.

Throughout his later lectures and writing, Fuller also championed some of the revolutionary ideas which he realized would not come to fruition during his lifetime but which he believed were seeds for the future work of others. One of the most dramatic of those projects was the "floating city."

During the early 1960s, Fuller had accepted a challenge from a wealthy Japanese patron, Matsutaro Shoriki, to design what amounted to a gigantic houseboat community that was to be floated in Tokyo Bay and would alleviate some of Japan's overcrowding. Always concerned with maximum efficiency rather than "practical aspects," Fuller proposed constructing a floating nine-thousand-foot-high tetrahedronal building. The structure was to house 300,000 families in spacious apartments, each containing two thousand square feet and overlooking the water. Since the World's tallest humanly fabricated structure at that time was a two-thousand-foot television tower and the World's tallest building was the Empire State Building at 1472 feet, Fuller's proposal appeared to be more fantasy than reality. However, after examining his engineering calculations, many engineering and architectural experts agreed that the idea, although radical, was feasible.[48]

In an effort to appease those who felt that something less formidable would be more practical during the initial stages of the project, Bucky redesigned the structure so that it could be built in sections, with the first segment housing only one thousand occupants. He also planned to construct the floating city from prefabricated modules, so that obsolete or

worn-out apartments, schools, or stores could simply be removed from the
main structure and their materials recycled. The open spaces in the framework created by such a change would then be filled with whatever new
modules were required. Thus, it would be a flexible structure that could be
anchored wherever it was needed.[49]

Although the idea was carefully studied by everyone, including the
Japanese financier's engineers, and proved to be valid, Shoriki died while
the project was still in the drawing and proposal phase. It was, however,
picked up on a much smaller scale by the United States Department of
Housing and Urban Development during the late 1960s and was chris-

Fig. 17–2   The Triton City, floating self-contained community model.

tened Triton City. Triton City was to be a floating neighborhood of no more than five thousand people, embodying all the positive amenities of any neighborhood, including parks strategically placed within the massive courtyard area of the structure, stores, and parking. Since Fuller felt that the flow and storage of automobiles were major problems for urban dwellers, he relegated all motorized-vehicle access and parking to unseen, lower-level locations. Above the parking areas, pedestrians would move about safely on elevators and escalators, all of which would be enclosed in glass for increased visibility, security, and ambience.[50]

After examining Fuller's models, drawings, and specifications, the Department of the Navy certified the seaworthiness of Triton City as well as the feasibility of building it at Fuller's estimated cost by using construction facilities which were, at the time, devoted to aircraft carriers. As President Johnson's administration was deeply interested in social service programs for the poor, the fact that apartments could be realistically rented at a rate so low that people living just above the poverty level could afford them was particularly appealing.[51]

The City of Baltimore was seriously considering acquiring the first Triton City and installing it in Baltimore's harbor, when politics once again resulted in the demise of a Fuller project. When the Nixon administration was elected in 1968, social service programs were no longer a high priority, and the Democrat-controlled city of Baltimore lost its national political advantage. Thus, work on its Triton City ground to an abrupt halt, and another of Fuller's futuristic ideas was relegated to obscurity until the moment it is needed and will be rediscovered.[52]

Although Triton City and its larger Japanese version were designed for protected harbors, Fuller also envisioned mooring similar structures throughout the World's oceans. He felt that cities anchored at sea could be supported by enormous pontoons submerged below the water's surface where there was no turbulence. Structural columns would then rise from those pontoons to support the floating city high above the surface of the sea. With such a system, even during the most furious storm, the city would remain as motionless as any structure built on land.[53]

Similarly, Bucky proposed building fully submerged cities which could be anchored at sea. The only portion of such structures which would extend above the surface of the water was to be a large vertical tower through which air would flow. On top of that tower, Fuller proposed installing platforms for landing helicopters or other aircraft which do not require runways. The primary means of transporting goods and people to submerged cities was, however, to be submarines. Fuller felt that long-range atomic submarines would become the future trucks and buses of the World, linking modern oceangoing colonies both above and below the sea. Such colonies could also serve as "ports" for repairing and resupplying the submarines.[54]

Even more radical than his proposals for structures at sea were Fuller's ideas for structures floating in the sky. During the late 1950s, he hypothesized that the weight of construction materials was decreasing so radically, while the strength of those same materials was increasing, that he would soon be able to design and construct enormous geodesic spheres which weighed very little.[55] Bucky then began calculating further possibilities and found that a tensegrity-supported geodesic sphere one hundred feet in diameter would enclose 7 tons of air. The structure itself, however, would weigh only 3½ tons.When the diameter of the structure was doubled, the weight of the structure doubled to 7 tons, but the mass and weight of the enclosed air increased to a total of 56 tons. The ratio of air to structure weight became 8 to 1 rather than 2 to 1, as was true of the smaller structure.[56]

Doubling the diameter again to four hundred feet, Fuller found that the same phenomenon continued to hold true, so that the ratio of air to structure weight increased to 33 to 1. He continued hypothetically expanding the size of the geodesic spheres until he was discussing a structure one-half mile in diameter in which the air-to-structure weight ratio would be approximately 1000 to 1. In other words, the weight of the structure would be negligible in comparison to the weight of the air it would enclose.

He also found that when the sun shone on such an open-frame aluminum geodesic sphere, the light passing through would reflect off the concave interior and would heat the inside air slightly. That air would be so heavy in comparison to the structure that even a change of one degree would become significant. Bucky calculated that a single-degree increase in the temperature of the internal air would cause it to expand and force a quantity of air weighing more than the structure itself out of the sphere. The result of such a change would be that the total weight of the interior air plus the structure itself would become less than the total weight of an equivalent volume of atmospheric air. Hence, the structure, with its enclosed thinner air, would rise to a level where the atmosphere was of an equal density. In other words, the structure would rise into the sky and float.[57]

Fuller felt that any geodesic spheres larger than one-half mile could become "floating clouds," and he christened those theoretical structures "Cloud Nines." He also calculated that the air in his Cloud Nine structures would become denser as cooler air rushed in, or during the evening when the sun's light was not available. To remedy that problem, he proposed that large sections of the surface be covered with lightweight polyethylene "curtains" which could be closed, retarding the rate at which air could enter the structure. Thus, by regulation of the opening of the curtains, a Cloud Nine structure could be maintained floating at a specific altitude. It could then be left to float at will or, more practically, could be anchored above specific sites to provide a new form of shelter. Such shelters could also be moved to disaster sites in a matter of days to handle emergencies.[58]

Like the weight of the structure itself, the weight of humans occupying a Cloud Nine and their belongings, including buildings, would be negligible in comparison to the weight of the displaced air. With that ratio maintained, the structures could float above the Earth's surface indefinitely, providing a unique new shelter option.[59]

Always more concerned with individuals than organizations, Fuller also devoted a great deal of his time to examining and developing life support which focused on individuals. Chief among his interests was the "autonomous black box," or living-technology packet, which Bucky forsaw as an ideal companion to small geodesic domes of the future.[60]

Having observed changing technology for decades, in the 1960s Bucky was certain that the same technology which permitted human beings to survive in and explore space would soon be applied to survival on Earth. Specifically, he believed that the life-support packages, which weighed hundreds of pounds and sustained several people in space for days, were becoming increasingly smaller and more efficient and would someday be available to support human beings living on Earth.[61]

In Fuller's estimation, such a package would eventually be small enough for an average person to carry and, like any major appliance, could be rented for a price he calculated to be about two hundred dollars per year. The package would provide an individual with everything he or she needed in terms of energy, water treatment, sewage recycling, and communication, as well as the other services which now compel most people to live in communities where their homes are connected to the public corporations providing such services. With such an autonomous black box, an individual would no longer be limited by the dictates of

410   CHAPTER 17

society, and it could be incorporated in an easily constructed geodesic-
dome home to create a quality, inexpensive living environment almost
anywhere on Earth.[62]

Should such a black box become available, it is unlikely that Fuller
will receive any credit for considering the invention years ago. He will,
however, receive a great deal of recognition if and when a project he
initiated during the early 1970s is realized. The project was to be constructed in East St. Louis, Illinois, a poor, predominantly black community directly across the Mississippi from St. Louis, Missouri. East St. Louis
had always been a destitute neighborhood in that region, but the people of
that community had survived working for the heavy industry located
there. However, when American heavy industry began closing plants in
the 1960s, East St. Louis became a truly impoverished community.

As East St. Louis was near his Carbondale headquarters, Fuller was
asked by a friend, the well-known black dancer Katherine Dunham, if he

Fig. 17–3   An early model of the Old Man River City, without its geodesic dome covering.

Fig. 17–4 A later model of the Old Man River city in which the geodesic dome has been replaced by a clear covering.

might consider using his design science strategy to create a practical program which would help that community.[63] Fuller felt that the town of seventy thousand was an ideal location for again demonstrating the practicality of his ideas, and in 1971, he made his initial proposal to the community.[64]

Bucky suggested creating a form of the floating city on land. His proposal was to build an enormous self-contained community which would house twenty-five thousand families within a giant crater-like structure that was to be one-half mile from rim to rim. The structure's environment was to be controlled by covering it with a massive, clear geodesic dome one mile in diameter.[65]

Within the structure, individual families would each live in apartments of two thousand five hundred square feet, which would provide views both outwardly and inwardly. The interior view was to be of the public area, which would occupy the lower interior tiers of the structure as well as its ground level. Those lower areas were to be filled with recreational facilities and stores. As in the floating cities as well as the small communities of the past whose living arrangements had served as a model for Fuller's visions, the residents of the complex would be able to work and go about their daily business without ever leaving the complex. All necessary services would be provided internally.[66]

When Fuller's ideas were welcomed by the East St. Louis citizens, he helped form a team to facilitate detailed design and construction of the complex. Since he was constantly traveling, the primary responsibility for the operation fell to his good friend Professor James Fitzgibbon, the head of Washington University's School of Architecture. Fuller himself donated his time and ideas to the project, and two or three times each year, he would return to the community, meeting with and advising the team and interested citizens.[67]

Bucky's only stipulation for his unpaid work on the project reflected the same distrust which had hampered so many of his other endeavors. He demanded that the design solutions be free of any compromises which might be required to qualify for private foundation or government financing.[68] By agreeing to that single requirement, the residents and the design team condemned Old Man River to a fate similar to that of Fuller's Dymaxion Car and House. The fact that it had to remain pristine and

Fig. 17–5   Fuller with John Denver at Bucky's eightieth-birthday celebration, hosted by Denver at his Windstar Foundation in 1975.

Fig. 17–6   Bucky and Anne in front of the Dymaxion Car and the Fly's-Eye Geodesic Dome at the eightieth-birthday celebration.

uncompromised by the traditional societal restrictions is one of the elements which has kept Old Man River from moving beyond the dream of a local community. Largely unfinanced and without the national or international support such a massive undertaking requires, Old Man River's progress has been slow. Still, because the project was not completely dominated by Fuller as was the case with his other endeavors, the initiative he inspired during the early 1970s endures among several East St. Louis citizens. Thus, Old Man River remains a small but viable project championed by local residents of East St. Louis who continue working on the undertaking. Even though the idea is represented only by a model, some design drawings, a small office, and the dreams of a few individuals, those people continue to envision its completion in the year 2004.

Even though the 1981 publication of *Critical Path* essentially marked the culmination of Fuller's work, as old age began to catch up with him, he continued his extensive series of lectures. The last of his in-depth series took place during the summer of 1981, when he devoted six full days to sharing "everything he knew" with an assembled group of 150 people who were just as eager to listen to his thinking out loud as had been the college students of Black Mountain College thirty-three years earlier. And Bucky did not disappoint that audience either.

Although he continued expounding on the same topics that characterized most of his talks and that have been discussed throughout these pages, he did so with an enthusiasm for life that tended to cause people to forget he was eighty-six years old. Still, his age could not be denied, and it finally slowed him down in late 1981, when, for the first time in decades, he was forced to spend several weeks in one place while he recovered from a hip replacement operation. Not that the time was wasted idly lying around. In need of constantly sharing his latest thoughts, even before *Critical Path*'s publication, Bucky was working on the book's sequel, *Grunch of Giants*, which was published in 1983.

He was also planning one last attempt at experientially sharing his vision of humanity's status on Spaceship Earth. That attempt was designed to utilize all of his previously gained understanding, and it resulted in the production of the largest portable World map in history. That innovation, which became known simply as the "Big Map," was a Dymaxion Map the size of a basketball court, and Bucky presented several of his last lectures in auditoriums where the Map was displayed on the floor. Using the Big Map, Fuller created a presentation in which one hundred participants, each representing 1 percent of the World's population, were positioned in their correct geographic location. Then, with well over a dozen people crowded onto both India and China, while only six people occupied all of North America, items representing resources produced and consumed were distributed.

It was not long before the entire audience was aware of obvious discrepancies, as the people representing India and the East were left holding only a small amount of food, while those in North America possessed more than they could carry. Global warfare was also explored as participants experienced the vast military overproduction of the United States and the Soviet Union. That enormous presentation was extremely successful, and Fuller even managed to stage it for members of the United States Congress. Still, he was very aware of the limitation of his age as well as the toll Anne's health was taking on him.

In late 1982, she underwent an operation for a cancer which had spread through her internal organs. Following that, she was constantly in and out of the hospital with complications. Bucky balanced his hectic schedule to be with her as often as possible, but his commitment to humanity also drove him to work even more furiously on sharing his message whenever possible.

In early June of 1983, that schedule once again caught up with him. He was trapped on a Boston subway-station escalator with one of his young associates when it stopped running. Caught within a mob of pushing humanity, the eighty-seven-year-old Fuller had no choice but to go along with the crowd as it pushed him rapidly up the escalator stairs. Upon reaching the upper landing, he collapsed. Paramedics were called to the scene, and although he appeared fine after a short rest, they insisted that he be examined at a local hospital. Doctors at that hospital pronounced him in excellent physical condition for a man his age, and he was soon back on the road.

Anne was not so fortunate. On May 13 she underwent another operation for cancer, and she spent most of May and June confined to home or a Los Angeles hospital. Although Bucky made every effort to be with her whenever possible, he also continued the frenetic pace which had characterized most of his life, traveling at breakneck speed in order to meet all his commitments most efficiently.

A sample of his itinerary even during those last days of his life shows him in Los Angeles on May 30, in Philadelphia on May 31 and June 1, and taping a National Public Radio interview in Washington, DC, on the afternoon of June 2. Late that same night, Bucky appeared on the Larry King radio show from midnight to 3 A.M., and he spoke at "The Frontiers of Healing" Conference at 8 P.M. the next evening.

Following that speech, he flew to New York, where the very next day (Saturday, June 4) he spent seven hours presenting an Integrity Day talk. The next two days were devoted to meetings and receptions throughout New York, followed by two days of meetings and work at his Philadelphia office and a one-day trip to Boston where he attended the Harvard commencement ceremonies.

On June 10, Bucky returned to his Pacific Palisades home for five days of interviews, work, and talks throughout the Los Angeles area before leaving for San Francisco and a speech at the opening of the new headquarters of the North Face, a major geodesic tent and outdoor equipment manufacturer. He then flew to Philadelphia, where he worked at his office during the afternoon of June 18 prior to flying to London for four days of meetings and interviews.

Fig. 17–7   Fuller receives the Medal of Freedom (the highest civilian award given by the United States government) from President Reagan in 1983, shortly before his death. The citation reads: "A true renaissance man, and one of the greatest minds of our times, Richard Buckminster Fuller's contribution as a geometrician, educator, and architect-designer are benchmarks of accomplishment in their fields. Among his most notable inventions and discoveries are synergetic geometry, geodesic structures and tensegrity structures. Mr. Fuller reminds us all that America is a land of pioneers, haven for innovation and the free expression of ideas."

June 23 found him flying back to Washington, DC, where he made a presentation at the White House in the afternoon and boarded another plane to Chicago to attend an evening dinner dance honoring people with Chicago connections who had been subjects of *Time* magazine covers. After a picture-taking session in Chicago the following day, Bucky flew back to his Pacific Palisades home for his final public appearance.

The next day, Saturday, June 25, he presented an all-day, thinking out loud Integrity Day in Huntington Beach. During the course of that talk, he spoke of Anne's health as well as his own mortality and mentioned that he felt she was lingering and waiting for him.

His conviction was fulfilled just six days later. After a morning meeting with coauthor of *Humans in Universe* Anwar Dil on July 1, just eleven days short of his eighty-eighth birthday, Bucky visited Anne at Good Samaritan Hospital. She had lapsed into a coma, and that morning the doctors informed him that it was unlikely she would regain consciousness.

After sitting with her, Bucky and his grandson, Jaime Snyder, went out for lunch. Jaime then left Bucky at the hospital to be with Anne, and sitting at his wife's bedside in the intensive care unit, Bucky suffered a massive heart attack. Even though he was in the one location which would maximize his possibility of being revived, Bucky died that afternoon. Anne never regained consciousness and died thirty-six hours later.

A few days hence, they were buried side by side in Milton, Massachusetts, under a double tombstone which characteristically reflected Bucky's wishes exclusively. It reads: "CALL ME TRIMTAB."[69]

Within weeks, his daughter, Allegra, and grandson, Jaime, had closed Fuller's Philadelphia office and had begun the arduous process of packing the enormous archive which had served Bucky, his associates, and anyone else who requested information or assistance. The staff moved on to new challenges, and the material itself was soon rehoused in a Los Angeles vault so that it would be closer to the family. There it remains, a vast informational resource which may someday once again be made available to the public and to scholars.

Buckminster Fuller left a legacy of which any individual would be proud. He spread seeds of ideas throughout the World, and they have enriched the lives of millions of individuals. More importantly, those seeds which have yet to germinate may well provide the essence of solutions to the issues that face humanity now and will face us in the future.

Fuller's legacy to each of us also includes his very life. Devoting himself to living as an experiment to determine what one individual could accomplish on behalf of humanity, Fuller clearly proved that an average individual does possess capabilities far beyond what most of us realize. During the fifty-six years of his personal experiment between 1927 and 1983, he demonstrated and documented that even without the advantages modern technology affords individuals today, one person could make a major difference. And, as time passes and people around the World increasingly come together to consider and solve global issues, his true legacy may well become even more apparent.

We are one people living on the tiny sphere Fuller named Spaceship Earth, and because of his decision to work on behalf of each and every member of the crew of that Spaceship for the majority of his life, the Planet and its inhabitants are better off. Yet, he did not leave us stranded without the spirit of his ongoing inspiration. As he reminded us in his epic poem entitled *How Little I Know:*

. . . the prime code
Or angle and frequency modulated signal
Could have been transmitted
From a remote stellar location.
It seems more likely . . .
That the inanimate structural pattern integrity,
Which we call human being,
Was a frequency modulated code message
Beamed at Earth from remote location.
Man as prime organizing
"Principle" construct pattern integrity
Was radiated here from the stars—
Not as primal cell, but as
A fully articulated high order being,
Possibly as the synergetic totality
Of all the gravitation
And radiation effects
Of all the stars
In our galaxy
And from all the adjacent galaxies
With some weak effects
And some strong effects
And from all time
And pattern itself being weightless,
The life integrities are apparently
Inherently immortal.

You and I
Are essential functions
of Universe
We are exquisite syntropy.

I'll be seeing you!
Forever.[70]

Chronology of the Life of R. Buckminster Fuller

The following listing provides a historical context for examining Fuller's life. It is not intended to be a complete listing. Rather, it presents an overview of the environment in which Fuller matured, worked, and developed his ideas. It also includes events and innovations which Fuller himself felt were important in humanity's evolution.

1895   Buckminster Fuller is born, July 12, in Milton, Massachusetts.
First American gasoline-engine-powered automobile is designed.
X-rays are discovered.
Tallest fabricated structure in the World is the Eiffel Tower, at 984 feet.
Grover Cleveland is president of the United States.

1896   William McKinley is inaugurated as president.

1897   Electron discovered.

1898   Spanish-American War begins.

1899   Fuller enters kindergarten, where he builds the first octet truss out of dried peas and toothpicks.
Fuller is diagnosed as nearsighted and receives his first glasses; for the first time in his life he can see objects clearly.

1901   Marconi completes first transatlantic radio transmission.

President McKinley is assassinated and Theodore Roosevelt becomes president.

1902 Mount Pelée on the island of Martinique erupts, killing thirty thousand people. This devastation leaves a powerful impression on young Fuller.

1903 The Wright brothers successfully fly the first gas-engine propelled airplane.
Ford Motor Company is founded and begins mass production of automobiles.

1904 Fuller enters Milton Academy, lower school.

1905 Einstein's theory of relativity is published.

1906 Fuller enters Milton Academy, upper school.
Fuller receives one of the first appendectomies ever performed.
San Francisco is devastated by earthquake and fire.

1908 William Howard Taft is elected president.

1909 North Pole is discovered by Admiral Peary.

1910 Fuller's father dies.

1911 South Pole is discovered by Amundsen.

1912 Woodrow Wilson is elected president.
Titanic sinks from collision with an iceberg.

1913 Fuller graduates from Milton Academy and enters Harvard University as a member of the class of 1917.

1914 Fuller is expelled from Harvard and reinstated after an intensive internship as an apprentice millwright in a Canadian cotton mill.
Panama Canal is completed and opened.

1915 Fuller is expelled from Harvard for the second time and takes a job with Armour and Company in New York City working 3 A.M.–5 P.M. daily, six days per week.
First transcontinental telephone is put into operation.
Woolworth Building in New York City is completed and becomes the tallest occupied structure in the World at sixty stories.

1916 Fuller becomes engaged to Anne Hewlett.

1917 Fuller enlists in United States Navy Reserve.
Fuller and Anne Hewlett are married, July 12.

United States enters World War I.

Russian Revolution begins.

1918   Fuller is assigned to special short course for officer training at United States Naval Academy, Annapolis.

Fuller is promoted to lieutenant (junior grade) in United States Navy.

Fuller is assigned to active war zone in Atlantic as aide to admiral commanding cruiser and transport forces of Atlantic fleet.

Fullers' first child, Alexandra, is born, December 12.

1919   Fuller is assigned as ensign to temporary duty on USS *George Washington,* the ship that carried President Wilson to France for the signing of the Versailles Treaty ending World War I.

First two-way transatlantic radio telephone conversation occurs between President Wilson on the USS *George Washington* in France and Washington, DC. As the *George Washington*'s communication officer, Fuller is intimately involved.

Word War I ends.

Alexandra Fuller contracts infantile paralysis and spinal meningitis, and Fuller resigns his commission in the navy to be with her and his wife, Anne.

Fuller takes a job as assistant export manager for Armour and Company in New York City.

Prohibition of alcoholic beverages in the United States begins.

1920   Warren Harding is elected president.

Women obtain the right to vote in the United States.

1921   Fuller resigns from Armour and Company to become National Accounts Sales Manager for Kelley-Springfield Truck Company.

1922   The Fullers' daughter, Alexandra, dies prior to her fourth birthday. Bucky is devastated by this incident and his inability to prevent the death of his daughter. He feels that better shelter and environmental controls might have saved her, and he begins a lifelong exploration of these issues.

Fuller loses his position with Kelley-Springfield and begins a career as an independent entrepreneur. His first enterprise is the founding of the Stockade Corporation, which manufactures buildings using a revolutionary technology invented by Fuller's father-in-law, noted architect J. M. Hewlett.

The first practical self-starting automobile is put into production.

1924   Calvin Coolidge is elected president.

1925   First commercial passenger airline route is established in America between Detroit and Chicago.

1926   Fuller fails to make a profit running the Stockade Corporation, loses all the money his friends have invested in the company, and is fired from his position as president when the company is sold to Celotex Company.

Electric refrigerator is adapted for use in ordinary households.

1927   Fuller considers himself a failure and seriously contemplates suicide. He decides that he might be of value to Universe, realizes that he does not have the right to end his own life, and dedicates himself as an individual to the service of all humanity.

Fuller writes and privately publishes his first book, *4D Timelock.*

Fuller invents the Dymaxion House as a part of his concept of air-deliverable, mass-produced dwelling units based on his strategy of anticipatory design science.

Fuller founds the 4D Company for research, development, and patent protection of his Dymaxion House and Car.

Fuller creates Energetic-Synergetic geometry.

The Fullers' second child, Allegra, is born in Chicago.

Lindbergh is the first to fly nonstop from New York to Paris.

1928   Herbert Hoover is elected president.

1929   The word *Dymaxion* is created by public relations people working for Marshall Field Department Store, where Fuller is to display a model of what he then called his 4D House. The word is a combination of syllables from the words *dynamic, maximum,* and *ion.* It is copyrighted in Fuller's name by Marshall Field and becomes his trademark over the years.

The Fullers move from Chicago to New York City.

Stock market crashes and the Great Depression begins.

1930   Fuller cashes in his life insurance policies to finance taking over *T-Square* magazine in Philadelphia. He becomes owner, publisher, and editor and changes the name to *Shelter*.

Fuller also accepts a position as Assistant to Director of Research at Pierce Foundation, a division of American Standard Sanitary Manufacturing Company. There, he begins two years of work on mass-produced kitchens and bathrooms.

1932   Franklin Roosevelt is elected president.

Fuller discontinues publication of *Shelter* magazine after the election of Franklin Roosevelt as president of the United States because he feels that Roosevelt will correct some of the construction problems Fuller founded *Shelter* to editorialize about. Fuller also feels that it is time to stop talking and to focus more attention on doing.

Fortune magazine publishes an article entitled "The Industry That Industry Missed," which cites Fuller's Dymaxion House as the prototype of a new mass-production housing industry.

1933   Fuller founds Dymaxion Corporation, through which he builds and successfully demonstrates the first prototype Dymaxion Car in Bridgeport, Connecticut. It is designed as the first stage of an experimental omnimedium, wingless transport vehicle to be propelled and maneuvered by rocket jet stilts, which Fuller anticipates will be invented in the future.

Prohibition of alcoholic beverages ends in the United States.

Adolf Hitler becomes chancellor of Germany.

United States bank failures reach a peak of five thousand banks in one day, and President Roosevelt declares a bank moratorium.

Congress gives the president the power to control money in the United States.

1934   Fuller's mother dies.

1935   Fuller completes writing the book *Nine Chains to the Moon*.

Dymaxion Car prototypes numbers two and three are completed and displayed at the Chicago World's Fair.

1936   Fuller meets with Albert Einstein, and Einstein acknowledges that Fuller's interpretation of Einstein's work as written in *Nine Chains to the Moon* is correct. Einstein also expresses amazement that Fuller could conceive of Einstein's theories as having practical application.

Fuller joins Phelps Dodge Corporation as assistant to the director of research. There, he continues his study of plumbing and the self-contained bathroom. At Phelps Dodge, he finalizes invention of the Dymaxion Bathroom as well as a new system of automobile brakes. He also continues his study of World resources and focuses his attention on copper. In that study, he discovers the regular recirculation pattern of metals.

Fuller is a frequent guest on experimental CBS television broadcasts to one hundred sets throughout New York.

1938   *Nine Chains to the Moon* is published.

Fuller travels throughout the United States with Christopher Morley.

Fuller joins *Fortune* magazine staff as its science and technology consultant.

1939   World War II begins.

Einstein and others warn President Roosevelt that the Germans might develop an atomic bomb, and Roosevelt immediately authorizes research into an American atomic bomb.

1940   Fuller leaves *Fortune* to work on the design and construction of Dymaxion Deployment Units in conjunction with the Butler Manufacturing Company of Kansas City. These are housing units developed from grain bins which Butler manufactures. They are employed extensively during World War II to protect delicate new radar equipment and troops in remote locations.

1941   Japan attacks Pearl Harbor and the United States enters World War II.

Fuller gives up drinking and smoking.

1942   Fuller joins the United States Board of Economic Warfare as Director of Mechanical Engineering.

1943   A full-color rendition of Fuller's Dymaxion Sky-Ocean World Map is published in *Life* magazine resulting in the largest printing of that magazine thus far.

1944   Fuller resigns his government position to complete designs of and supervise manufacturing of the first prototype Dymaxion House in Wichita, Kansas, in conjunction with the Beech Aircraft Company. For the project, Fuller moves to Wichita through 1946 while his wife and daughter remain in New York.

1945   President Roosevelt dies and Harry S. Truman succeeds him as president.
First atomic bomb is exploded.
Word War II ends.

1946   Fuller is awarded the first cartographic projection patent since the turn of the century by the United States Patent Office for the Dymaxion Map.
The first all-electric, general-purpose, digital computer is completed.

1947   Fuller invents the geodesic dome.
Fuller is appointed professor at Black Mountain College in North Carolina for the first of three summer sessions.

1948   Fuller accepts an appointment to teach at MIT.
Fuller works on several prototype geodesic structures and invents the geodesic dome.
President Truman wins close reelection over Thomas Dewey.
Gandhi is assassinated in India.

1949   Fuller teaches at Chicago Institute of Design in addition to Black Mount College and MIT.
Russia explodes its first atomic bomb.

1950   Fuller begins his extensive travels in responding to invitations to work and lecture at many universities and corporations around the World.
Korean War begins.

1952   Fuller begins work on the Ford Motor Company dome for its River Rouge headquarters.
Fuller and a group of Cornell University students construct the first Geoscope.
Fuller receives Award of Merit, American Institute of Architects, New York Chapter.
Dwight Eisenhower is elected president.
First commercial jetliner service is initiated.

1953   Fuller completes Ford dome. It is the first successful industrial
         acceptance of Fuller's concepts and the first practical ap-
         plication of the geodesic dome.
         Fuller directs construction of a fifty-foot tensegrity structure at
         Princeton University.
         Polio vaccine is discovered and developed.

1954   Fuller receives patent for geodesic dome.
         Fuller receives the first of his forty-seven honorary degrees,
         doctor of design, North Carolina State University.
         Fuller is awarded the Gran Premio, Triennale di Milano, Italy (a
         prestigious design award).
         Marine Corps airlifts and delivers a family-house-sized geodesic
         dome at sixty knots by helicopter. A picture of this opera-
         tion is featured on the front page of *The New York Times*.
         Geodesic domes are adopted as the structures to be used for all
         United States Marine Corps advance bases.
         Walter O'Malley, owner of Brooklyn Dodgers, asks Fuller to
         develop a geodesic dome to be installed over his baseball
         stadium.
         Color television is introduced.
         Supreme Court of United States orders desegregation of all
         schools.

1955   Fuller begins a development project for Brooklyn Dodgers Sta-
         dium geodesic dome at Princeton University School of Ar-
         chitecture. Because it is the first such large stadium dome
         proposed, the project receives a great deal of newspaper
         publicity.
         Fuller is awarded Centennial Award and Citation by Michigan
         State College and Award of Merit by the United States
         Marine Corps.
         DEW Line geodesic radomes are first installed in the Arctic to
         protect radar.
         United Nations Food and Agriculture Organization announces
         that its scientists have conceded that Malthus was wrong
         and that there could be enough food for all of humanity,
         even though a high percentage of people are starving. They
         attribute this inequity in global resource distribution to the
         political intervention of all nations.

1956   Fuller receives an appointment as a visiting lecturer at Southern Illinois University. In 1959 his position at SIU becomes permanent.

A geodesic dome is first used to house the United States display at an International Trade Fair. In this instance a one-hundred-foot-diameter dome is flown to Kabul, Afghanistan, in a single DC-4 airplane and erected in forty-eight hours by inexperienced Afghan workers supervised by the one American engineer Fuller was allowed to send along.

First transatlantic telephone cable installed.

1957   Fuller designs and supervises construction of the largest clear-span structural enclosure in history. This is the 384-foot-diameter Union Tank Car Company geodesic dome in Baton Rouge, Louisiana. Prior to this time, the largest such structure was St. Peter's Cathedral dome in Rome, which is 150 feet in diameter.

Fuller is awarded the Gran Premio, Triennale di Milano, Italy.

The first dome built by Kaiser Aluminum is erected of prefabricated sections in Hawaii in less than twenty-two hours. The Hawaii Symphony plays a hastily scheduled concert to a capacity audience in that dome twenty-four hours after the dome sections have arrived on the island.

Manufacturing begins on playdomes for playground use.

The Dodgers move to Los Angeles, and the New York stadium dome project is canceled.

The year is declared "International Geophysical Year" by many countries working together on scientific projects.

First satellite, Russian *Sputnik,* is launched by Earthians.

1958   Fuller makes the first of his many annual circuits traveling around the World fulfilling his regular university appointments in countries such as South Africa, India, Japan, England, and the United States.

Fuller is awarded the Gold Medal, Scarab by the National Architectural Society.

Fuller's Energetic-Synergetic Geometry is examined by nuclear physicists and molecular biologists, and they find that it mathematically explains Nature's fundamental structuring at the atomic nucleus and virus levels.

United States nuclear submarine *Nautilus* makes the first crossing of the Arctic Ocean and the North Pole submerged below the polar ice cap.

1959 Fuller is appointed by the State Department to visit the Soviet Union as the United States Representative for Engineering in a protocol exchange. At a dinner given by the Soviets in Fuller's honor, they tell him that they have been following his work for twenty-nine years.

Fuller is appointed as university professor (research) and awarded an honorary doctor of arts degree at Southern Illinois University. He and Anne erect, and move into, one of the first geodesic dome houses near the university. He also sets up an office as his global headquarters at SIU.

Fuller meets with Judge Roy Hofheinz of Houston, Texas, who wants to acquire a major league baseball franchise for his city. An agreement is reached, and Fuller begins work on what will eventually become the Astrodome.

Fuller works with Kaiser Aluminum to produce a gold-anodized, aluminum geodesic dome which is used for the United States International Exhibit in Moscow. Premier Khrushchev and other Russians are so thrilled by this dome that Russia buys it and leaves it as a permanent structure in Moscow's Sokolniki Park.

One-year outdoor garden exhibit of Fuller's geodesic domes, octet truss, and tensegrity structures opens at Museum of Modern Art in New York City.

Global jet-aircraft passenger-service network is established.

Alaska and Hawaii are admitted as states.

The St. Lawrence Seaway opens mid-America to ocean ships.

1960 Fuller publishes *The Dymaxion World of Buckminster Fuller*, which is written with Robert W. Marks.

Fuller is awarded the following: Frank P. Brown Medal by Philadelphia's Franklin Institute; Gold Medal of the American Institute of Architects (Philadelphia Chapter); Roy Edwin Crummer Award by the Crummer School of Finance and Business Administration.

Fuller continues intensive work on development of the stadium dome for Houston.

Geodesic dome is delivered fully erected to Ford Motor Company by helicopter. It is 114 feet in diameter and contains ten thousand square feet of floor space.

John Kennedy is elected president.

Peace Corps is established.

1961   Fuller speaks to two thousand architects at the International Union of Architects' World Congress and proposes initiation of Phase I of a Design Science Decade. In this phase, the World will be put on notice that the success of all humanity is not a political responsibility. Rather, such success depends on an invention initiative and a design revolution.

Fuller is granted patent for the octet truss.

Over two thousand geodesic domes are produced by over one hundred companies which have been licensed by Fuller. These domes are primarily delivered by air and rapidly installed in forty countries around the World and the North and South Polar regions.

DNA is deciphered as the genetic code and design control for all life.

Soviet cosmonaut Yuri Gagarin is the first human to orbit Earth in space.

1962   Fuller is appointed by Harvard University as the Charles Eliot Norton Professor of Poetry, a one-year appointment.

Fuller establishes his "Inventory of World Resources, Human Trends, and Needs" at Southern Illinois University as an accounting system to examine human evolution.

Fuller is awarded a United States patent for tensegrity.

On the eve of signing a contract with Fuller's company, Synergetics, to build an efficient aluminum dome for his Houston stadium, Judge Hofheinz backs out. Instead, he signs a contract with Roof Structures.

One Telstar communications relay satellite weighing only one-quarter ton replaces and outperforms transatlantic cables weighting seventy-five thousand tons.

1963   Fuller publishes the books *No More Secondhand God, Ideas and Integrities,* and *Education Automation.*

Fuller is appointed to NASA's Advanced Structures Research

Team which adopts his octet truss and geodesic dome as primary space structures.

Fuller is awarded the American Institute of Architects' Allied Professions Gold Medal and the Plomado de Oro Award of the Mexican Society of Architects.

Fuller is the subject of five half-hour television shows on National Educational Television.

Fuller is honored in a recognition day at the University of Colorado.

World Congress of Virologists announces the discovery of the protein shells of viruses, which they publicly acknowledge was anticipated by Fuller in his formulas of geodesic structures.

President Kennedy is assassinated, and Lyndon Johnson succeeds him.

Zip code is introduced into the United States mail system.

1964   Fuller is subject of cover story in *Time* magazine.

Fuller is commissioned as the architect of the United States pavilion for the 1967 Montreal World's Fair.

Fuller is a member of the meeting of "Leading Citizens of USSR–USA" assembled by USSR's Academy of Sciences.

Fuller's *Design Science Decade: World Inventory, Human Trends and Needs* is published.

Fuller's series of articles, "Prospects of Humanity," appears in *Saturday Review*.

Fuller is subject of BBC's first science program on their new, wide-range Channel Two network.

Fuller is awarded Gold Key Laureate, Delta Phi Delta, National Fine Arts Honor Society.

United States Civil Rights Act becomes law.

Free-speech movement begins at University of California, Berkeley.

Khrushchev is ousted as leader of Soviet Union.

Verrazano Narrows Bridge, New York, is completed and at 4,260 feet becomes the longest suspension bridge in the World.

1965   Fuller is given Creative Achievement Award of Brandeis University.

First commercial intercontinental relay satellite placed in Earth orbit.

First space walks by both United States and Soviet astronauts occur.

Westward mobility of population is evident as California becomes the most populous state.

Massive electric-power failure blacks out most of northeastern United States and two provinces of Canada.

United States forces in South Vietnam reach 184,000, and U.S. begins bombing North Vietnam.

1966   Fuller completes design for the USA Pavilion at the 1967 Montreal World's Fair. It consists of a 250-foot-diameter geodesic sphere christened a *skybreak bubble*.

Fuller receives First Award of Excellence from the Industrial Designers Society of America.

Fuller inaugurates World Game at Southern Illinois University to demonstrate how to make the World work so that all humanity may achieve physical and economic success without individuals' or groups' interfering with one another.

Fuller lectures scientists and engineers at Cape Kennedy, explaining how fallout from space technology into domestic economy will bring about the first scientifically designed house, and how that house could catalyze the physical success of all humanity.

Fuller is appointed as Fellow of the Graham Foundation.

Both the Soviet Union and United States make successful landings of instrument packages on the Moon.

Time magazine runs "God is Dead" story.

1967   Fuller is featured in cover story of *Saturday Review*.

Fuller's Montreal Expo '67 Dome draws 5.3 million people in six months, setting a World's Fair attendance record.

Fuller is awarded the following: Appreciation Award by Mensa; Appreciation Award by Young President's Organization, Western Area Conference; Alumnus Honoris Causa by Asbestos-Danville-Shipton High School; Centennial Award by Boston Society of Architects; Honorary Citizen by Hot Springs, Arkansas; Leadership Award by Alice Lloyd College; Order of Lincoln Medal by Lincoln Academy of Illinois; Fifth Thanksgiving Award by Clarke College; Golden Eagle Award for Excellence in Film by the Council on International Non-theatrical Events.

Fuller is elected to honorary membership in Alpha Chapter of Phi Beta Kappa (Harvard University) on the occasion of Fuller's class of 1917 fifty-year reunion.

Fuller is commissioned by United States Department of Housing and Urban Development to complete design drawings, economic analyses, and model of tetrahedronal floating city.

Fuller is granted patent for "star tensegrity."

Fuller delivers ninety public lectures, including the centennial address at American University, Beirut, Lebanon; keynote address to Austrian Architects Association; World Congress of Architectural Students, Barcelona, Spain; Fiftieth anniversary address, American Planners Association; First World Congress of Engineers and Architects, Tel Aviv, Israel.

Fuller receives honorary doctor of engineering, Clarkson College of Technology, bringing his total number of honorary doctorates to thirteen.

Fuller differentiates between human brain and mind while speaking as Harvey Cushing Orator to annual congress of two thousand members of American Association of Neuro-Surgeons.

Arab–Israeli Six-Day War is fought.

Human bone fragments 2.5 million years old are discovered in Kenya.

First human heart transplant is performed.

1968   Fuller elected to National Academy of Design and World Academy of Arts and Sciences.

Fuller appointed as Distinguished University Professor at Southern Illinois University, one of only three so honored in the ninety-nine year history of that institution.

Fuller is awarded the following: British Royal Gold Medal of Architecture by Her Majesty the Queen on the recommendation of the Royal Institute of British Architects; Award for Excellence by the American Institute of Steel Construction (for Expo '67 Dome); Boss of the Year by the National Secretaries Association; Citation by Buffalo Western New York Chapter of AIA; First Architectural Design Award by the American Institute of Architects (for Expo '67 Dome); Gold Medal (Architecture) by the National Institute of Arts and Letters; President's Cabinet Award by the University of Detroit; Fellowship Award by the Building Research Institute of the National Academy of Science.

Fuller receives honorary doctorates from Dartmouth College, University of Rhode Island, Ripon College, and New England College.

Fuller's restored Dymaxion car number two is displayed at the Museum of Modern Art in New York.

Richard Nixon is elected president.

United States first orbits men around the moon.

Direct airline service between the United States and the USSR is initiated.

Martin Luther King, Jr., and Robert Kennedy are assassinated.

Police and antiwar demonstrators clash at Democratic National Convention in Chicago.

The Soviet Union invades Czechoslovakia.

1969 Fuller leads the first public World Game workshop (June 12–July 31).

Fuller testifies on World Game before United States Senate Subcommittee on Intergovernmental Relations.

Fuller delivers Jawaharlal Nehru Memorial Lecture on "Planetary Planning" in New Delhi, India.

Fuller publishes *Operating Manual for Spaceship Earth* and *Utopia or Oblivion: The Prospects for Humanity*.

Fuller is awarded the following: Humanist of the Year Citation by the American Association of Humanists; Citation of Merit by the United States Department of Housing and Urban Development; Master Designer, Product Engineering, McGraw-Hill Award; Diploma of Achievement by Dean Institute of Technology.

Fuller is appointed as Hoyt Fellow at Yale University.

Fuller receives honorary degrees from University of Wisconsin, Boston College, and Bates College.

First humans land on moon. This landing is viewed live by 600,000 people in forty-nine countries.

First remote sensing satellite is launched. It senses global temperature patterns and revolutionizes weather forecasting.

The Beatles become a global musical success.

Woodstock Music and Art Festival draws 300,000 young people.

DeGaulle resigns as leader of France.

Strategic Arms Limitation Talks (SALT) begin.

Integrated circuits, fiber optics, microwaves, and fluidics all emerge as new technologies.

1970　Fuller publishes *I Seem to Be a Verb.*

Fuller is installed as Master Architect for Life by the National Chapter of Alpha Rho Chi Architectural Fraternity.

Fuller is awarded the following: Award of Merit by the Cliff Dwellers; Eighth Lively Arts Award for Architecture by A.W.N.Y.; Gold Medal by California Polytechnic College; Gold Medal by the American Institute of Architects.

Fuller is visiting professor at International University of Art, Venice and Florence, Italy.

Fuller makes eighty-six major public lectures, which include the following: Marshall McLuhan Executive Seminar, Bahamas; Dartmouth College Bicentennial; Oxford University, England; International Association of Machinists, Washington, DC; Plenary Session of the United Nations; General Assembly of the World Society for Ekistics, Athens, Greece; Association for Human Resources, Boston; Architects and Engineers Conference on the Planning of Jerusalem, Jerusalem, Israel.

Fuller receives honorary degrees from Minneapolis School of Art, Park College, Brandeis University, Columbia College, and Wilberforce University.

Fuller receives honorary ''donship'' as fellow of St. Peter's College, Oxford University.

Detroit's oldest architectural firm endows the Buckminster Fuller Chair at the School of Architecture, University of Detroit.

First ''Earth Day'' celebrated around the World.

First commercial flight of Boeing 747 jumbo jet occurs. It carries four hundred passengers and is constructed from over 500,000 precision machined parts.

One computer exists for every four thousand people in the United States (twice the ratio of any other country).

109 million telephones in the United States carry 141 billion calls per year.

231 million televisions and 620 million radios are in use around the World.

United States voting age is lowered to eighteen.

National Guardsmen kill four students at Kent State University during protest of United States invasion of Cambodia.

1971 Fuller presents his proposal for "Old Man River City" project to the East St. Louis community, and it is enthusiastically accepted.

Fuller heads "Project Toronto" proposal for future design of that city.

Fuller is awarded the following: Hero for the Nuclear Age Citation by the Center for Teaching about Peace and War at Wayne State University; President's Cabinet Award by the University of Detroit; Salmagundi Medal and Honorable Life Membership by the Salmagundi Club; Distinguished Research Professor by the Institute for Behavioral Research Experimental College.

Fuller-designed geoscope is dedicated at Southern Illinois University.

Fuller publishes major articles in *Life, Christian Science Monitor,* and *Rolling Stone.*

Fuller receives honorary doctorate from Southeastern Massachusetts University.

NBC broadcasts a one-hour national program on Fuller entitled "Buckminster Fuller on Spaceship Earth."

World Trade Center towers in New York become the tallest building in the World. Fifteen buildings in the United States are now taller than the tallest building in 1915.

Governor Nelson Rockefeller orders riot at Attica Prison in New York quelled, resulting in the death of forty-two people.

1972 Fuller becomes "World Fellow in Residence" at a consortium of Philadelphia area institutions (University of Pennsylvania; Bryn Mawr, Haverford, and Swarthmore Colleges; and the University City Science Center).

Fuller publishes *Intuition* and *Buckminster Fuller to Children of Earth.*

Fuller receives honorary doctorates from Grinnell College, University of Maine (Orono), and Emerson College.

Fuller is awarded Founders Medal by Austin College and Outstanding Lecturer and Author Award by the City of Anchorage, Alaska.

Fuller is consultant to Design Science Institute and adviser to Earth Metabolic Design.

Fuller delivers over 120 major lectures, including the following: University of Washington (John Danz Lecturer); University of California at Santa Barbara ("Future of Man" series); Massachusetts Institute of Technology (Lecture Series on World Peace); National Association of Student Personnel Administrators Annual Conference; University of Colorado Environmental Action Conference; University of Minnesota Earth Week '72; American Philosophical Society Annual Conference; IBM's Watson Research Center in New York; Unitarian Universalist Church of Akron (one-hundredth anniversary); University of Pennsylvania, Fine Arts; Monsanto Co. International Division, Executive Session.

Playboy publishes an extensive interview with Fuller.

Fortieth anniversary issue of *Architectural Forum* is dedicated to Fuller.

England's *Architectural Design* magazine devotes an entire issue to "Buckminster Fuller Retrospective."

In "Operation Deep Freeze," the United States installs an entire research station of several buildings beneath a large stainless steel and aluminum geodesic dome at the exact South Pole.

President Nixon visits China and Russia.

Black September terrorists assassinate Israeli Olympic athletes at Munich, Germany.

1973   Fuller establishes his publications and research office in Philadelphia at University City Center as part of being "World Fellow in Residence" there.

Fuller is granted patents on "floating breakwater" and "tensegrity dome."

Fuller completes design project for airports of New Delhi, Bombay, and Madras, India, in conjunction with Shoji Sadao.

Fuller delivers 124 major public lectures, including the following: keynote address at the Milton S. Eisenhower Symposium at Johns Hopkins University; University of Oregon, Eugene; National Conference on Managing the Environment, Washington, DC; Hiroshima Chamber of Commerce and Industry, Japan; University of Southern California (keynote address, World Man/World Environment Symposium); Association of University Architects; National Center for State Courts; Milton Academy (175th anniversary), Milton, Massachusetts; International Meditation Society and Maharishi International University (World Plan Week-USA), St. Louis, Missouri; Harvard Law School Forum.

Fuller receives the following awards: Award of Merit from the Philadelphia Art Alliance; Citation of Honorary Citizenship from the City of Philadelphia; First Annual DVM Award from the Delaware Valley Mensa Association; Honorary Citizenship from Saskatchewan, Canada; Brockington Visitor from Queens University, Canada.

Fuller receives honorary doctorates from Nasson College, Rensselaer Polytechnic Institute, and Beaver College.

The Fullers move to Philadelphia.

Chicago Museum of Science and Industry creates an exhibition of Fuller artifacts which travels for two years to Minneapolis Institute of Art, Ontario Science Center, Franklin Institute in Philadelphia, Bronfman Center in Montreal, California Museum of Science and Industry in Los Angeles, Des Moines Center of Science and Industry, and East St. Louis Senior High School.

Japan constructs two weather geodesic radomes atop Mount Fuji and issues memorial stamp, "Pearl in the Crown of Fuji-San."

Watergate scandal; congressional hearings begin.

Energy crisis and fuel shortages develop in United States.

United States launches orbiting space station.

1974   Fuller makes his thirty-seventh complete circuit of Earth as he delivers 150 major public addresses including the following: centennial speaker, Notre Dame University; Prentiss M. Brown Visiting Lectureship, Albion College; Reynolds

Lecturer, Davidson College; Symposium on the Science of Creative Intelligence, Tulane University; Annual Meeting of American Academy of Political and Social Science; "Perspectives on Man" Symposium, University of Houston; American Psychiatric Association Annual Meeting; Edinburgh Festival, Edinburgh, Scotland; Mensa International Committee, London, England; University of Nottingham, England.

Fuller receives honorary doctorates from University of Notre Dame, St. Joseph's College, Pratt Institute, University of Pennsylvania, and McGill University.

Fuller is granted a New York State architect's license.

Fuller is awarded the following: Annual Award by the Boston Montessori Schoolhouse; Appreciation Medal by Harvard Business School Club of New York; Educator of the Year by Montessori Institute of America; Appreciation Award by Wabash Valley Association Inc.; Citation by Drexel University (Nesbitt College); Commendation by California Museum of Science and Industry.

Fuller becomes Consultant to Team 3 Architects International, Penang, Malaysia.

Astronauts spend eighty-five days in space aboard United States space station *Skylab*.

Nixon becomes first American president to resign, and Gerald Ford becomes president.

Unemployment in United States rises to 6.5 percent.

India becomes the sixth country to explode a nuclear device.

Americans are given the right to own gold bullion after a forty-year prohibition, and the United States Treasury auctions off most of its gold, weakening the World gold market.

Oil prices quadruple.

1975   Fuller publishes *Synergetics,* the result of a half century of work on the geometry of Nature's coordinate system.

Fuller is appointed professor emeritus at Southern Illinois University and University of Pennsylvania.

Fuller becomes international president of the World Society for Ekistics.

Fuller becomes a member of Advisory Committee for Windworks, Mukwonago, Wisconsin.

Fuller receives patent for "Non-Symmetrical, Tension-Integrity Structures."

Fuller becomes Tutor in Design Science at International College, Los Angeles.

Fuller is awarded the following: Planetary Citizens Award by the United Nations; Distinguished Service Award Medal by Beech Aircraft; Certificate of Honor by the City and County of San Francisco; Honorary Diploma by Stowe School; Marshal Visitor Award by International House; Reward for Peace by World Unification Movement.

Fuller is invested as a fellow in the American Institute of Architects.

Fuller receives honorary degree from Hobart and William Smith College.

Medard Gabel and the World Game Workshop publish *Energy, Earth and Everyone* with introduction by Fuller.

The largest and richest American companies have become international operations.

Cracks found in Illinois atomic reactor cause shutdown of twenty-three similar reactors.

American and Russian astronauts meet in orbit during cooperative space linkup.

The marketing of home videorecorders begins.

Major figures in Watergate scandal found guilty.

Unemployment in United States rises to 9 percent.

New York City narrowly avoids bankruptcy as $3 billion in debt comes due.

1976   Fuller conceives and designs synergetics exhibit for opening of Smithsonian/Cooper-Hewitt Museum of Design.

Fuller participates in convention to draft and sign a Declaration of Principles and Rights for American Children.

Fuller proposes constructing Geoscope using steel frame of Montreal Expo '67 Dome, which remains structurally undamaged after fire burns off vinyl skin.

Fuller publishes *And It Came to Pass—Not to Stay*.

Fuller is First Distinguished Lecturer, College of Engineering, Villanova University.

Fuller receives the following awards: Messing Award from St. Louis University Library Association; Development of

Consciousness Award from the International Meditation Society; America's Tricentennial Resolution for Utopia over Oblivion and founding of the Buckminster Fuller Matrix Society at Von Braun Civic Center, Huntsville, Alabama; award for sixteen years of service from Southern Illinois University, Carbondale; Gold Medal of Honor from the National Arts Club; International New Thought Alliance Humanitarian Award; Recognition Award from Oakland Community College; Bicentennial Award from the Golden Slipper Club, Philadelphia.

Fuller completes work on the World's first tetrahedronal book, *Tetrascroll,* which is conceived, illustrated, and written by him and published in limited edition.

North and South Vietnam officially reunite, with Hanoi as capital.

First synthetic gene constructed.

Twenty-six people die from "legionnaire's disease" at American Legion Convention in Philadelphia.

Jimmy Carter is elected president.

700,000 people die in Chinese Earthquake, which is worst disaster in recorded history.

1977   Fuller designs and develops two prototype geodesic domes, "Pinecone Dome" and "Fly's Eye Dome."

Fuller's limited-edition book *Tetrascroll* is exhibited at the Museum of Modern Art in New York.

Fuller delivers over one hundred major lectures, including the following: Distinguished Half Century Lecturer, University of Houston; Third Annual Ezra Pound Lecture, University of Idaho; Grauer Lecture, University of British Columbia; Creative Presentations Seminar with Werner Erhard; American Management Association, New York; Habitat Conference, Southwest Texas State University; American Association of Humanistic Psychology; UNESCO Round Table Conference on "Challenge of the Year 2000," Paris, France; United Nations University; United States Air Force Academy.

Fuller is honored at "Buckminster Fuller Day," proclaimed by governor of Massachusetts and mayors of Boston and Cambridge.

Fuller receives the following awards: First Annual Heald Award from Illinois Institute of Technology; Eleanor Roosevelt Humanitarian Award from League for the Hard of Hearing; Stevens Honor Award from Stevens Institute of Technology; "The Golden Plate Award" from the American Academy of Achievement; Planetary Citizens Award from the United Nations; Engineering and Science Award from Drexel University; Eli Whitney Award from Connecticut Patent Law Association; Honorary Citizen from Park Forest, Illinois; Distinguished Speaker from the United States Air Force Academy; Achievers Award in Architecture from the Samsonite Corporation.

Fuller travels on lecture tour of the Far East sponsored by the United States Information Agency and the State Department.

Fuller continues as Philadelphia's "World Fellow in Residence," University Professor Emeritus at Southern Illinois University and University of Pennsylvania, and consultant to Architects 3 International and Tutor in Design Science at International University.

First oil flows through the Alaska pipeline.

Twenty-five-hour blackout leads to extensive looting and vandalism in New York City.

Department of Energy added to president's Cabinet.

President Carter pardons Vietnam war draft evaders as one of his first acts in office.

United States Census Bureau announces that, for the first time, a majority of Americans live in the Sunbelt states of the South and the West.

1978   Fuller's *Synergetics Folio* is published.

Fuller makes one of his numerous presentations before a United States Senate Committee and describes how satellites could be used for taking daily inventories of everything from World resources to public opinion.

Fuller appears in Honda advertisement.

Fuller receives citation from the Regents of the University of Minnesota and the Medal of Merit from the Lotus Club.

Fuller continues his numerous speaking engagements, which include the following: World Congress of the New Age in Florence, Italy; "Wholistic Health" Conference in Washington, DC; Aspen Healing Arts Conference; keynote speaker at Annual Conference of the Society for College and University Planning; keynote speaker for Colorado Energyfest; Peter Goldmark Memorial Lecture at Electromedia Conference in Copenhagen, Denmark; Vikram Sarabai Memorial Lecture at Nehru Foundation in India; Guggenheim Lecture, National Air and Space Museum.

Fuller is scholar in residence at the University of Massachusetts and attends an informal gathering of World leaders at D'Arros in the Seychelles.

Fuller is honored at "Buckminster Fuller Day," jointly proclaimed by governor of Minnesota and mayors of Minneapolis and St. Paul.

Fuller becomes senior partner in the New York Architectural firm of Fuller & Sadao PC.

World population growth rate begins to decline, confirming Fuller's earlier prognostications.

Scientists propose a "global electrical circuit" which would double the amount of electricity available for practical use on Earth. This strategy was first publicly proposed by Fuller in 1969 in an address to a World Game workshop.

Largest snowstorm in the history of northeastern United States; $1.4-million roof of Hartford Coliseum collapses hours after audience of five thousand leaves under weight of only 4.8 inches of wet snow. At that time Fuller's geodesic domes are handling much greater Arctic snow loads of over three hundred pounds per square foot.

Last case of smallpox in the World is eradicated. This is the first instance of complete disease eradication in recorded history.

World's first test-tube baby is born in England.

Two Soviet cosmonauts break record by spending 140 days orbiting the Earth in space station.

President Carter mediates peace talks between President Sadat of Egypt and Prime Minister Begin of Israel.

Pope Paul VI dies, and newly elected Pope John Paul I dies after a reign of only thirty-four days. He is succeeded by Pope John Paul II, the first non-Italian Pope in 456 years.

1979   Fuller makes an extensive visit to People's Republic of China and finds people anxious to industrialize using his "Design-Science Revolution" strategy.

Fuller publishes *Synergetics 2,* which amplifies and expands on the material in *Synergetics.*

Fuller is invested into the Order of Knights of St. John of Jerusalem and Priory of King Valdemar the Great in Copenhagen, Denmark.

Fuller publishes *Buckminster Fuller on Education.*

Fuller receives the following awards: Quest Medal of St. Edward's University; First Annual Humanities Award from the World Symposium on Humanity; American-Israel Arts, Sciences and Humanities Award from the State of Israel Bonds; First Lifelong Learning Award from Metropolitan State University; Raymond A. Dart Award ("The Steel Brain") from the United Steelworkers Association; John Scott Award from the City of Philadelphia.

Fuller continues intensive speaking engagements, including the following: keynote address at Mind Child Architecture Conference at New Jersey Institute of Technology; Flag Day Ceremony at Betsy Ross House and Radio City Music Hall; Distinguished Lecturer at the University of Calgary.

Fuller becomes chairman of the board of R. Buckminster Fuller, Sadao & Zung Architects, Inc., an Ohio architectural firm.

Fuller becomes senior partner of Buckminster Fuller Associates, London, England.

Fuller is awarded honorary degrees by International College, Southern Illinois University, and Alaska Pacific University.

Medard Gabel and The World Game Laboratory publish *Ho-Ping: Food for Everyone.*

Gossamer Albatross is first human-powered vehicle to fly across English Channel. This is possible because scientific technology has developed materials which accomplish much more while using fewer resources.

Radiation accident and near meltdown at Three-Mile Island nuclear power plant leads to reevaluation of nuclear power program.

Egypt and Israel sign peace treaty.

Chrysler Corporation asks federal government for $4 billion in aid to avoid bankruptcy.

Iranian hostage crisis begins.

1980   Fuller publishes *Critical Path.*

Fuller patents "tensegrity truss."

Fuller is honored with exhibition and gold medal from American Academy of Arts and Letters.

Fuller travels to Brazil and views implementation of industrialization strategies he first described in 1942.

Fuller is awarded honorary degrees by Roosevelt University, Georgian Court College, and Newport University.

Fuller delivers over ninety major lectures, including the following: Sam Rayburn Public Affairs Symposium, East Texas State University; Distinguished Visitors Series, Columbus, Indiana; International Dome Symposium; American Teachers of Mathematics Conference; Symposium on Hunger sponsored by The Hunger Project; Windstar Foundation, Snowmass, Colorado; Digital Equipment Corporation; Chamber of Commerce, Budapest, Hungary; Science of Mind Institute Seminar; Belgium Television Communications Conference, Liège.

Fuller is appointed to presidential commission to further develop the "Global 2000 Report" outlining the crisis in energy and the global environment. He is also appointed to the Congressional Committee on the Future.

Fuller is honored with recognition days by the State of Illinois and the City of Buffalo, New York.

Fuller receives the following awards: Academician from the American Academy and the Institute of Arts and Letters; Special Award from Erie County, New York; First Dean's Award from the State University of New York at Buffalo Architectural School; Honorary Royal Designer for Industry from the Royal Society of Arts (England); adjunct professor of humanities from Texas Wesleyan University.

The Fullers move to Pacific Palisades, California.

City of Los Angeles leases a fifty-foot-diameter Fly's-Eye Dome as the theme building for its bicentennial celebration.

Grip-Kitrick edition of Fuller's Dymaxion Map issued. It is the largest, most accurate whole Earth map in human history.

Reagan is elected president of United States.

Mount St. Helen's volcano erupts.

United States and fifty other countries boycott Moscow Olympic Games, primarily in protest of Russia's invasion of Afghanistan.

$1.5-billion federally guaranteed loan granted to Chrysler Corporation.

1981 Fuller becomes chairman of Fuller-Patterson Corporation and Buckminster Fuller Research Park.

Fuller continues as World Fellow in Residence, University City Science Center, Philadelphia; university professor emeritus, Southern Illinois University and University of Pennsylvania; senior partner, Fuller & Sadao PC; chairman of the board, R. Buckminster Fuller, Sadao & Zung Architects, Inc; senior partner, Buckminster Fuller Associates; and consultant to Team 3 Architects International.

Fuller is awarded honorary degree by Texas Wesleyan University, bringing his total number of honorary degrees to forty-seven.

Fuller is honored at Recognition Day in Austin, Texas, and with "Resolution of Honor" from the State of California.

Fuller is inducted into the Housing Hall of Fame.

Fuller is awarded Nesbitt College Honor Citation, Drexel University.

American hostages in Iran are freed minutes after Reagan is inaugurated as president.

First reusable space shuttle is launched into orbit around the Earth and completes a successful two-day mission.

Sandra Day O'Connor becomes first woman Associate Justice of the United States Supreme Court.

1982 Fuller designs and develops tensional Dymaxion hanging bookshelf and deresonated tensegrity dome.

Fuller publishes *Grunch of Giants*.

Fuller designs and supervises construction of Dymaxion "Big Map," a basketball-court-sized Dymaxion Map on which global resources and policies can be graphically displayed. He displays this map and its capabilities to members of the United States Congress.

Fuller receives the following awards: President's Fellow Award

from Rhode Island School of Design; Distinguished Service Award from the World Future Society; Regents Lecturer From the University of California at Los Angeles.

Fuller delivers over seventy major lectures, including the following: World Game Workshop; Touche Ross & Com-
pany; New Alchemy Institute; American Society for Training and Development; Harvard University; Wellness Conference, Institute for the Advancement of Human Behavior; American Society of Interior Designers; World Affairs
Conference; Insights into the Future Symposium; Smithso-
nian Institute; University of Pennsylvania.

Fuller has the star Tau Persei located on Charles C. Gates
Planetarium Dome dedicated to him.

Fuller is inducted into the Engineering and Science Hall of
Fame.

Fuller publishes *Inventions* and *Humans in Universe* (with Dr.
Anwar Dil).

Fuller is awarded the Medal of Freedom by President Reagan,
the highest civilian award given by the United States
government.

Fuller patents "Hanging Storage Shelf Unit."

Aluminum 415-foot-diameter dome is erected to house Howard
Hughs's *Spruce Goose* airplane in Long Beach, California.
This becomes the largest clear-span dome in the World.

Walt Disney World erects a 165-foot-diameter steel-and-alumi-
num geodesic sphere using the theme "Spaceship Earth"
at its EPCOT Center in Orlando, Florida.

AT&T agrees to break up into smaller companies in response to
a thirteen-year lawsuit.

The United States unemployment rate reaches 10.8 percent,
which is the highest rate since the Great Depression.

First permanent artificial heart is installed in a human.

1983   Fuller continues as World Fellow in Residence, University City
Science Center, Philadelphia: university professor emeritus, Southern Illinois University and University of Pennsylvania; senior partner, Fuller & Sadao PC; chairman of
the board, R. Buckminster Fuller, Sadao & Zung Architects Inc.; senior partner, Buckminster Fuller Associates;
consultant to Team 3 International Architects; and chair-

man, Fuller-Patterson Corporation and Buckminster Fuller Research and Development Park.

Fuller is honored in "Integrity Day" declarations by Cities of Los Angeles, Long Beach, and Santa Cruz; Orange and Tulare Counties; and the State of California. The declarations honor his last series of public-speaking engagements, called "Integrity Days," which were founded on his experience that personal integrity was ultimately the most important issue with which any individual has to deal.

R. Buckminster Fuller dies on July 1, 1983, in Los Angeles after suffering a massive heart attack while visiting his comatose wife, Anne, at Good Samaritan Hospital. He has been evaluated in excellent health just three weeks earlier by a doctor in Boston. Anne never recovers from the coma and dies thirty-six hours later.

Source Notes

Throughout my years of intently studying Buckminster Fuller, I have learned that his seemingly complex and diverse message, like Nature itself, was actually quite simple. Although he was constantly updating his thoughts and ideas on almost every subject, the essence of that public message remained constant from his first published writings of 1928 through his final public lecture in June of 1983.

Fuller generally recounted similar communications at each of the hundreds of lectures and conversations he presented from the late 1940s through his death. Those same few messages are also evident in the volumes of writing he produced.

Fuller served humanity by making his thoughts and ideas as publicly available as those of any public person, without relying on the auspices of a bureaucracy to diseminate that information. Hence, the following notes could reference literally thousands of sources. I do not claim that any of these references presents the "best" or only source of a particular item of information. Nor do I claim that these references represent the only time I encountered an item. Rather, the following list presents for each piece of information one source from the thousands of hours of audiotape and the multitude of pages of written material I utilized to synthesize this book.

CHAPTER 1

1. R. Buckminster Fuller and Robert Marks, *The Dymaxion World of Buckminster Fuller* (Garden City, N.Y.: Anchor Books, 1973; originally published by Southern Illinois Univer-

sity Press, 1960), p. 11.

2. Alden Hatch, *Buckminster Fuller at Home in the Universe* (New York: Crown Publishers, Inc., 1974), p. 8.

3. Sidney Rosen, *Wizard of the Dome* (Boston: Little Brown, 1969), p. 9.

4. R. Buckminster Fuller and Robert Marks, *The Dymaxion World of Buckminster Fuller* (Garden City, NY: Anchor Books, 1973; originally published by Southern Illinois University Press, 1960), p. 11.

5. Hugh Kenner, *Bucky: A Guided Tour of Buckminster Fuller* (New York: William Morrow & Company, Inc., 1973), p. 73.

6. R. Buckminster Fuller, *No More Secondhand God and Other Writings* (Carbondale: Southern Illinois University Press, 1963), p. 4.

7. Alden Hatch, *Buckminster Fuller at Home in the Universe* (New York: Crown Publishers, Inc., 1974), p. 7.

8. Ibid.

9. Robert Snyder, ed., *Buckminster Fuller: Autobiographical Monologue Scenario* (New York: St. Martin's Press, 1980), p. 11.

10. Lecture Series by R. Buckminster Fuller, University of Oregon, Eugene, July 1–12, 1962.

11. World Game Lecture Series by R. Buckminster Fuller, New York University, New York, 14–21 July 1979.

12. Lecture by R. Buckminster Fuller, *Buckminster Fuller's Personal Odyssey,* New Dimensions Foundation, San Francisco.

13. Lecture Series by R. Buckminster Fuller, *The Future of Business,* Lake Tahoe, CA, 17–21 August 1981.

14. "The Dymaxion American," *Time Magazine* (10 January 1984), p. 47.

15. Alden Hatch, *Buckminster Fuller at Home in the Universe* (New York: Crown Publishers, Inc., 1974), p. 17.

16. Ibid., p. 11.

17. Lecture Series by R. Buckminster Fuller, *The Future of Business,* Lake Tahoe, CA, 17–21 August 1981.

18. Lecture Series by R. Buckminster Fuller, University of Oregon, Eugene, 1–12 July 1962.

19. Ibid.

20. Nathan Aeseng, *More with Less: The Future World of Buckminster Fuller* (Minneapolis: Lerner Publications, 1986), p. 15.

21. Hugh Kenner, *Bucky: A Guided Tour of Buckminster Fuller* (New York: William Morrow & Company, Inc., 1973), p. 57.

22. Ibid.

23. Robert Snyder, ed., *Buckminster Fuller: Autobiographical Monologue Scenario* (New York: St. Martin's Press, 1980), p. 7.

24. Ibid.

25. Hugh Kenner, *Bucky: A Guided Tour of Buckminster Fuller* (New York: William Morrow & Company, Inc., 1973), p. 52.

26. Lecture Series by R. Buckminster Fuller, *The Future of Business,* Lake Tahoe, CA, 17–21 August 1981.

27. R. Buckminster Fuller, *Critical Path* (New York: St. Martin's Press, 1981), p. 134.

28. Lecture by R. Buckminster Fuller, *Buckminster Fuller's Personal Odyssey,* New Dimensions Foundation, San Francisco.

29. Hugh Kenner, *Bucky: A Guided Tour of Buckminster Fuller* (New York: William Morrow & Company, Inc., 1973), p. 68.

30. John McHale, ed., "Richard Buckminster Fuller," *Architectural Design* (July 1961), p. 293.

31. R. Buckminster Fuller, *Ideas and Integrities* (New York: Prentice-Hall, Inc., 1963), p. 19.

32. Ibid.

33. Athena V. Lord, *Pilot for Spaceship Earth* (New York: Macmillan Publishing Co., Inc., 1978), p. 3.

34. John McHale, ed., "Richard Buckminster Fuller," *Architectural Design* (July 1961), p. 17.

35. Robert Snyder, ed., *Autobiographical Monologue Scenario* (New York: St. Martin's Press, 1980), p. 17.

36. World Game Lecture Series by R. Buckminster Fuller, Washington, DC, 16–18 July 1982.

37. Lecture Series by R. Buckminster Fuller, University of Oregon, Eugene, 1–12 July 1962.

38. World Game Lecture Series by R. Buckminster Fuller, New York University, New York, 14–21 July 1979.

39. Ibid.

40. Hugh Kenner, *Bucky: A Guided Tour of Buckminster Fuller* (New York: William Morrow & Company, Inc., 1973), p. 18.

41. Athena V. Lord, *Pilot for Spaceship Earth* (New York: Macmillan Publishing Co., Inc., 1978), p. 12.

42. Ibid., p. 13.

43. Alden Hatch, *Buckminster Fuller at Home in the Universe* (New York: Crown Publishers, Inc., 1974), p. 24.

44. Lecture Series by R. Buckminster Fuller, University of Oregon, Eugene, 1–12 July 1962.

45. Robert Snyder, ed., *Buckminster Fuller: Autobiographical Monologue Scenario* (New York: St. Martin's Press, 1980), p. 22.

46. Ibid.

47. Alden Hatch, *Buckminster Fuller at Home in the Universe* (New York: Crown Publishers, Inc., 1974), p. 32.

48. Sidney Rosen, *Wizard of the Dome* (Boston: Little Brown, 1969), p. 7.

49. Calvin Tomkins, "Profiles, In the Outlaw Area," *The New Yorker* (6 January 1966), p. 52.

50. Amy Edmondson, *A Fuller Explanation: The Synergetic Geometry of R. Buckminster Fuller* (Boston: Birkhauser, 1987), p. 6.

51. Calvin Tomkins, "Profiles, in the Outlaw Area," *The New Yorker* (6 January 1966), p. 53.

52. Lecture Series by R. Buckminster Fuller, University of California, Santa Barbara, December 1967.

53. Ibid.

54. Lecture Series by R. Buckminster Fuller, *The Future of Business*, Lake Tahoe, CA, 17–21 August 1981.

55. Lecture Series, R. Buckminster Fuller, University of California, Santa Barbara, December 1967.

56. Alden Hatch, *Buckminster Fuller at Home in the Universe* (New York: Crown Publishers, Inc., 1974), p. 28.

57. Ibid.

58. Sidney Rosen, *Wizard of the Dome* (Boston: Little Brown, 1969), p. 6.

59. R. Buckminster Fuller, *Critical Path* (New York: St. Martin's Press, 1981), p. 129.

60. Hugh Kenner, *Bucky: A Guided Tour of Buckminster Fuller* (New York: William Morrow & Company, Inc., 1973), p. 75.

61. R. Buckminster Fuller, *Critical Path* (New York: St. Martin's Press, 1981), p. 130.

62. Alden Hatch, *Buckminster Fuller at Home in the Universe* (New York: Crown Publishers, Inc., 1974), p. 30.

63. Hugh Kenner, *Bucky: A Guided Tour of Buckminster Fuller* (New York: William Morrow & Company, Inc., 1973), p. 74.

64. Alden Hatch, *Buckminster Fuller at Home in the Universe* (New York: Crown Publishers, Inc., 1974), p. 32.

65. Robert Snyder, ed., *Buckminster Fuller: Autobiographical Monologue Scenario* (New York: St. Martin's Press, 1980), p. 23.

66. Lecture Series by R. Buckminster Fuller, University of Oregon, Eugene, 1–12 July 1962.

67. Lecture by R. Buckminster Fuller, Burklyn Business School, Kirkwood Meadows, CA, 1 August 1982.

68. Robert Snyder, ed., *Buckminster Fuller: Autobiographical Monologue Scenario* (New York: St. Martin's Press, 1980), p. 24.

69. Ibid.

70. Lecture Series by R. Buckminster Fuller, *The Future of Business*, Lake Tahoe, CA, 17–21 August 1981.

71. Alden Hatch, *Buckminster Fuller at Home in the Universe* (New York: Crown Publishers, Inc., 1974), p. 34.

72. Ibid.

73. Lecture, R. Buckminster Fuller, *Vision 65*, World Conference on Communication Arts, Southern Illinois University, Carbondale, 12 October 1965 (reprinted, R. Buckminster Fuller, *Utopia or Oblivion*, Chapter 4, Bantam, 1969).

74. Calvin Tomkins, "Profiles, in the Outlaw Area," *The New Yorker* (6 January 1966), p. 53.

75. Athena V. Lord, *Pilot for Spaceship Earth* (New York: Macmillan Publishing Co., Inc., 1978), p. 21.

76. Lecture Series by R. Buckminster Fuller, University of Oregon, Eugene, 1–12 July 1962.

77. Athena V. Lord, *Pilot for Spaceship Earth* (New York: Macmillan Publishing Co., Inc., 1978), p. 21.

78. Ibid., p. 22.

79. John McHale, ed., "Richard Buckminster Fuller," *Architectural Design* (July 1961), p. 291.

80. Ibid.

81. Alden Hatch, *Buckminster Fuller at Home in the Universe* (New York: Crown Publishers, Inc., 1974), p. 35.

82. John McHale, ed., "Richard Buckminster Fuller," *Architectural Design* (July 1961), p. 292.

83. Ibid., p. 293.

84. Athena V. Lord, *Pilot for Spaceship Earth* (New York: Macmillan Publishing Co., Inc., 1978), p. 22.

85. Calvin Tompkins, "Profiles, in the Outlaw Area," *The New Yorker* (6 January 1966), p. 52.

86. Lecture Series by R. Buckminster Fuller, *The Future of Business*, Lake Tahoe, CA, 17–21 August 1981.

87. John McHale, ed., "R. Buckminster Fuller," *Architectural Design* (July 1961), p. 294.

88. Alden Hatch, *Buckminster Fuller at Home in the Universe* (New York: Crown Publishers, Inc., 1974), p. 39.

89. Lecture Series by R. Buckminster Fuller, University of Oregon, Eugene, 1–12 July 1962.

90. John McHale, ed., "R. Buckminster Fuller" *Architectural Design* (July 1961), p. 294.

CHAPTER 2

1. Lecture by R. Buckminster Fuller, *How to Make Our World Work,* New Dimensions Foundation, San Francisco.
2. Calvin Tomkins, "Profiles, in the Outlaw Area," *The New Yorker* (6 January 1966), p. 68.
3. Lecture Series by R. Buckminster Fuller, University of Oregon, Eugene, 1–12 July 1962.
4. Lecture Series, R. Buckminster Fuller, University of California, Santa Barbara, December 1967.
5. Lecture by R. Buckminster Fuller, *How to Make Our World Work,* New Dimensions Foundation, San Francisco.
6. World Game Lecture Series by R. Buckminster Fuller, New York University, New York, 14–21 July 1979.
7. World Game Lecture Series by R. Buckminster Fuller, Drexel University, Philadelphia, 1977.
8. Lecture Series by R. Buckminster Fuller, *The Future of Business,* Lake Tahoe, CA, 17–21 August 1981.
9. Ibid.
10. Ibid.
11. Lecture Series, R. Buckminster Fuller University of California, Santa Barbara, December 1967.
12. Ibid.
13. Ibid.
14. Lecture Series by R. Buckminster Fuller, University of Oregon, Eugene, 1–12 July 1962.
15. Ibid.
16. World Game Lecture Series, by R. Buckminster Fuller, Washington, DC, 16–18 July 1982.
17. Ibid.
18. World Game Lecture Series, by R. Buckminster Fuller, University of Pennsylvania, Philadelphia, 1976.
19. Lecture Series by R. Buckminster Fuller, University of Oregon, Eugene, 1–12 July 1962.
20. Ibid.
21. Athena V. Lord, *Pilot for Spaceship Earth* (New York: Macmillan Publishing Co., Inc., 1978), p. 14.
22. Lecture Series by R. Buckminster Fuller, *The Future of Business,* Lake Tahoe, CA, 17–21 August 1981.
23. Lecture Series by R. Buckminster Fuller, University of Oregon, Eugene, 1–12 July 1962.
24. Lecture by R. Buckminster Fuller, Burklyn Business School, Kirkwood Meadows, CA, 1 August 1982.
25. Lecture Series by R. Buckminster Fuller, University of Oregon, Eugene, 1–12 July 1962.
26. Lecture Series by R. Buckminster Fuller, *The Future of Business,* Lake Tahoe, CA, 17–21 August 1981.
27. World Game Lecture Series by R. Buckminster Fuller, University of Pennsylvania, Philadelphia, 1976.
28. Amy Edmondson, *A Fuller Explanation: The Synergetic Geometry of R. Buckminster Fuller* (Boston: Birkhauser, 1987), p. 30.
29. Lecture Series by R. Buckminster Fuller, University of Oregon, Eugene, 1–12 July 1962.
30. Ibid.

31. World Game Lecture Series by R. Buckminster Fuller, Drexel University, Philadelphia, 1977.

32. Lecture Series by R. Buckminster Fuller, University of Oregon, Eugene, 1–12 July 1962.

33. Lecture Series, R. Buckminster Fuller University of California, Santa Barbara, December 1967.

34. Lecture by R. Buckminster Fuller, *Education Tomorrow,* New Dimensions Foundation, San Francisco.

35. Ibid.

36. Lecture Series by R. Buckminster Fuller, University of Oregon, Eugene, 1–12 July 1962.

37. Ibid.

38. World Game Lecture Series by R. Buckminster Fuller, University of Pennsylvania, Philadelphia, 1976.

39. Ibid.

40. Ibid.

41. Ibid.

42. Lecture Series by R. Buckminster Fuller, University of Oregon, Eugene, 1–12 July 1962.

43. Lecture Series, R. Buckminster Fuller, University of California, Santa Barbara, December 1967.

44. World Game Lecture Series by R. Buckminster Fuller, University of Pennsylvania, Philadelphia, 1976.

45. Lecture by R. Buckminster Fuller, *Education Tomorrow,* New Dimensions Foundation, San Francisco.

46. Lecture Series by R. Buckminster Fuller, University of Oregon, Eugene, 1–12 July 1962.

47. Hugh Kenner, *Bucky: A Guided Tour of Buckminster Fuller* (New York: William Morrow & Company, Inc., 1973), p. 35.

48. Ibid.

49. R. Buckminster Fuller, *Synergetics* (New York: Macmillan Publishing Co., Inc., 1975), p. 442.

50. Ibid., p. 443.

51. Amy Edmondson, *A Fuller Explanation: The Synergetic Geometry of R. Buckminster Fuller* (Boston: Birkhauser, 1987), p. 29.

52. Lecture Series, R. Buckminster Fuller, University of California, Santa Barbara, December 1967.

53. Lecture Series by R. Buckminster Fuller, University of Oregon, Eugene, 1–12 July 1962.

54. Ibid.

55. Ibid.

56. R. Buckminster Fuller, *Synergetics* (New York: Macmillan Publishing Co., Inc., 1975), p. 446.

57. World Game Lecture Series by R. Buckminster Fuller, University of Pennsylvania, Philadelphia, 1976.

58. Lecture Series, R. Buckminster Fuller, University of California, Santa Barbara, December 1967.

59. World Game Lecture Series by R. Buckminster Fuller, New York University, New York, 14–21 July 1979.

60. Robert Snyder, ed., *Buckminster Fuller: Autobiographical Monologue Scenario* (New York: St. Martin's Press, 1980), p. 51.

61. Ibid.

62. R. Buckminster Fuller, *Synergetics* (New York: Macmillan Publishing Co., Inc., 1975), p. 443.

63. World Game Lecture Series by R. Buckminster Fuller, University of Pennsylvania, Philadelphia, 1976.

64. Amy Edmondson, *A Fuller Explanation: The Synergetic Geometry of R. Buckminster Fuller* (Boston: Birkhauser, 1987), p. 29.

65. Lecture Series by R. Buckminster Fuller, University of Oregon, Eugene, 1–12 July 1962.

66. Lecture Series by R. Buckminster Fuller, *The Future of Business*, Lake Tahoe, CA, 17–21 August 1981.

67. Lecture Series, R. Buckminster Fuller, University of California, Santa Barbara, December 1967.

68. World Game Lecture Series by R. Buckminster Fuller, Amherst, MA, 1978.

CHAPTER 3

1. Alden Hatch, *Buckminster Fuller at Home in the Universe* (New York: Crown Publishers, Inc., 1974), p. 39.

2. Athena V. Lord, *Pilot for Spaceship Earth* (New York: Macmillan Publishing Co., Inc., 1978), p. 26.

3. Alden Hatch, *Buckminster Fuller at Home in the Universe* (New York: Crown Publishers, Inc., 1974), p. 42.

4. Ibid., p. 41.

5. Ibid., p. 42.

6. Ibid.

7. Lecture Series by R. Buckminster Fuller, University of Oregon, Eugene, 1–12 July 1962.

8. Ibid.

9. Lecture Series, R. Buckminster Fuller, University of California, Santa Barbara, December 1967.

10. Athena V. Lord, *Pilot for Spaceship Earth* (New York: Macmillan Publishing Co., Inc., 1978), p. 25.

11. Alden Hatch, *Buckminster Fuller at Home in the Universe* (New York: Crown Publishers, Inc., 1974), p. 44.

12. Ibid., p. 46.

13. Lecture Series by R. Buckminster Fuller, *The Future of Business*, Lake Tahoe, CA, 17–21 August 1981.

14. Lecture Series by R. Buckminster Fuller, University of Oregon, Eugene, 1–12 July 1962.

15. Lecture Series, R. Buckminster Fuller, University of California, Santa Barbara, December 1967.

16. Alden Hatch, *Buckminster Fuller at Home in the Universe* (New York: Crown Publishers, Inc., 1974), p. 47.

17. Ibid.

18. Sidney Rosen, *Wizard of the Dome* (Boston: Little Brown, 1969), p. 21.

19. Ibid.

20. Alden Hatch, *Buckminster Fuller at Home in the Universe* (New York: Crown Publishers, Inc., 1974), p. 51.

21. Ibid., p. 52.
22. Ibid., p. 53.
23. Ibid.
24. Athena V. Lord, *Pilot for Spaceship Earth* (New York: Macmillan Publishing Co., Inc., 1978), p. 29.
25. Lecture Series by R. Buckminster Fuller, University of Oregon, Eugene, 1–12 July 1962.
26. Alden Hatch, *Buckminster Fuller at Home in the Universe* (New York: Crown Publishers, Inc., 1974), p. 59.
27. Athena V. Lord, *Pilot for Spaceship Earth* (New York: Macmillan Publishing Co., Inc., 1978), p. 29.
28. Ibid., p. 30.
29. Ibid.
30. Sidney Rosen, *Wizard of the Dome* (Boston: Little, Brown, 1969), p. 22.
31. Athena V. Lord, *Pilot for Spaceship Earth* (New York: Macmillan Publishing Co., Inc., 1978), p. 31.
32. Sidney Rosen, *Wizard of the Dome* (Boston: Little, Brown, 1969), p. 23.
33. Lecture Series by R. Buckminster Fuller, University of Oregon, Eugene, 1–12 July 1962.
34. Athena V. Lord, *Pilot for Spaceship Earth* (New York: Macmillan Publishing Co., Inc., 1978), p. 31.
35. Hugh Kenner, *Bucky: A Guided Tour of Buckminster Fuller* (New York: William Morrow & Company, Inc., 1973), p. 102.
36. Ibid.
37. Lecture Series, R. Buckminster Fuller, University of California, Santa Barbara, December 1967.
38. Alden Hatch, *Buckminster Fuller at Home in the Universe* (New York: Crown Publishers, Inc., 1974), p. 62.
39. Athena V. Lord, *Pilot for Spaceship Earth* (New York: Macmillan Publishing Co., Inc., 1978), p. 32.
40. Lecture Series, R. Buckminster Fuller, University of California, Santa Barbara, December 1967.
41. World Game Lecture Series by R. Buckminster Fuller, University of Pennsylvania, Philadelphia, 1976.
42. Ibid.
43. John McHale, ed., "R. Buckminster Fuller," *Architectural Design* (July 1961), p. 298.
44. World Game Lecture Series by R. Buckminster Fuller, University of Pennsylvania, Philadelphia, 1976.
45. Ibid.
46. Ibid.
47. Ibid.
48. Lecture Series by R. Buckminster Fuller, University of Oregon, Eugene, 1–12 July 1962.
49. World Game Lecture Series by R. Buckminster Fuller, University of Pennsylvania, Philadelphia, 1976.
50. Lecture Series by R. Buckminster Fuller, *The Future of Business*, Lake Tahoe, CA, 17–21 August 1981.
51. Interview with R. Buckminster Fuller, Andrew D. Basiago, Pacific Palisades, CA, 20–21 October 1981.
52. John McHale, ed., "R. Buckminster Fuller," *Architectural Design* (July 1961), p. 291.

53. R. Buckminster Fuller, *Critical Path* (New York: St. Martin's Press, 1981), p. 62.
54. Lecture Series, R. Buckminster Fuller, University of California, Santa Barbara, December 1967.
55. John McHale, ed., "R. Buckminster Fuller," *Architectural Design* (July 1961), p. 291.
56. World Game Lecture Series by R. Buckminster Fuller, University of Pennsylvania, Philadelphia, 1976.
57. Ibid.
58. Ibid.
59. Lecture Series by R. Buckminster Fuller, University of Oregon, Eugene, 1–12 July 1962.
60. Ibid.
61. World Game Lecture Series by R. Buckminster Fuller, Amherst, MA, 1978.
62. Ibid.
63. Ibid.
64. World Game Lecture Series by R. Buckminster Fuller, New York University, New York, 14–21 July 1979.
65. Ibid.
66. Ibid.
67. Calvin Tomkins, "Profiles, in the Outlaw Area," *The New Yorker* (6 January 1966), p. 78.
68. World Game Lecture Series by R. Buckminster Fuller, New York University, New York, 14–21 July 1979.
69. Calvin Tomkins, "Profiles, in the Outlaw Area," The New Yorker (6 January 1966), p. 80.
70. Ibid.
71. Interview with R. Buckminster Fuller, Andrew D. Basiago, Pacific Palisades, CA, 20–21 October 1981.
72. Ibid.
73. Calvin Tomkins, "Profiles, in the Outlaw Area," *The New Yorker* (6 January 1966), p. 81.
74. Lecture Series by R. Buckminster Fuller, *The Future of Business*, Lake Tahoe, CA, 17–21 August 1981.
75. Ibid.
76. World Game Lecture Series by R. Buckminster Fuller, Washington, DC, 16–18 July 1982.
77. Lecture Series by R. Buckminster Fuller, University of Oregon, Eugene, 1–12 July 1962.
78. Ibid.
79. Ibid.
80. Athena V. Lord, *Pilot for Spaceship Earth* (New York: Macmillan Publishing Co., Inc., 1978), p. 35.
81. World Game Lecture Series by R. Buckminster Fuller, Washington, DC, 16–18 July 1982.
82. Ibid.

CHAPTER 4

1. Alden Hatch, *Buckminster Fuller at Home in the Universe* (New York: Crown Publishers, Inc., 1974), p. 64.
2. Ibid.
3. Lecture Series, R. Buckminster Fuller, University of California, Santa Barbara, December 1967.

4. Lecture Series by R. Buckminster Fuller, *The Future of Business,* Lake Tahoe, CA, 17–21 August 1981.
5. Alden Hatch, *Buckminster Fuller at Home in the Universe* (New York: Crown Publishers, Inc., 1974), p. 65.
6. Lecture Series by R. Buckminster Fuller, *The Future of Business,* Lake Tahoe, CA, 17–21 August 1981.
7. Athena V. Lord, *Pilot for Spaceship Earth* (New York: Macmillan Publishing Co., Inc., 1978), p. 33.
8. World Game Lecture Series by R. Buckminster Fuller, Drexel University, Philadelphia, 1977.
9. World Game Lecture Series by R. Buckminster Fuller, New York University, New York, 14–21 July 1979.
10. Ibid.
11. World Game Lecture Series by R. Buckminster Fuller, University of Pennsylvania, Philadelphia, 1976.
12. World Game Lecture Series by R. Buckminster Fuller, Washington, DC, 16–18 July 1982.
13. Ibid.
14. Ibid.
15. World Game Lecture Series by R. Buckminster Fuller, University of Pennsylvania Museum, Philadelphia, 1975.
16. Athena V. Lord, *Pilot for Spaceship Earth* (New York: Macmillan Publishing Co., Inc., 1978), p. 36.
17. Sidney Rosen, *Wizard of the Dome* (Boston: Little, Brown, 1969), p. 25.
18. Ibid.
19. Alden Hatch, *Buckminster Fuller at Home in the Universe* (New York: Crown Publishers, Inc., 1974), p. 67.
20. Ibid.
21. Athena V. Lord, *Pilot for Spaceship Earth* (New York: Macmillan Publishing Co., Inc., 1978), p. 36.
22. Ibid.
23. Alden Hatch, *Buckminster Fuller at Home in the Universe* (New York: Crown Publishers, Inc., 1974), p. 68.
24. Robert Snyder, ed., *Buckminster Fuller: Autobiographical Monologue Scenario* (New York: St. Martin's Press, 1980), p. 32.
25. Ibid., p. 33.
26. Ibid.
27. Lecture Series by R. Buckminster Fuller, University of Oregon, Eugene, 1–12 July 1962.
28. Ibid.
29. Lecture Series, R. Buckminster Fuller, University of California, Santa Barbara, December 1967.
30. Ibid.
31. Robert Snyder, ed., *Buckminster Fuller: Autobiographical Monologue Scenario* (New York: St. Martin's Press, 1980), p. 33.
32. Hugh Kenner, *Bucky: A Guided Tour of Buckminster Fuller* (New York: William Morrow & Company, Inc., 1973), p. 25.
33. Lecture by R. Buckminster Fuller, Burklyn Business School, Kirkwood Meadows, CA, 1 August 1982.

34. Hugh Kenner, *Bucky: A Guided Tour of Buckminster Fuller* (New York: William Morrow & Company, Inc., 1973), p. 25.
35. Lecture by R. Buckminster Fuller, Burklyn Business School, Kirkwood Meadows, CA, 1 August 1982.
36. Lecture Series, R. Buckminster Fuller, University of California, Santa Barbara, December 1967.
37. Alden Hatch, *Buckminster Fuller at Home in the Universe* (New York: Crown Publishers, Inc., 1974), p. 69.
38. Ibid.
39. Ibid., p. 70.
40. Lecture Series by R. Buckminster Fuller, University of Oregon, Eugene, 1–12 July 1962.
41. Ibid.
42. Lecture Series, R. Buckminster Fuller, University of California, Santa Barbara, December 1967.
43. Ibid.
44. Athena V. Lord, *Pilot for Spaceship Earth* (New York: Macmillan Publishing Co., Inc., 1978), p. 37.
45. Alden Hatch, *Buckminster Fuller at Home in the Universe* (New York: Crown Publishers, Inc., 1974), p. 72.
46. Lecture Series by R. Buckminster Fuller, University of Oregon, Eugene, 1–12 July 1962.
47. Alden Hatch, *Buckminster Fuller at Home in the Universe* (New York: Crown Publishers, Inc., 1974), p. 72.
48. Lecture Series, R. Buckminster Fuller, University of California, Santa Barbara, December 1967.
49. Athena V. Lord, *Pilot for Spaceship Earth* (New York: Macmillan Publishing Co., Inc., 1978), p. 37.
50. Ibid., p. 38.
51. Ibid., p. 40.
52. Lecture Series by R. Buckminster Fuller, *The Future of Business,* Lake Tahoe, CA, 17–21 August 1981.
53. Alden Hatch, *Buckminster Fuller at Home in the Universe* (New York: Crown Publishers, Inc., 1974), p. 76.
54. Ibid.
55. Lecture Series by R. Buckminster Fuller, *The Future of Business,* Lake Tahoe, CA, 17–21 August 1981.
56. Ibid.
57. Ibid.
58. Alden Hatch, *Buckminster Fuller at Home in the Universe* (New York: Crown Publishers, Inc., 1974), p. 78.
59. Ibid.
60. Ibid., p. 79.
61. Ibid.
62. Lecture Series, R. Buckminster Fuller, University of California, Santa Barbara, December 1967.
63. Ibid.
64. Lecture Series by R. Buckminster Fuller, University of Oregon, Eugene, 1–12 July 1962.

65. World Game Lecture Series by R. Buckminster Fuller, University of Pennsylvania, Philadelphia, 1976.

66. Ibid.

67. Sidney Rosen, *Wizard of the Dome* (Boston: Little, Brown, 1969), p. 39.

68. World Game Lecture Series by R. Buckminster Fuller, University of Pennsylvania, Philadelphia, 1976.

69. Athena V. Lord, *Pilot for Spaceship Earth* (New York: Macmillan Publishing Co., Inc., 1978), p. 40.

70. World Game Lecture Series by R. Buckminster Fuller, University of Pennsylvania, Philadelphia, 1976.

71. Robert Snyder, ed., *Buckminster Fuller: Autobiographical Monologue Scenario* (New York: St. Martin's Press, 1980), p. 31.

72. World Game Lecture Series by R. Buckminster Fuller, University of Pennsylvania, Philadelphia, 1976.

73. Ibid.

74. Lecture Series by R. Buckminster Fuller, *The Future of Business,* Lake Tahoe, CA, 17–21 August 1981.

75. Lecture Series, R. Buckminster Fuller, University of California, Santa Barbara, December 1967.

76. Athena V. Lord, *Pilot for Spaceship Earth* (New York: Macmillan Publishing Co., Inc., 1978), p. 43.

77. "The Dymaxion American," *Time Magazine* (10 January 1984), p. 48.

78. Ibid.

79. Robert Snyder, ed., *Buckminster Fuller: Autobiographical Monologue Scenario* (New York: St. Martin's Press, 1980), p. 34.

80. Alden Hatch, *Buckminster Fuller at Home in the Universe* (New York: Crown Publishers, Inc., 1974), p. 84.

81. World Game Lecture Series by R. Buckminster Fuller, University of Pennsylvania, Philadelphia, 1976.

82. Ibid.

83. Lecture Series by R. Buckminster Fuller, *The Future of Business,* Lake Tahoe, CA, 17–21 August 1981.

84. World Game Lecture Series by R. Buckminster Fuller, University of Pennsylvania, Philadelphia, 1976.

85. World Game Lecture Series by R. Buckminster Fuller, Washington, DC, 16–18 July 1982.

86. Ibid.

87. Sidney Rosen, *Wizard of the Dome* (Boston: Little, Brown, 1969), p. 40.

88. Ibid., p. 41.

89. World Game Lecture Series by R. Buckminster Fuller, University of Pennsylvania, Philadelphia, 1976.,

90. Ibid.

91. Lecture Series by R. Buckminster Fuller, University of Pennsylvania, Philadelphia, 1976.

90. Ibid.

91. Lecture Series by R. Buckminster Fuller, *The Future of Business,* Lake Tahoe, CA, 17–21 August 1981.

92. Ibid.

CHAPTER 5

1. Alden Hatch, *Buckminster Fuller at Home in the Universe* (New York: Crown Publishers, Inc., 1974), p. 87.
2. Lecture Series, R. Buckminster Fuller, University of California, Santa Barbara, December 1967.
3. Ibid.
4. Robert Snyder, ed., *Buckminster Fuller: Autobiographical Monologue Scenario* (New York: St. Martin's Press, 1980), p. 35.
5. Ibid.
6. Athena V. Lord, *Pilot for Spaceship Earth* (New York: Macmillan Publishing Co., Inc., 1978), p. 46.
7. Alden Hatch, *Buckminster Fuller at Home in the Universe* (New York: Crown Publishers, Inc., 1974), p. 92.
8. Lecture Series by R. Buckminster Fuller, *The Future of Business,* Lake Tahoe, CA, 17–21 August 1981.
9. Lecture Series by R. Buckminster Fuller, University of Oregon, Eugene, 1–12 July 1962.
10. Ibid.
11. Ibid.
12. Lecture Series by R. Buckminster Fuller, *The Future of Business,* Lake Tahoe, CA, 17–21 August 1981.
13. R. Buckminster Fuller and Robert Marks, *The Dymaxion World of Buckminster Fuller* (Garden City, NY: Anchor Books, 1973; originally published by Southern Illinois University Press, 1960), p. 11.
14. Lecture Series by R. Buckminster Fuller, *The Future of Business,* Lake Tahoe, CA, 17–21 August 1981.
15. Ibid.
16. R. Buckminster Fuller, *No More Secondhand God and Other Writings* (Carbondale: Southern Illinois University Press, 1963), p. v.
17. Alden Hatch, *Buckminster Fuller at Home in the Universe* (New York: Crown Publishers, Inc., 1974), p. 89.
18. Lecture Series by R. Buckminster Fuller, *The Future of Business,* Lake Tahoe, CA, 17–21 August 1981.
19. R. Buckminster Fuller, *Inventions: The Patented Works of R. Buckminster Fuller* (New York: St. Martin's Press, 1983), p. xv.
20. Lecture Series by R. Buckminster Fuller, *The Future of Business,* Lake Tahoe, CA, 17–21 1981.
21. R. Buckminster Fuller, *Inventions: The Patented Works of R. Buckminster Fuller* (New York: St. Martin's Press, 1983), p. xv.
22. Lecture Series by R. Buckminster Fuller, University of Oregon, Eugene, 1–12 July 1962.
23. Lecture Series by R. Buckminster Fuller, *The Future of Business,* Lake Tahoe, CA, 17–21 August 1981.
24. Lecture Series by R. Buckminster Fuller, University of Oregon, Eugene, 1–12 July 1962.
25. Lecture Series by R. Buckminster Fuller, *The Future of Business,* Lake Tahoe, CA, 17–21 August 1981.

26. Alden Hatch, *Buckminster Fuller at Home in the Universe* (New York: Crown Publishers, Inc., 1974), p. 90.

27. Lecture Series by R. Buckminster Fuller, University of Oregon, Eugene, 1–12 July 1962.

28. Alden Hatch, *Buckminster Fuller at Home in the Universe* (New York: Crown Publishers, Inc., 1974), p. 91.

29. Ibid.

30. Interview with R. Buckminster Fuller, Andrew D. Basiago, Pacific Palisades, CA, 20–21 October 1981.

31. Sidney Rosen, *Wizard of the Dome* (Boston: Little, Brown, 1969), p. 50.

32. Ibid., p. 48.

33. Reprint of talk by R. Buckminster Fuller given in October 1958 *Journal of the Royal Institute of Architects* (5 June 1958).

34. Ibid.

35. Lecture Series by R. Buckminster Fuller, University of Oregon, Eugene, 1–12 July 1962.

36. Lecture Series by R. Buckminster Fuller, *The Future of Business,* Lake Tahoe, CA, 17–21 1981.

37. Lecture Series, R. Buckminster Fuller, University of California, Santa Barbara, December 1967.

38. Lecture Series by R. Buckminster Fuller, *The Future of Business,* Lake Tahoe, CA, 17–21 August 1981.

39. Lecture Series, R. Buckminster Fuller, University of California, Santa Barbara, December 1967.

40. Lecture Series by R. Buckminster Fuller, *The Future of Business,* Lake Tahoe, CA, 17–21 August 1981.

41. World Game Lecture Series by R. Buckminster Fuller, Amherst, MA, 1978.

42. Lecture by R. Buckminster Fuller, *How to Make Our World Work,* New Dimensions Foundation, San Francisco.

43. Lecture Series by R. Buckminster Fuller, University of Oregon, Eugene, 1–12 July 1962.

44. Ibid.

45. Hugh Kenner, *Bucky: A Guided Tour of Buckminster Fuller* (New York: William Morrow & Company, Inc., 1973), p. 21.

46. Lecture Series, R. Buckminster Fuller, University of California, Santa Barbara, December 1967.

47. Alden Hatch, *Buckminster Fuller at Home in the Universe* (New York: Crown Publishers, Inc., 1974), p. 90.

48. Lecture Series, R. Buckminster Fuller, University of California, Santa Barbara, December 1967.

49. Lecture Series by R. Buckminster Fuller, University of Oregon, Eugene, 1–12 July 1962.

50. R. Buckminster Fuller and Robert Marks, *The Dymaxion World of Buckminster Fuller* (Garden City, NY: Anchor Books, 1973; originally published by Southern Illinois University Press, 1960), p. 39.

51. Lecture Series by R. Buckminster Fuller, University of Oregon, Eugene, 1–12 July 1962.

52. Lecture Series, R. Buckminster Fuller, University of California, Santa Barbara, December 1967.

53. Lecture Series by R. Buckminster Fuller, University of Oregon, Eugene, 1–12 July 1962.

54. Ibid.

55. "The Dymaxion American," *Time* (10 January 1984), p. 51.

56. Lecture Series by R. Buckminster Fuller, University of Oregon, Eugene, 1–12 July 1962.

57. Lecture Series, R. Buckminster Fuller, University of California, Santa Barbara, December 1967.

58. Ibid.

59. Robert Snyder, ed., *Buckminster Fuller: Autobiographical Monologue Scenario* (New York: St. Martin's Press, 1980), p. 47.

60. Ibid.

61. World Game Lecture Series by R. Buckminster Fuller, University of Pennsylvania Museum, Philadelphia, 1975.

62. Hugh Kenner, *Bucky: A Guided Tour of Buckminster Fuller* (New York: William Morrow & Company, Inc., 1973), p. 22.

63. Ibid.

64. Lecture Series, R. Buckminster Fuller, University of California, Santa Barbara, December 1967.

65. Lecture Series by R. Buckminster Fuller, University of Oregon, Eugene, 1–12 July 1962.

66. R. Buckminster Fuller, *No More Secondhand God and Other Writings* (Carbondale: Southern Illinois University Press, 1963), p. 103.

67. Ibid.

68. World Game Lecture Series by R. Buckminster Fuller, University of Pennsylvania Museum, Philadelphia, 1975.

69. Ibid.

70. World Game Lecture Series by R. Buckminster Fuller, New York University, New York, 14–21 July 1979.

71. Sidney Rosen, *Wizard of the Dome* (Boston: Little, Brown, 1969), p. 163.

72. World Game Lecture Series by R. Buckminster Fuller, University of Pennsylvania, Philadelphia, 1976.

73. R. Buckminster Fuller and Robert Marks, *The Dymaxion World of Buckminster Fuller* (Garden City, NY: Anchor Books, 1973; originally published by Southern Illinois University Press, 1960), p. 58.

74. World Game Lecture Series by R. Buckminster Fuller, University of Pennsylvania, Philadelphia, 1976.

75. World Game Lecture Series by R. Buckminster Fuller, Washington, DC, 16–18 July 1982.

76. Lecture Series by R. Buckminster Fuller, University of Oregon, Eugene, 1–12 July 1962.

77. Ibid.

78. Lecture Series, R. Buckminster Fuller, University of California, Santa Barbara, December 1967.

79. Robert Snyder, ed., *Buckminster Fuller: Autobiographical Monologue Scenario* (New York: St. Martin's Press, 1980), p. 47.

80. Ibid.

81. Hugh Kenner, *Bucky: A Guided Tour of Buckminster Fuller* (New York: William Morrow & Company, Inc., 1973), p. 19.

82. Lecture Series by R. Buckminster Fuller, University of Oregon, Eugene, 1–12 July 1962.

83. Hugh Kenner, *Bucky: A Guided Tour of Buckminster Fuller* (New York: William Morrow & Company, Inc., 1973), p. 23.

84. Lecture Series by R. Buckminster Fuller, University of Oregon, Eugene, 1–12 July 1962.

85. Lecture Series by R. Buckminster Fuller, *The Future of Business,* Lake Tahoe, CA, 17–21 August 1981.

86. World Game Lecture Series by R. Buckminster Fuller, University of Pennsylvania Museum, Philadelphia, 1975.

87. Lecture Series by R. Buckminster Fuller, University of Oregon, Eugene, 1–12 July 1962.
88. Ibid.
89. Ibid.
90. World Game Lecture Series by R. Buckminster Fuller, University of Pennsylvania Museum, Philadelphia, 1975.
91. Lecture Series by R. Buckminster Fuller, *The Future of Business,* Lake Tahoe, CA, 17–21 August 1981.
92. World Game Lecture Series by R. Buckminster Fuller, University of Pennsylvania Museum, Philadelphia, 1975.
93. Lecture Series by R. Buckminster Fuller, University of Oregon, Eugene, 1–12 July 1962.
94. Lecture Series by R. Buckminster Fuller, *The Future of Business,* Lake Tahoe, CA, 17–21 August 1981.
95. Ibid.
96. Lecture Series, R. Buckminster Fuller, University of California, Santa Barbara, December 1967.
97. Lecture Series by R. Buckminster Fuller, *The Future of Business,* Lake Tahoe, CA, 17–21 August 1981.
98. Lecture Series, R. Buckminster Fuller, University of California, Santa Barbara, December 1967.
99. Lecture Series by R. Buckminster Fuller, University of Oregon, Eugene, 1–12 July 1962.
100. Lecture Series, R. Buckminster Fuller, University of California, Santa Barbara, December 1967.
101. World Game Lecture Series by R. Buckminster Fuller, Drexel University, Philadelphia, 1977.
102. Lecture by R. Buckminster Fuller, *Principles of the Universe,* New Dimensions Foundation, San Francisco.
103. Ibid.
104. Ibid.
105. Ibid.
106. World Game Lecture Series by R. Buckminster Fuller, Drexel University, Philadelphia, 1977.
107. Ibid.
108. Lecture by R. Buckminster Fuller, Burklyn Business School, Kirkwood Meadows, CA, 1 August 1982.
109. Ibid.
110. Ibid.
111. Ibid.
112. World Game Lecture Series by R. Buckminster Fuller, University of Pennsylvania, Philadelphia, 1976.
113. Ibid.
114. Lecture Series by R. Buckminster Fuller, *The Future of Business,* Lake Tahoe, CA, 17–21 August 1981.
115. Ibid.
116. Ibid.
117. World Game Lecture Series by R. Buckminster Fuller, University of Pennsylvania, Philadelphia, 1976.
118. Hugh Kenner, *Bucky: A Guided Tour of Buckminster Fuller* (New York: William Morrow & Company, Inc., 1973), p. 19.

119. Ibid., p. 25.
120. Lecture Series by R. Buckminster Fuller, *The Future of Business*, Lake Tahoe, CA, 17–21 August 1981.
121. World Game Lecture Series by R. Buckminster Fuller, Washington, DC, 16–18 July 1982.
122. Lecture Series by R. Buckminster Fuller, *The Future of Business*, Lake Tahoe, CA, 17–21 August 1981.
123. World Game Lecture Series by R. Buckminster Fuller, Amherst, MA, 1978.
124. Ibid.
125. World Game Lecture Series by R. Buckminster Fuller, Washington, DC, 16–18 July 1982.
126. Lecture Series by R. Buckminster Fuller, University of Oregon, Eugene, 1–12 July 1962.
127. World Game Lecture Series by R. Buckminster Fuller, Washington, DC, 16–18 July 1982.
128. Lecture Series by R. Buckminster Fuller, University of Oregon, Eugene, 1–12 July 1962.

CHAPTER 6

1. Lecture Series by R. Buckminster Fuller, University of Oregon, Eugene, 1–12 July 1962.
2. Lecture Series, R. Buckminster Fuller, University of California, Santa Barbara, December 1967.
3. Lecture Series by R. Buckminster Fuller, University of Oregon, Eugene, 1–12 July 1962.
4. Ibid.
5. Lecture Series, R. Buckminster Fuller, University of California, Santa Barbara, December 1967.
6. World Game Lecture Series by R. Buckminster Fuller, University of Pennsylvania, Philadelphia, 1976.
7. Robert Snyder, ed., *Buckminster Fuller: Autobiographical Monologue Scenario* (New York: St. Martin's Press, 1980), p. 55.
8. World Game Lecture Series by R. Buckminster Fuller, University of Pennsylvania, Philadelphia, 1976.
9. Robert Snyder, ed., *Buckminster Fuller: Autobiographical Monologue Scenario* (New York: St. Martin's Press, 1980), p. 55.
10. Lecture Series, R. Buckminster Fuller, University of California, Santa Barbara, December 1967.
11. Ibid.
12. R. Buckminster Fuller and Robert Marks, *The Dymaxion World of Buckminster Fuller* (Garden City, NY: Anchor Books, 1973; originally published by Southern Illinois University Press, 1960), p. 22.
13. Ibid.
14. Sidney Rosen, *Wizard of the Dome* (Boston: Little, Brown, 1969), p. 42.
15. Ibid., p. 44.
16. World Game Lecture Series by R. Buckminster Fuller, University of Pennsylvania, Philadelphia, 1976.
17. R. Buckminster Fuller and Robert Marks, *The Dymaxion World of Buckminster Fuller* (Garden City, NY: Anchor Books, 1973; originally published by Southern Illinois University Press, 1960), p. 77.
18. World Game Lecture Series by R. Buckminster Fuller, University of Pennsylvania, Philadelphia, 1976.

19. World Game Lecture Series by R. Buckminster Fuller, University of Pennsylvania Museum, Philadelphia, 1975.
20. R. Buckminster Fuller and Robert Marks, *The Dymaxion World of Buckminster Fuller* (Garden City, NY: Anchor Books, 1973; originally published by Southern Illinois University Press, 1960), p. 77.
21. World Game Lecture Series by R. Buckminster Fuller, University of Pennsylvania Museum, Philadelphia, 1975.
22. Ibid.
23. Ibid.
24. Athena V. Lord, *Pilot for Spaceship Earth* (New York: Macmillan Publishing Co., Inc., 1978), p. 56.
25. Ibid.
26. Sidney Rosen, *Wizard of the Dome* (Boston: Little, Brown, 1969), p. 58.
27. Ibid.
28. World Game Lecture Series by R. Buckminster Fuller, University of Pennsylvania Museum, Philadelphia, 1975.
29. R. Buckminster Fuller and Robert Marks, *The Dymaxion World of Buckminster Fuller* (Garden City, NY: Anchor Books, 1973; originally published by Southern Illinois University Press, 1960), p. 77.
30. Ibid., p. 89.
31. Alden Hatch, *Buckminster Fuller at Home in the Universe* (New York: Crown Publishers, Inc., 1974), p. 103.
32. R. Buckminster Fuller and Robert Marks, *The Dymaxion World of Buckminster Fuller* (Garden City, NY: Anchor Books, 1973; originally published by Southern Illinois University Press, 1960), p. 89.
33. Ibid.
34. Athena V. Lord, *Pilot for Spaceship Earth* (New York: Macmillan Publishing Co., Inc., 1978), p. 66.
35. World Game Lecture Series by R. Buckminster Fuller, University of Pennsylvania, Philadelphia, 1976.
36. Alden Hatch, *Buckminster Fuller at Home in the Universe* (New York: Crown Publishers, Inc., 1974), p. 103.
37. Sidney Rosen, *Wizard of the Dome* (Boston: Little, Brown, 1969), p. 64.
38. Athena V. Lord, *Pilot for Spaceship Earth* (New York: Macmillan Publishing Co., Inc., 1978), p. 64.
39. World Game Lecture Series by R. Buckminster Fuller, Drexel University, Philadelphia, 1977.
40. R. Buckminster Fuller, *4D Timelock* (Albuquerque: Lama Foundation, reprinted, 1970).
41. Lecture Series by R. Buckminster Fuller, *The Future of Business,* Lake Tahoe, CA, 17–21 August 1981.
42. Sidney Rosen, *Wizard of the Dome* (Boston: Little, Brown, 1969), p. 65.
43. R. Buckminster Fuller and Robert Marks, *The Dymaxion World of Buckminster Fuller* (Garden City, NY: Anchor Books, 1973; originally published by Southern Illinois University Press, 1960), p. 21.
44. Ibid.
45. Ibid.
46. Sidney Rosen, *Wizard of the Dome* (Boston: Little, Brown, 1969), p. 66.

47. Lecture Series by R. Buckminster Fuller, *The Future of Business,* Lake Tahoe, CA, 17–21 August 1981.
48. Lecture Series by R. Buckminster Fuller, University of Oregon, Eugene, 1–12 July 1962.
49. Robert Snyder, ed., *Buckminster Fuller: Autobiographical Monologue Scenario* (New York: St. Martin's Press, 1980), p. 60.
50. Ibid.
51. Hugh Kenner, *Bucky: A Guided Tour of Buckminster Fuller* (New York: William Morrow & Company, Inc., 1973), p. 3.
52. Ibid.
53. Athena V. Lord, *Pilot for Spaceship Earth* (New York: Macmillan Publishing Co., Inc., 1978), p. 68.
54. Alden Hatch, *Buckminster Fuller at Home in the Universe* (New York: Crown Publishers, Inc., 1974), p. 111.
55. Athena V. Lord, *Pilot for Spaceship Earth* (New York: Macmillan Publishing Co., Inc., 1978), p. 68.
56. Alden Hatch, *Buckminster Fuller at Home in the Universe* (New York: Crown Publishers, Inc., 1974), p. 114.
57. Lecture by R. Buckminster Fuller, *Buckminster Fuller's Personal Odyssey,* New Dimensions Foundation, San Francisco, CA.
58. Alden Hatch, *Buckminster Fuller at Home in the Universe* (New York: Crown Publishers, Inc., 1974), p. 115.
59. Lecture Series by R. Buckminster Fuller, University of Oregon, Eugene, 1–12 July 1962.
60. World Game Lecture Series by R. Buckminster Fuller, Amherst, MA, 1978.
61. World Game Lecture Series by R. Buckminster Fuller, University of Pennsylvania, Philadelphia, 1976.
62. World Game Lecture Series by R. Buckminster Fuller, Amherst, MA, 1978.
63. Ibid.
64. World Game Lecture Series by R. Buckminster Fuller, University of Pennsylvania, Philadelphia, 1976.
65. Ibid.
66. World Game Lecture Series by R. Buckminster Fuller, New York University, New York, 14–21 July 1979.
67. Ibid.
68. World Game Lecture Series by R. Buckminster Fuller, University of Pennsylvania Museum, Philadelphia, 1975.
69. Athena V. Lord, *Pilot for Spaceship Earth* (New York: Macmillan Publishing Co., Inc., 1978), p. 69.
70. Alden Hatch, *Buckminster Fuller at Home in the Universe* (New York: Crown Publishers, Inc., 1974), p. 118.
71. World Game Lecture Series by R. Buckminster Fuller, New York University, New York, 14–21 July 1979.
72. Lecture Series by R. Buckminster Fuller, University of Oregon, Eugene, 1–12 July 1962.
73. Alden Hatch, *Buckminster Fuller at Home in the Universe* (New York: Crown Publishers, Inc., 1974), p. 110.
74. Lecture Series by R. Buckminster Fuller, University of Oregon, Eugene, 1–12 July 1962.
75. Alden Hatch, *Buckminster Fuller at Home in the Universe* (New York: Crown Publishers, Inc., 1974), p. 118.

76. Athena V. Lord, *Pilot for Spaceship Earth* (New York: Macmillan Publishing Co., Inc., 1978), p. 73.
77. Alden Hatch, *Buckminster Fuller at Home in the Universe* (New York: Crown Publishers, Inc., 1974), p. 119.
78. Robert Synder, ed., *Buckminster Fuller: Autobiographical Monologue Scenario* (New York: St. Martin's Press, 1980), p. 66.
79. Lecture by R. Buckminster Fuller, *Buckminster Fuller's Personal Odyssey,* New Dimensions Foundation, San Francisco.
80. World Game Lecture Series by R. Buckminster Fuller, University of Pennsylvania Museum, Philadelphia, 1975.
81. Lecture Series by R. Buckminster Fuller, *The Future of Business,* Lake Tahoe, CA, 17–21 August 1981.
82. Ibid.
83. Ibid.
84. Lecture Series by R. Buckminster Fuller, University of Oregon, Eugene, 1–12 July 1962.
85. Alden Hatch, *Buckminster Fuller at Home in the Universe* (New York: Crown Publishers, Inc., 1974), p. 121.
86. Lecture Series by R. Buckminster Fuller, University of Oregon, Eugene, 1–12July 1962.
88. Lecture by R. Buckminster Fuller, *Education Tomorrow,* New Dimensions Foundation, San Francisco.
89. Alden Hatch, *Buckminster Fuller at Home in the Universe* (New York: Crown Publishers, Inc., 1974), p. 120.
90. Ibid.
91. Robert Synder, ed., *Buckminster Fuller: Autobiographical Monologue Scenario* (New York: St. Martin's Press, 1980), p. 67.
92. Athena V. Lord, *Pilot for Spaceship Earth* (New York: Macmillan Publishing Co., Inc., 1978), p. 77.

CHAPTER 7

1. Athena V. Lord, *Pilot for Spaceship Earth* (New York: Macmillan Publishing Co., Inc., 1978), p. 77.
2. Lecture Series by R. Buckminster Fuller, University of Oregon, Eugene, 1–12 July 1962.
3. Ibid.
4. J. Baldwin, "Dymaxion Transports," *Automobile Magazine* (July 1988), p. 109.
5. Sidney Rosen, *Wizard of the Dome* (Boston: Little, Brown, 1969), p. 26.
6. World Game Lecture Series by R. Buckminster Fuller, University of Pennsylvania Museum, Philadelphia, 1975.
7. World Game Lecture Series by R. Buckminster Fuller, Amherst, MA, 1978.
8. World Game Lecture Series by R. Buckminster Fuller, University of Pennsylvania Museum, Philadelphia, 1975.
9. Ibid.
10. World Game Lecture Series by R. Buckminster Fuller, Amherst, MA, 1978.
11. World Game Lecture Series by R. Buckminster Fuller, University of Pennsylvania Museum, Philadelphia, 1975.
12. Ibid.

13. Ibid.
14. Ibid.
15. World Game Lecture Series by R. Buckminster Fuller, Amherst, MA, 1978.
16. R. Buckminster Fuller and Robert Marks, *The Dymaxion World of Buckminster Fuller* (Garden City, NY: Anchor Books, 1973; originally published by Southern Illinois University Press, 1960), p. 28.
17. Ibid.
18. John McHale, ed., "Richard Buckminster Fuller," *Architectural Design* (July 1961), p. 293.
19. R. Buckminster Fuller and Robert Marks, *The Dymaxion World of Buckminster Fuller* (Garden City, NY: Anchor Books, 1973; originally published by Southern Illinois University Press, 1960), p. 28.
20. World Game Lecture Series by R. Buckminster Fuller, Amherst, MA, 1978.
21. Ibid.
22. World Game Lecture Series by R. Buckminster Fuller, University of Pennsylvania Museum, Philadelphia, 1975.
23. R. Buckminster Fuller and Robert Marks, *The Dymaxion World of Buckminster Fuller* (Garden City, NY: Anchor Books, 1973; originally published by Southern Illinois University Press, 1960), p. 25.
24. Ibid., p. 26.
25. World Game Lecture Series by R. Buckminster Fuller, University of Pennsylvania Museum, Philadelphia, 1975.
26. Ibid.
27. J. Baldwin, "Dymaxion Transports," *Automobile Magazine* (July 1988), p. 109.
28. World Game Lecture Series by R. Buckminster Fuller, University of Pennsylvania Museum, Philadelphia, 1975.
29. Ibid.
30. World Game Lecture Series by R. Buckminster Fuller, New York University, New York, 14–21 July 1979.
31. Ibid.
32. Ibid.
33. Lecture Series by R. Buckminster Fuller, *The Future of Business,* Lake Tahoe, CA, 17–21 August 1981.
34. Lecture Series by R. Buckminster Fuller, University of Oregon, Eugene, 1–12 July 1962.
35. Ibid.
36. Robert Synder, ed., *Buckminster Fuller: Autobiographical Monologue Scenario* (New York: St. Martin's Press, 1980), p. 71.
37. World Game Lecture Series by R. Buckminster Fuller, New York University, New York, 14–21 July 1979.
38. Lecture Series by R. Buckminster Fuller, University of Oregon, Eugene, 1–12 July 1962.
39. Athena V. Lord, *Pilot for Spaceship Earth* (New York: Macmillan Publishing Co., Inc., 1978), p. 86.
40. Lecture Series by R. Buckminster Fuller, University of Oregon, Eugene, 1–12 July 1962.
41. R. Buckminster Fuller and Robert Marks, *The Dymaxion World of Buckminster Fuller* (Garden City, NY: Anchor Books, 1973; originally published by Southern Illinois University Press, 1960), p. 25.
42. John McHale, ed., "Richard Buckminster Fuller," *Architectural Design* (July 1961), p. 294.

43. Robert Synder, ed., *Buckminster Fuller: Autobiographical Monologue Scenario* (New York: St. Martin's Press, 1980), p. 71.

44. Ibid.

45. J. Baldwin, "Dymaxion Transports," *Automobile Magazine* (July 1988), p. 110.

46. Lecture by R. Buckminster Fuller, *Buckminster Fuller's Personal Odyssey,* New Dimensions Foundation, San Francisco.

47. Ibid.

48. Athena V. Lord, *Pilot for Spaceship Earth* (New York: Macmillan Publishing Co., Inc., 1978), p. 80.

49. Lecture Series by R. Buckminster Fuller, University of Oregon, Eugene, 1–12 July 1962.

50. J. Baldwin, "Dymaxion Transports," *Automobile Magazine* (July 1988), p. 115.

51. Alden Hatch, *Buckminster Fuller at Home in the Universe* (New York: Crown Publishers, Inc., 1974), p. 123.

52. Athena V. Lord, *Pilot for Spaceship Earth* (New York: Macmillan Publishing Co., Inc., 1978), p. 79.

53. Lecture Series by R. Buckminster Fuller, *The Future of Business,* Lake Tahoe, CA, 17–21 August 1981.

54. Ibid.

55. Alden Hatch, *Buckminster Fuller at Home in the Universe* (New York: Crown Publishers, Inc., 1974), p. 133.

56. Lecture Series by R. Buckminster Fuller, University of Oregon, Eugene, 1–12 July 1962.

57. Alden Hatch, *Buckminster Fuller at Home in the Universe* (New York: Crown Publishers, Inc., 1974), p. 124.

58. Lecture Series by R. Buckminster Fuller, *The Future of Business,* Lake Tahoe, CA, 17–21 August 1981.

59. R. Buckminster Fuller and Robert Marks, *The Dymaxion World of Buckminster Fuller* (Garden City, NY: Anchor Books, 1973; originally published by Southern Illinois University Press, 1960), p. 27.

60. Alden Hatch, *Buckminster Fuller at Home in the Universe* (New York: Crown Publishers, Inc., 1974), p. 126.

61. Lecture Series by R. Buckminster Fuller, *The Future of Business,* Lake Tahoe, CA, 17–21 August 1981.

62. Athena V. Lord, *Pilot for Spaceship Earth* (New York: Macmillan Publishing Co., Inc., 1978), p. 79.

63. Lecture Series by R. Buckminster Fuller, University of Oregon, Eugene, 1–12 July 1962.

64. Athena V. Lord, *Pilot for Spaceship Earth* (New York: Macmillan Publishing Co., Inc., 1978), p. 81.

65. World Game Lecture Series by R. Buckminster Fuller, New York University, New York, 14–21 July 1979.

66. Ibid.

67. Lecture Series by R. Buckminster Fuller, University of Oregon, Eugene, 1–12 July 1962.

68. Ibid.

69. Alden Hatch, *Buckminster Fuller at Home in the Universe* (New York: Crown Publishers, Inc., 1974), p. 127.

70. "The Dymaxion American," *Time* (10 January 1984), p. 49.

71. Ibid.

72. Sidney Rosen, *Wizard of the Dome* (Boston: Little, Brown, 1969), p. 78.

73. Ibid.

74. Lecture by R. Buckminster Fuller, Burklyn Business School, Kirkwood Meadows, CA, 1 August 1982.

75. Alden Hatch, *Buckminster Fuller at Home in the Universe* (New York: Crown Publishers, Inc., 1974), p. 127.

76. Lecture Series by R. Buckminster Fuller, *The Future of Business,* Lake Tahoe, CA, 17–21 August 1981.

77. Alden Hatch, *Buckminster Fuller at Home in the Universe* (New York: Crown Publishers, Inc., 1974), p. 127.

78. J. Baldwin, "Dymaxion Transports," *Automobile Magazine* (July 1988), p. 110.

79. Ibid.

80. Lecture by R. Buckminster Fuller, Burklyn Business School, Kirkwood Meadows, CA, 1 August 1982.

81. Ibid.

82. World Game Lecture Series by R. Buckminster Fuller, New York University, New York, 14–21 July 1979.

83. Ibid.

84. Athena V. Lord, *Pilot for Spaceship Earth* (New York: Macmillan Publishing Co., Inc., 1978), p. 81.

85. Hugh Kenner, *Bucky: A Guided Tour of Buckminster Fuller* (New York: William Morrow & Company, Inc., 1973), p. 51.

86. Lecture Series by R. Buckminster Fuller, University of Oregon, Eugene, 1–12 July 1962.

87. Lecture Series by R. Buckminster Fuller, *The Future of Business,* Lake Tahoe, CA, 17–21 August 1981.

88. Ibid.

89. World Game Lecture Series by R. Buckminster Fuller, New York University, New York 14–21 July 1979.

90. Ibid.

91. Ibid.

92. Athena V. Lord, *Pilot for Spaceship Earth* (New York: Macmillan Publishing Co., Inc., 1978), p. 85.

CHAPTER 8

1. J. Baldwin, "Dymaxion Transports," *Automobile Magazine* (July 1988), p. 113.

2. Lecture Series by R. Buckminster Fuller, *The Future of Business,* Lake Tahoe, CA, 17–21 August 1981.

3. J. Baldwin, "Dymaxion Transports," *Automobile Magazine* (July 1988), p. 111.

4. Athena V. Lord, *Pilot for Spaceship Earth* (New York: Macmillan Publishing Co., Inc., 1978), p. 83.

5. Lecture Series, R. Buckminster Fuller, University of California, Santa Barbara, December 1967.

6. Lecture Series by R. Buckminster Fuller, *The Future of Business,* Lake Tahoe, CA, 17–21 August 1981.

7. Ibid.

8. Lecture Series, R. Buckminster Fuller, University of California, Santa Barbara, December 1967.

9. Ibid.

10. J. Baldwin, "Dymaxion Transports," *Automobile Magazine* (July 1988), p. 113.

11. Lecture Series by R. Buckminster Fuller, University of Oregon, Eugene, 1–12 July 1962.

12. Ibid.

13. Lecture Series, R. Buckminster Fuller, University of California, Santa Barbara, December 1967.

14. Ibid.

15. Lecture Series by R. Buckminster Fuller, University of Oregon, Eugene, 1–12 July 1962.

16. Ibid.

17. Ibid.

18. Athena V. Lord, *Pilot for Spaceship Earth* (New York: Macmillan Publishing Co., Inc., 1978), p. 85.

19. Ibid., p. 86.

20. Alden Hatch, *Buckminster Fuller at Home in the Universe* (New York: Crown Publishers, Inc., 1974), p. 132.

21. Sidney Rosen, *Wizard of the Dome* (Boston: Little, Brown, 1969), p. 80.

22. Lecture Series by R. Buckminster Fuller, University of Oregon, Eugene, 1–12 July 1962.

23. Ibid.

24. R. Buckminster Fuller and Robert Marks, *The Dymaxion World of Buckminster Fuller* (Garden City, NY: Anchor Books, 1973; originally published by Southern Illinois University Press, 1960), p. 29.

25. Lecture Series by R. Buckminster Fuller, University of Oregon, Eugene, 1–12 July 1962.

26. Ibid.

27. R. Buckminster Fuller and Robert Marks, *The Dymaxion World of Buckminster Fuller* (Garden City, NY: Anchor Books, 1973; originally published by Southern Illinois University Press, 1960), p. 30.

28. Lecture Series, R. Buckminster Fuller University of California, Santa Barbara, December 1967.

29. J. Baldwin, "Dymaxion Transports," *Automobile Magazine* (July 1988), p. 114.

30. Lecture Series, R. Buckminster Fuller, University of California, Santa Barbara, December 1967.

31. Ibid.

32. J. Baldwin, "Dymaxion Transports," *Automobile Magazine* (July 1988), p. 114.

33. Lecture Series, R. Buckminster Fuller, University of California, Santa Barbara, December 1967.

34. Ibid.

35. Alden Hatch, *Buckminster Fuller at Home in the Universe* (New York: Crown Publishers, Inc., 1974), p. 132.

36. Lecture Series by R. Buckminster Fuller, University of Oregon, Eugene, 1–12 July 1962.

37. Lecture Series, R. Buckminster Fuller University of California, Santa Barbara, December 1967.

38. Athena V. Lord, *Pilot for Spaceship Earth* (New York: Macmillan Publishing Co., Inc., 1978), p. 85.

39. Ibid.

40. J. Baldwin, "Dymaxion Transports," *Automobile Magazine* (July 1988), p. 111.

41. Lecture Series by R. Buckminster Fuller, University of Oregon, Eugene, 1–12 July 1962.

42. Calvin Tomkins, "Profiles, in the Outlaw Area, *The New Yorker* (6 January 1966), p. 68.

43. Lecture Series by R. Buckminster Fuller, University of Oregon, Eugene, 1–12 July 1962.
44. Ibid.
45. Ibid.
46. Alden Hatch, *Buckminster Fuller at Home in the Universe* (New York: Crown Publishers, Inc., 1974), p. 133.
47. Lecture by R. Buckminster Fuller, *Buckminster Fuller's Personal Odyssey,* New Dimensions Foundation, San Francisco.
48. Alden Hatch, *Buckminster Fuller at Home in the Universe* (New York: Crown Publishers, Inc., 1974), p. 133.
49. World Game Lecture Series by R. Buckminster Fuller, New York University, New York, 14–21 July 1979.
50. World Game Lecture Series by R. Buckminster Fuller, University of Pennsylvania, Philadelphia, 1976.
51. Lecture Series by R. Buckminster Fuller, *The Future of Business,* Lake Tahoe, CA, 17–21 August 1981.
52. Lecture Series by R. Buckminster Fuller, University of Oregon, Eugene, 1–12 July 1962.
53. Ibid.
54. Athena V. Lord, *Pilot for Spaceship Earth* (New York: Macmillan Publishing Co., Inc., 1978), p. 88.
55. J. Baldwin, "Dymaxion Transports," *Automobile Magazine* (July 1988), p. 115.
56. World Game Lecture Series by R. Buckminster Fuller, University of Pennsylvania, Philadelphia, 1976.
57. Lecture Series by R. Buckminster Fuller, *The Future of Business,* Lake Tahoe, CA, 17–21 August 1981.
58. Ibid.
59. Ibid.
60. Ibid.
61. R. Buckminster Fuller and Robert Marks, *The Dymaxion World of Buckminster Fuller* (Garden City, NY: Anchor Books, 1973; originally published by Southern Illinois University Press, 1960), p. 29.
62. Ibid.
63. Alden Hatch, *Buckminster Fuller at Home in the Universe* (New York: Crown Publishers, Inc., 1974), p. 130.
64. J. Baldwin, "Dymaxion Transports," *Automobile Magazine* (July 1988), p. 113.
65. R. Buckminster Fuller and Robert Marks, *The Dymaxion World of Buckminster Fuller* (Garden City, NY: Anchor Books, 1973; originally published by Southern Illinois University Press, 1960), p. 29.
66. Lecture Series, R. Buckminster Fuller, University of California, Santa Barbara, December 1967.
67. Athena V. Lord, *Pilot for Spaceship Earth* (New York: Macmillan Publishing Co., Inc., 1978), p. 88.
68. Lecture Series by R. Buckminster Fuller, *The Future of Business,* Lake Tahoe, CA, 17–21 August 1981.
69. Ibid.
70. Lecture Series by R. Buckminster Fuller, University of Oregon, Eugene, 1–12 July 1962.
71. Ibid.
72. Ibid.

73. Ibid.
74. Lecture Series, R. Buckminster Fuller, University of California, Santa Barbara, December 1967.
75. Ibid.
76. Alden Hatch, *Buckminster Fuller at Home in the Universe* (New York: Crown Publishers, Inc., 1974), p. 166.
77. J. Baldwin, "Dymaxion Transports," *Automobile Magazine* (July 1988), p. 115.
78. Lecture Series by R. Buckminster Fuller, University of Oregon, Eugene, 1–12 July 1962.
79. Alden Hatch, *Buckminster Fuller at Home in the Universe* (New York: Crown Publishers, Inc., 1974), p. 169.
80. Ibid.
81. Ibid.
82. World Game Lecture Series by R. Buckminster Fuller, University of Pennsylvania Museum, Philadelphia, 1975.
83. Ibid.
84. R. Buckminster Fuller and Robert Marks, *The Dymaxion World of Buckminster Fuller* (Garden City, NY: Anchor Books, 1973; originally published by Southern Illinois University Press, 1960), p. 113.
85. John McHale, "Buckminster Fuller," *Architectural Review* (July 1956), p. 19.
86. Alden Hatch, *Buckminster Fuller at Home in the Universe* (New York: Crown Publishers, Inc., 1974), p. 170.
87. Lecture Series by R. Buckminster Fuller, University of Oregon, Eugene, 1–12 July 1962.
88. Ibid.
89. Alden Hatch, *Buckminster Fuller at Home in the Universe* (New York: Crown Publishers, Inc., 1974), p. 171.
90. Athena V. Lord, *Pilot for Spaceship Earth* (New York: Macmillan Publishing Co., Inc., 1978), p. 87.
91. Alden Hatch, *Buckminster Fuller at Home in the Universe* (New York: Crown Publishers, Inc., 1974), p. 136.
92. Robert Synder, ed., *Buckminster Fuller: Autobiographical Monologue Scenario* (New York: St. Martin's Press, 1980), p. 76.
93. Lecture Series by R. Buckminster Fuller, University of Oregon, Eugene, 1–12 July 1962.

CHAPTER 9

1. Lecture Series by R. Buckminster Fuller, University of Oregon, Eugene, 1–12 July 1962.
2. Robert Synder, ed., *Buckminster Fuller: Autobiographical Monologue Scenario* (New York: St. Martin's Press, 1980), p. 76.
3. Alden Hatch, *Buckminster Fuller at Home in the Universe* (New York: Crown Publishers, Inc., 1974), p. 140.
4. Lecture Series by R. Buckminster Fuller, University of Oregon, Eugene, 1–12 July 1962.
5. Robert Snyder, ed., *Buckminster Fuller: Autobiographical Monologue Scenario* (New York: St. Martin's Press, 1980), p. 69.
6. Lecture Series by R. Buckminster Fuller, University of Oregon, Eugene, 1–12 July 1962.
7. R. Buckminster Fuller and Robert Marks, *The Dymaxion World of Buckminster Fuller*

(Garden City, NY: Anchor Books, 1973; originally published by Southern Illinois University Press, 1960), p. 7.

8. Ibid., p. 8.

9. Athena V. Lord, *Pilot for Spaceship Earth* (New York: Macmillan Publishing Co., Inc., 1978), p. 92.

10. Alden Hatch, *Buckminster Fuller at Home in the Universe* (New York: Crown Publishers, Inc., 1974), p. 142.

11. R. Buckminster Fuller and Robert Marks, *The Dymaxion World of Buckminster Fuller* (Garden City, NY: Anchor Books, 1973; originally published by Southern Illinois University Press, 1960), p. 8.

12. Calvin Tomkins, "Profiles, in the Outlaw Area," *The New Yorker* (6 January 1966), p. 36.

13. Ibid.

14. R. Buckminster Fuller, *Nine Chains to the Moon* (Carbondale: Southern Illinois University Press, 1938), p. xiv.

15. Lecture Series by R. Buckminster Fuller, *The Future of Business,* Lake Tahoe, CA, 17–21 August 1981.

16. Alden Hatch, *Buckminster Fuller at Home in the Universe* (New York: Crown Publishers, Inc., 1974), p. 140.

17. Ibid., p. 144.

18. Ibid., p. 145.

19. Lecture Series by R. Buckminster Fuller, *The Future of Business,* Lake Tahoe, CA, 17–21 August 1981.

20. Sidney Rosen, *Wizard of the Dome* (Boston: Little, Brown, 1969), p. 85.

21. World Game Lecture Series by R. Buckminster Fuller, University of Pennsylvania Museum, Philadelphia, 1975.

22. Sidney Rosen, *Wizard of the Dome* (Boston: Little, Brown, 1969), p. 86.

23. World Game Lecture Series by R. Buckminster Fuller, University of Pennsylvania Museum, Philadelphia, 1975.

24. Alden Hatch, *Buckminster Fuller at Home in the Universe* (New York: Crown Publishers, Inc., 1974), p. 146.

25. World Game Lecture Series by R. Buckminster Fuller, Washington, DC, 16–18 July 1982.

26. Sidney Rosen, *Wizard of the Dome* (Boston: Little, Brown, 1969), p. 86.

27. World Game Lecture Series by R. Buckminster Fuller, University of Pennsylvania Museum, Philadelphia, 1975.

28. R. Buckminster Fuller and Robert Marks, *The Dymaxion World of Buckminster Fuller* (Garden City, NY: Anchor Books, 1973; originally published by Southern Illinois University Press, 1960), p. 33.

29. Athena V. Lord, *Pilot for Spaceship Earth* (New York: Macmillan Publishing Co., Inc., 1978), p. 89.

30. Ibid., p. 90.

31. World Game Lecture Series by R. Buckminster Fuller, University of Pennsylvania Museum, Philadelphia, 1975.

32. Sidney Rosen, *Wizard of the Dome* (Boston: Little, Brown, 1969), p. 6.

33. Ibid., p. 60.

34. World Game Lecture Series by R. Buckminster Fuller, University of Pennsylvania Museum, Philadelphia, 1975.

35. Ibid.

36. World Game Lecture Series by R. Buckminster Fuller, Amherst, MA, 1978.
37. Ibid.
38. Ibid.
39. World Game Lecture Series by R. Buckminster Fuller, New York University, New York, 14–21 July 1979.
40. Athena V. Lord, *Pilot for Spaceship Earth* (New York: Macmillan Publishing Co., Inc., 1978), p. 93.
41. Ibid.
42. Ibid.
43. Eric Burgess, "Fuller Insists World Can End Poverty within 25 Years," *Christian Science Monitor* (13 August 1969).
44. Ibid.
45. Calvin Tomkins, "Profiles, in the Outlaw Area," *The New Yorker* (6 January 1966), p. 88.
46. Ibid.
47. Ibid.
48. Alden Hatch, *Buckminster Fuller at Home in the Universe* (New York: Crown Publishers, Inc., 1974), p. 150.
49. Lecture Series by R. Buckminster Fuller, *The Future of Business*, Lake Tahoe, CA, 17–21 August 1981.
50. Ibid.
51. Alden Hatch, *Buckminster Fuller at Home in the Universe* (New York: Crown Publishers, Inc., 1974), p. 151.
52. World Game Lecture Series by R. Buckminster Fuller, University of Pennsylvania Museum, Philadelphia, 1975.
53. World Game Lecture Series by R. Buckminster Fuller, Washington, DC, 16–18 July 1982.
54. Lecture Series by R. Buckminster Fuller, University of Oregon, Eugene, 1–12 July 1962.
55. Ibid.
56. Alden Hatch, *Buckminster Fuller at Home in the Universe* (New York: Crown Publishers, Inc., 1974), p. 156.
57. Athena V. Lord, *Pilot for Spaceship Earth* (New York: Macmillan Publishing Co., Inc., 1978), p. 95.
58. World Game Lecture Series by R. Buckminster Fuller, Amherst, MA, 1978.
59. Ibid.
60. Sidney Rosen, *Wizard of the Dome* (Boston: Little, Brown, 1969), p. 93.
61. Athena V. Lord, *Pilot for Spaceship Earth* (New York: Macmillan Publishing Co., Inc., 1978), p. 95.
62. Alden Hatch, *Buckminster Fuller at Home in the Universe* (New York: Crown Publishers, Inc., 1974), p. 159.
63. Sidney Rosen, *Wizard of the Dome* (Boston: Little, Brown, 1969), p. 94.
64. Ibid., p. 98.
65. Ibid.
66. World Game Lecture Series by R. Buckminster Fuller, Amherst, MA, 1978.
67. Alden Hatch, *Buckminster Fuller at Home in the Universe* (New York: Crown Publishers, Inc., 1974), p. 159.
68. Athena V. Lord, *Pilot for Spaceship Earth* (New York: Macmillan Publishing Co., Inc., 1978), p. 96.
69. Sidney Rosen, *Wizard of the Dome* (Boston: Little, Brown, 1969), p. 96.

70. World Game Lecture Series by R. Buckminster Fuller, Amherst, MA, 1978.

71. Ibid.

72. Ibid.

73. World Game Lecture Series by R. Buckminster Fuller, University of Pennsylvania, Philadelphia, 1976.

74. Alden Hatch, *Buckminster Fuller at Home in the Universe* (New York: Crown Publishers, Inc., 1974), p. 159.

75. World Game Lecture Series by R. Buckminster Fuller, University of Pennsylvania, Philadelphia, 1976.

76. Sidney Rosen, *Wizard of the Dome* (Boston: Little, Brown, 1969), p. 98.

77. Alden Hatch, *Buckminster Fuller at Home in the Universe* (New York: Crown Publishers, Inc., 1974), p. 162.

78. Ibid.

79. Robert Synder, ed., *Buckminster Fuller: Autobiographical Monologue Scenario* (New York: St. Martin's Press, 1980), p. 78.

80. World Game Lecture Series by R. Buckminster Fuller, University of Pennsylvania, Philadelphia, 1976.

81. Athena V. Lord, *Pilot for Spaceship Earth* (New York: Macmillan Publishing Co., Inc., 1978), p. 97.

82. Alden Hatch, *Buckminster Fuller at Home in the Universe* (New York: Crown Publishers, Inc., 1974), p. 162.

83. Sidney Rosen, *Wizard of the Dome* (Boston: Little, Brown, 1969), p. 99.

84. Alden Hatch, *Buckminster Fuller at Home in the Universe* (New York: Crown Publishers, Inc., 1974), p. 162.

85. Sidney Rosen, *Wizard of the Dome* (Boston: Little, Brown, 1969), p. 100.

86. Ibid.

87. Athena V. Lord, *Pilot for Spaceship Earth* (New York: Macmillan Publishing Co., Inc., 1978), p. 99.

CHAPTER 10

1. World Game Lecture Series by R. Buckminster Fuller, University of Pennsylvania Museum, Philadelphia, 1975.

2. World Game Lecture Series by R. Buckminster Fuller, Amherst, MA, 1978.

3. R. Buckminster Fuller and Robert Marks, *The Dymaxion World of Buckminster Fuller* (Garden City, NY: Anchor Books, 1973; originally published by Southern Illinois University Press, 1960), p. 50.

4. World Game Lecture Series by R. Buckminster Fuller, University of Pennsylvania Museum, Philadelphia, 1975.

5. Lecture Series by R. Buckminster Fuller, University of Oregon, Eugene, 1–12 July 1962.

6. Ibid.

7. Lecture Series by R. Buckminster Fuller, *The Future of Business,* Lake Tahoe, CA, 17–21 August 1981.

8. R. Buckminster Fuller *Inventions: The Patented Works of R. Buckminster Fuller* (New York: St. Martin's Press, 1983), p. 90.

9. World Game Lecture Series by R. Buckminster Fuller, New York University, New York, 14–21 July 1979.

10. World Game Lecture Series by R. Buckminster Fuller, University of Pennsylvania Museum, Philadelphia, 1975.

11. World Game Lecture Series by R. Buckminster Fuller, Drexel University, Philadelphia 1977.

12. World Game Lecture Series by R. Buckminster Fuller, University of Pennsylvania Museum, Philadelphia, 1975.

13. World Game Lecture Series by R. Buckminster Fuller, Drexel University, Philadelphia 1977.

14. World Game Lecture Series by R. Buckminster Fuller, University of Pennsylvania Museum, Philadelphia, 1975.

15. World Game Lecture Series by R. Buckminster Fuller, Drexel University, Philadelphia 1977.

16. Ibid.

17. Interview with R. Buckminster Fuller, Andrew D. Basiago, Pacific Palisades, CA, 20–21 October 1981.

18. World Game Lecture Series by R. Buckminster Fuller, University of Pennsylvania Museum, Philadelphia, 1975.

19. Ibid.

20. Lecture Series, R. Buckminster Fuller University of California, Santa Barbara, December 1967.

21. Ibid.

22. Ibid.

23. World Game Lecture Series by R. Buckminster Fuller, Washington, DC, 16–18 July 1982.

24. World Game Lecture Series by R. Buckminster Fuller, University of Pennsylvania Museum, Philadelphia, 1975.

25. World Game Lecture Series by R. Buckminster Fuller, Washington, DC, 16–18 July 1982.

26. Lecture Series by R. Buckminster Fuller, University of Oregon, Eugene, 1–12 July 1982.

27. R. Buckminster Fuller, *Inventions: The Patented Works of R. Buckminster Fuller* (New York: St. Martin's Press, 1983), p. 94.

28. Lecture by R. Buckminster Fuller, *On the Future,* New Dimensions Foundation, San Francisco.

29. Lecture Series by R. Buckminster Fuller, University of Oregon, Eugene, 1–12 July 1962.

30. World Game Lecture Series by R. Buckminster Fuller, Washington, DC, 16–18 July 1982.

31. Lecture Series by R. Buckminster Fuller, University of Oregon, Eugene, 1–12 July 1962.

32. World Game Lecture Series by R. Buckminster Fuller, Washington, DC, 16–18 July 1982.

33. R. Buckminster Fuller and Robert Marks, *The Dymaxion World of Buckminster Fuller* (Garden City, NY: Anchor Books, 1973; originally published by Southern Illinois University Press, 1960), p. 50.

34. Lecture Series by R. Buckminster Fuller, University of Oregon, Eugene, 1–12 July 1962.

35. Ibid.

36. World Game Lecture Series by R. Buckminster Fuller, Washington, DC, 16–18 July 1982.

37. R. Buckminster Fuller and Robert Marks, *The Dymaxion World of Buckminster Fuller* (Garden City, NY: Anchor Books, 1973; originally published by Southern Illinois University Press, 1960), p. 50.

38. Ibid.

39. World Game Lecture Series by R. Buckminster Fuller, Washington, DC, 16–18 July 1982.

40. R. Buckminster Fuller and Robert Marks, *The Dymaxion World of Buckminster Fuller* (Garden City, NY: Anchor Books, 1973; originally published by Southern Illinois University Press, 1960), p. 51.

41. Ibid.
42. Lecture Series, R. Buckminster Fuller, University of California, Santa Barbara, December 1967.
43. Ibid.
44. R. Buckminster Fuller and Robert Marks, *The Dymaxion World of Buckminster Fuller* (Garden City, NY: Anchor Books, 1973; originally published by Southern Illinois University Press, 1960), p. 51.
45. Lecture Series by R. Buckminster Fuller, University of Oregon, Eugene, 1–12 July 1962.
46. Lecture Series by R. Buckminster Fuller, *The Future of Business,* Lake Tahoe, CA, 17–21 August 1981.
47. Lecture Series by R. Buckminster Fuller, University of Oregon, Eugene, 1–12 July 1962.
48. World Game Lecture Series by R. Buckminster Fuller, New York University, New York, 14–21 July 1979.
49. Lecture Series by R. Buckminster Fuller, University of Oregon, Eugene, 1–12 July 1962.
50. Lecture Series, R. Buckminster Fuller University of California, Santa Barbara, December 1967.
51. World Game Lecture Series by R. Buckminster Fuller, New York University, New York, 14–21 July 1979.
52. Lecture Series by R. Buckminster Fuller, University of Oregon, Eugene, 1–12 July 1962.
53. Ibid.
54. Lecture Series by R. Buckminster Fuller, *The Future of Business,* Lake Tahoe, CA, 17–21 August 1981.
55. Lecture Series by R. Buckminster Fuller, University of Oregon, Eugene, 1–12 July 1962.
56. World Game Lecture Series by R. Buckminster Fuller, New York University, New York, 14–21 July 1979.
57. Lecture Series by R. Buckminster Fuller, University of Oregon, Eugene, 1–12 July 1962.
58. Ibid.
59. Ibid.
60. Lecture by R. Buckminster Fuller, Burklyn Business School, Kirkwood Meadows, CA, 1 August 1982.
61. World Game Lecture Series by R. Buckminster Fuller, University of Pennsylvania Museum, Philadelphia, 1975.
62. Ibid.
63. Lecture Series, R. Buckminster Fuller University of California, Santa Barbara, December 1967.
64. Ibid.
65. Ibid.
66. World Game Lecture Series by R. Buckminster Fuller, Amherst, MA, 1978.
67. Ibid.
68. Ibid.
69. World Game Lecture Series by R. Buckminster Fuller, University of Pennsylvania, Philadelphia, 1976.
70. Ibid.
71. Lecture Series by R. Buckminster Fuller, University of Oregon, Eugene, 1–12 July 1962.
72. Lecture Series by R. Buckminster Fuller, *The Future of Business,* Lake Tahoe, CA, 17–21 August 1981.
73. Alden Hatch, *Buckminster Fuller at Home in the Universe* (New York: Crown Publishers, Inc., 1974), p. 167.

74. Lecture Series by R. Buckminster Fuller, *The Future of Business,* Lake Tahoe, CA, 17–21 August 1981.
75. Alden Hatch, *Buckminster Fuller at Home in the Universe* (New York: Crown Publishers, Inc., 1974), p. 167.
76. Lecture Series by R. Buckminster Fuller, *The Future of Business,* Lake Tahoe, CA, 17–21 August 1981.
77. R. Buckminster Fuller *Inventions: The Patented Works of R. Buckminster Fuller* (New York: St. Martin's Press, 1983), p. 90.
78. Alden Hatch, *Buckminster Fuller at Home in the Universe* (New York: Crown Publishers, Inc., 1974), p. 167.
79. Lecture Series by R. Buckminster Fuller, *The Future of Business,* Lake Tahoe, CA, 17–21 August 1981.
80. Ibid.
81. World Game Lecture Series by R. Buckminster Fuller, New York University, New York, 14–21 July 1979.
82. Ibid.
83. Ibid.
84. Lecture Series by R. Buckminster Fuller, University of Oregon, Eugene, 1–12 July 1962.
85. Lecture Series, R. Buckminster Fuller University of California, Santa Barbara, December 1967.
86. Ibid.
87. Ibid.
88. Ibid.
89. Lecture, R. Buckminster Fuller *Children in the Universe,* University of California, Santa Barbara, 7 February 1973.
90. Ibid.

CHAPTER 11

1. World Game Lecture Series by R. Buckminster Fuller, University of Pennsylvania, Philadelphia, 1976.
2. Ibid.
3. Lecture by R. Buckminster Fuller *On the Future,* New Dimensions Foundation, San Francisco.
4. World Game Lecture Series by R. Buckminster Fuller, University of Pennsylvania, Philadelphia, 1976.
5. Ibid.
6. Lecture by R. Buckminster Fuller, *How to Make Our World Work,* New Dimensions Foundation, San Francisco.
7. Ibid.
8. World Game Lecture Series by R. Buckminster Fuller, New York University, New York, 14–21 July 1979.
9. Lecture by R. Buckminster Fuller, *On the Future,* New Dimensions Foundation, San Francisco.
10. World Game Lecture Series by R. Buckminster Fuller, University of Pennsylvania, Philadelphia, 1976.

11. R. Buckminster Fuller and Robert Marks, *The Dymaxion World of Buckminster Fuller* (Garden City, NY: Anchor Books, 1973; originally published by Southern Illinois University Press, 1960), p. 159.

12. Lecture by R. Buckminster Fuller, *On the Future,* New Dimensions Foundation, San Francisco.

13. World Game Lecture Series by R. Buckminster Fuller, University of Pennsylvania, Philadelphia, 1976.

14. World Game Lecture Series by R. Buckminster Fuller, New York University, New York, 14–21 July 1979.

15. Ibid.

16. Lecture Series by R. Buckminster Fuller, *The Future of Business,* Lake Tahoe, CA, 17–21 August 1981.

17. World Game Lecture Series by R. Buckminster Fuller, New York University, New York, 14–21 July 1979.

18. Lecture Series by R. Buckminster Fuller, *The Future of Business,* Lake Tahoe, CA, 17–21 August 1981.

19. Ibid.

20. World Game Lecture Series by R. Buckminster Fuller, New York University, New York, 14–21 July 1979.

21. Ibid.

22. Lecture Series by R. Buckminster Fuller, University of Oregon, Eugene, 1–12 July 1962.

23. Ibid.

24. World Game Lecture Series by R. Buckminster Fuller, New York University, New York, 14–21 July 1979.

25. Ibid.

26. Lecture Series by R. Buckminster Fuller, University of Oregon, Eugene, 1–12 July 1962.

27. World Game Lecture Series by R. Buckminster Fuller, New York University, New York, 14–21 July 1979.

28. R. Buckminster Fuller *Critical Path* (New York: St. Martin's Press, 1981), p. 22.

29. Calvin Tomkins, ''Profiles, in the Outlaw Area,'' *The New Yorker* (6 January 1966), p. 91.

30. Ibid.

31. Lecture Series by R. Buckminster Fuller, University of Oregon, Eugene, 1–12 July 1962.

32. Lecture by R. Buckminster Fuller, Burklyn Business School, Kirkwood Meadows, CA, 1 August 1982.

33. Ibid.

34. World Game Lecture Series by R. Buckminster Fuller, University of Pennsylvania Museum, Philadelphia, 1975.

35. Lecture Series by R. Buckminster Fuller, *The Future of Business,* Lake Tahoe, CA, 17–21 August 1981.

36. World Game Lecture Series by R. Buckminster Fuller, University of Pennsylvania Museum, Philadelphia, 1975.

37. Ibid.

38. World Game Lecture Series by R. Buckminster Fuller, Washington, DC, 16–18 July 1982.

39. Ibid.

40. Ibid.

41. Lecture, R. Buckminster Fuller, *Future of Man,* University of California, Santa Barbara, 11 January 1972.

42. World Game Lecture Series by R. Buckminster Fuller, New York University, New York, 14–21 July 1979.
43. Ibid.
44. R. Buckminster Fuller, *Critical Path* (New York: St. Martin's Press, 1981), p. 34.
45. Ibid., p. 46.
46. Lecture Series by R. Buckminster Fuller, University of Oregon, Eugene, 1–12 July 1962.
47. Lecture Series, R. Buckminster Fuller University of California, Santa Barbara, December 1967.
48. R. Buckminster Fuller and Robert Marks, *The Dymaxion World of Buckminster Fuller* (Garden City, NY: Anchor Books, 1973; originally published by Southern Illinois University Press, 1960), p. 158.
49. Interview with R. Buckminster Fuller, Andrew D. Basiago, Pacific Palisades, CA, 20–21 October 1981.
50. Lecture Series, R. Buckminster Fuller, University of California, Santa Barbara, December 1967.
51. Ibid.
52. Ibid.
53. Lecture by R. Buckminster Fuller, *On the Future,* New Dimensions Foundation, San Francisco.
54. Lecture Series, R. Buckminster Fuller University of California, Santa Barbara, December 1967.
55. World Game Lecture Series by R. Buckminster Fuller, Drexel University, Philadelphia, 1977.
56. Ibid.
57. Ibid.
58. Lecture, R. Buckminster Fuller *Future of Man,* University of California, Santa Barbara, 11 January 1972.
59. Ibid.
60. Ibid.
61. World Game Lecture Series by R. Buckminster Fuller, Washington, DC, 16–18 July 1982.
62. R. Buckminster Fuller, *No More Secondhand God and Other Writings* (Carbondale: Southern Illinois University Press, 1963), p. v.
63. Ibid.
64. Ibid.
65. Lecture Series by R. Buckminster Fuller, University of Oregon, Eugene, 1–12 July 1962.
66. Lecture by R. Buckminster Fuller, Burklyn Business School, Kirkwood Meadows, CA, 1 August 1982.
67. R. Buckminster Fuller *Critical Path* (New York: St. Martin's Press, 1981), p. 174.
68. World Game Lecture Series by R. Buckminster Fuller, New York University, New York, 14–21 July 1979.
69. Ibid.
70. Lecture Series by R. Buckminster Fuller, University of Oregon, Eugene, 1–12 July 1962.
71. Ibid.
72. Lecture Series by R. Buckminster Fuller, *The Future of Business,* Lake Tahoe, CA, 17–21 August 1981.
73. World Game Lecture Series by R. Buckminster Fuller, New York University, New York, 14–21 July 1979.

74. R. Buckminster Fuller *Critical Path* (New York: St. Martin's Press, 1981), p. 162.
75. Ibid.
76. Lecture Series, R. Buckminster Fuller, University of California, Santa Barbara, December 1967.
77. Ibid.
78. World Game Lecture Series by R. Buckminster Fuller, New York University, New York, 14–21 July 1979.
79. Ibid.
80. Ibid.
81. Athena V. Lord, *Pilot for Spaceship Earth* (New York: Macmillan Publishing Co., Inc., 1978), p. 120.
82. Lecture Series by R. Buckminster Fuller, University of Oregon, Eugene, 1–12 July 1962.
83. Ibid.
84. World Game Lecture Series by R. Buckminster Fuller, Washington, DC, 16–18 July 1982.
85. Ibid.
86. World Game Lecture Series by R. Buckminster Fuller, New York University, New York, 14–21 July 1979.
87. Ibid.
88. World Game Lecture Series by R. Buckminster Fuller, Drexel University, Philadelphia, 1977.
89. Lecture, R. Buckminster Fuller, *Vision 65,* World Conference on Communication Arts, Southern Illinois University, Carbondale, 12 October 1965 (reprinted, R. Buckminster Fuller, Utopia or Oblivion, Chapter 4, Bantam, 1969).
90. Ibid.
91. Ibid.
92. World Game Lecture Series by R. Buckminster Fuller, Drexel University, Philadelphia, 1977.
93. Ibid.
94. R. Buckminster Fuller, *Critical Path* (New York: St. Martin's Press, 1981), p. 175.
95. Lecture Series by R. Buckminster Fuller, University of Oregon, Eugene, 1–12 July 1962.
96. R. Buckminster Fuller, *Critical Path* (New York: St. Martin's Press, 1981), p. 183.
97. World Game Lecture Series by R. Buckminster Fuller, Drexel University, Philadelphia, 1977.
98. Ibid.

CHAPTER 12

1. Sidney Rosen, *Wizard of the Dome* (Boston: Little, Brown, 1969), p. 102.
2. World Game Lecture Series by R. Buckminster Fuller, Amherst, MA, 1978.
3. Alden Hatch, *Buckminster Fuller at Home in the Universe* (New York: Crown Publishers, Inc., 1974), p. 180.
4. Ibid., p. 173.
5. Lecture Series, R. Buckminster Fuller University of California, Santa Barbara, December 1967.

6. Alden Hatch, *Buckminster Fuller at Home in the Universe* (New York: Crown Publishers, Inc., 1974), p. 174.

7. Lecture Series, R. Buckminster Fuller University of California, Santa Barbara, December 1967.

8. Sidney Rosen, *Wizard of the Dome* (Boston: Little, Brown, 1969), p. 103.

9. World Game Lecture Series by R. Buckminster Fuller, Drexel University, Philadelphia, 1977.

10. Alden Hatch, *Buckminster Fuller at Home in the Universe* (New York: Crown Publishers, Inc., 1974), p. 168.

11. World Game Lecture Series by R. Buckminster Fuller, University of Pennsylvania, Philadelphia, 1976.

12. Alden Hatch, *Buckminster Fuller at Home in the Universe* (New York: Crown Publishers, Inc., 1974), p. 174.

13. World Game Lecture Series by R. Buckminster Fuller, University of Pennsylvania, Philadelphia, 1976.

14. Lecture Series by R. Buckminster Fuller, *The Future of Business,* Lake Tahoe, CA, 17–21 August 1981.

15. World Game Lecture Series by R. Buckminster Fuller, University of Pennsylvania, Philadelphia, 1976.

16. Lecture Series by R. Buckminster Fuller, *The Future of Business,* Lake Tahoe, CA, 17–21 August 1981.

17. World Game Lecture Series by R. Buckminster Fuller, University of Pennsylvania, Philadelphia, 1976.

18. Lecture Series by R. Buckminster Fuller, *The Future of Business,* Lake Tahoe, CA, 17–21 August 1981.

19. Lecture Series by R. Buckminster Fuller, University of Oregon, Eugene, 1–12 July 1962.

20. Ibid.

21. Sidney Rosen, *Wizard of the Dome* (Boston: Little, Brown, 1969), p. 103.

22. Athena V. Lord, *Pilot for Spaceship Earth* (New York: Macmillan Publishing Co., Inc., 1978), p. 110.

23. Ibid.

24. Alden Hatch, *Buckminster Fuller at Home in the Universe* (New York: Crown Publishers, Inc., 1974), p. 176.

25. Lecture Series by R. Buckminster Fuller, *The Future of Business,* Lake Tahoe, CA, 17–21 August 1981.

26. Lecture Series by R. Buckminster Fuller, University of Oregon, Eugene, 1–12 July 1962.

27. Ibid.

28. Alden Hatch, *Buckminster Fuller at Home in the Universe* (New York: Crown Publishers, Inc., 1974), p. 176.

29. Lecture Series by R. Buckminster Fuller, University of Oregon, Eugene, 1–12 July 1962.

30. Athena V. Lord, *Pilot for Spaceship Earth* (New York: Macmillan Publishing Co., Inc., 1978), p. 112.

31. Lecture Series by R. Buckminster Fuller, University of Oregon, Eugene, 1–12 July 1962.

32. Ibid.

33. Athena V. Lord, *Pilot for Spaceship Earth* (New York: Macmillan Publishing Co., Inc., 1978), p. 111.

34. Lecture Series by R. Buckminster Fuller, University of Oregon, Eugene, 1–12 July 1962.

35. Alden Hatch, *Buckminster Fuller at Home in the Universe* (New York: Crown Publishers, Inc., 1974), p. 179.
36. Lecture Series by R. Buckminster Fuller, University of Oregon, Eugene, 1–12 July 1962.
37. Alden Hatch, *Buckminster Fuller at Home in the Universe* (New York: Crown Publishers, Inc., 1974), p. 179.
38. World Game Lecture Series by R. Buckminster Fuller, New York University, New York, 14–21 July 1979.
39. Ibid.
40. Ibid.
41. Alden Hatch, *Buckminster Fuller at Home in the Universe* (New York: Crown Publishers, Inc., 1974), p. 180.
42. World Game Lecture Series by R. Buckminster Fuller, University of Pennsylvania, Philadelphia, 1976.
43. Ibid.
44. Ibid.
45. Alden Hatch, *Buckminster Fuller at Home in the Universe* (New York: Crown Publishers, Inc., 1974), p. 181.
46. Ibid.
47. Lecture Series by R. Buckminster Fuller, *The Future of Business,* Lake Tahoe, CA, 17–21 August 1981.
48. Ibid.
49. Ibid.
50. Lecture Series by R. Buckminster Fuller, University of Oregon, Eugene, 1–12 July 1962.
51. Alden Hatch, *Buckminster Fuller at Home in the Universe* (New York: Crown Publishers, Inc., 1974), p. 184.

CHAPTER 13

1. Lecture Series, R. Buckminster Fuller University of California, Santa Barbara, December 1967.
2. Ibid.
3. Lecture Series by R. Buckminster Fuller, *The Future of Business,* Lake Tahoe, CA, 17–21 August 1981.
4. Ibid.
5. Lecture Series, R. Buckminster Fuller University of California, Santa Barbara, December 1967.
6. Lecture Series by R. Buckminster Fuller, University of Oregon, Eugene, 1–12 July 1962.
7. Ibid.
8. Lecture Series by R. Buckminster Fuller, *The Future of Business,* Lake Tahoe, CA, 17–21 August 1981.
9. Ibid.
10. Ibid.
11. Lecture Series, R. Buckminster Fuller University of California, Santa Barbara, December 1967.

12. Lecture Series by R. Buckminster Fuller, *The Future of Business,* Lake Tahoe, CA, 17–21 August 1981.

13. Lecture Series, R. Buckminster Fuller University of California, Santa Barbara, December 1967.

14. Lecture Series by R. Buckminster Fuller, University of Oregon, Eugene, 1–12 July 1962.

15. Lecture Series by R. Buckminster Fuller, *The Future of Business,* Lake Tahoe, CA, 17–21 August 1981.

16. Ibid.

17. Lecture Series, R. Buckminster Fuller University of California, Santa Barbara, December 1967.

18. Lecture Series by R. Buckminster Fuller, University of Oregon, Eugene, 1–12 July 1962.

19. Hugh Kenner, *Bucky: A Guided Tour of Buckminster Fuller* (New York: William Morrow & Company, Inc., 1973), p. 35.

20. Lecture Series by R. Buckminster Fuller, University of Oregon, Eugene, 1–12 July 1962.

21. Lecture by R. Buckminster Fuller, *Buckminster Fuller's Personal Odyssey,* New Dimensions Foundation, San Francisco.

22. Ibid.

23. Lecture Series by R. Buckminster Fuller, University of Oregon, Eugene, 1–12 July 1962.

24. Ibid.

25. World Game Lecture Series by R. Buckminster Fuller, Drexel University, Philadelphia, 1977.

26. Ibid.

27. Lecture Series, R. Buckminster Fuller University of California, Santa Barbara, December 1967.

28. World Game Lecture Series by R. Buckminster Fuller, Drexel University, Philadelphia, 1977.

29. Lecture Series, R. Buckminster Fuller University of California, Santa Barbara, December 1967.

30. Ibid.

31. Lecture Series by R. Buckminster Fuller, University of Oregon, Eugene, 1–12 July 1962.

32. Ibid.

33. Ibid.

34. Lecture Series, R. Buckminster Fuller University of California, Santa Barbara, December 1967.

35. Lecture Series by R. Buckminster Fuller, University of Oregon, Eugene, 1–12 July 1962.

36. Ibid.

37. World Game Lecture Series by R. Buckminster Fuller, Washington, DC, 16–18 July 1982.

38. Ibid.

39. Ibid.

40. Lecture Series by R. Buckminster Fuller, *The Future of Business,* Lake Tahoe, CA, 17–21 August 1981.

41. Ibid.

42. Ibid.

43. Lecture Series, R. Buckminster Fuller University of California, Santa Barbara, December 1967.

44. Ibid.

45. Lecture Series by R. Buckminster Fuller, University of Oregon, Eugene, 1–12 July 1962.

46. Ibid.

47. Interview with R. Buckminster Fuller, Andrew D. Basiago, Pacific Palisades, CA, 20–21 October 1981.

48. World Game Lecture Series by R. Buckminster Fuller, University of Pennsylvania, Philadelphia, 1976.

49. Ibid.

50. Lecture by R. Buckminster Fuller, Burklyn Business School, Kirkwood Meadows, CA, 1 August 1982.

51. Ibid.

52. Ibid.

53. Ibid.

54. Lecture Series by R. Buckminster Fuller, University of Oregon, Eugene, 1–12 July 1962.

55. Ibid.

56. Ibid.

57. Ibid.

58. Ibid.

CHAPTER 14

1. Alden Hatch, *Buckminster Fuller at Home in the Universe* (New York: Crown Publishers, Inc., 1974), p. 184.

2. Athena V. Lord, *Pilot for Spaceship Earth* (New York: Macmillan Publishing Co., Inc., 1978), p. 115.

3. World Game Lecture Series by R. Buckminster Fuller, University of Pennsylvania Museum, Philadelphia, 1975.

4. World Game Lecture Series by R. Buckminster Fuller, Amherst, MA, 1978.

5. Ibid.

6. World Game Lecture Series by R. Buckminster Fuller, University of Pennsylvania Museum, Philadelphia, 1975.

7. Ibid.

8. Ibid.

9. Ibid.

10. World Game Lecture Series by R. Buckminster Fuller, Drexel University, Philadelphia, 1977.

11. Athena V. Lord, *Pilot for Spaceship Earth* (New York: Macmillan Publishing Co., Inc., 1978), p. 125.

12. Lecture Series by R. Buckminster Fuller, University of Oregon, Eugene, 1–12 July 1962.

13. Ibid.

14. Lecture Series by R. Buckminster Fuller, *The Future of Business,* Lake Tahoe, CA, 17–21 August 1981.

15. Alden Hatch, *Buckminster Fuller at Home in the Universe* (New York: Crown Publishers, Inc., 1974), p. 188.

16. Ibid., p. 189.

17. Robert Snyder, ed., *Buckminster Fuller: Autobiographical Monologue Scenario* (New York: St. Martin's Press, 1980), p. 85.

18. Ibid., p. 86.

19. Alden Hatch, *Buckminster Fuller at Home in the Universe* (New York: Crown Publishers, Inc., 1974), p. 191.
20. Lecture Series by R. Buckminster Fuller, University of Oregon, Eugene, 1–12 July 1962.
21. Ibid.
22. Alden Hatch, *Buckminster Fuller at Home in the Universe* (New York: Crown Publishers, Inc., 1974), p. 191.
23. World Game Lecture Series by R. Buckminster Fuller, Amherst, MA, 1978.
24. Ibid.
25. Ibid.
26. Alden Hatch, *Buckminster Fuller at Home in the Universe* (New York: Crown Publishers, Inc., 1974), p. 192.
27. World Game Lecture Series by R. Buckminster Fuller, New York University, New York, 14–21 July 1979.
28. Athena V. Lord, *Pilot for Spaceship Earth* (New York: Macmillan Publishing Co., Inc., 1978), p. 118.
29. Lecture Series by R. Buckminster Fuller, *The Future of Business*, Lake Tahoe, CA, 17–21 August 1981.
30. Ibid.
31. Ibid.
32. Hugh Kenner, *Bucky: A Guided Tour of Buckminster Fuller* (New York: William Morrow & Company, Inc., 1973), p. 87.
33. Ibid.
34. Lecture Series by R. Buckminster Fuller, *The Future of Business*, Lake Tahoe, CA, 17–21 August 1981.
35. Alden Hatch, *Buckminster Fuller at Home in the Universe* (New York: Crown Publishers, Inc., 1974), p. 195.
36. World Game Lecture Series by R. Buckminster Fuller, University of Pennsylvania Museum, Philadelphia, 1975.
37. Ibid.
38. Alden Hatch, *Buckminster Fuller at Home in the Universe* (New York: Crown Publishers, Inc., 1974), p. 196.
39. Athena V. Lord, *Pilot for Spaceship Earth* (New York: Macmillan Publishing Co., Inc., 1978), p. 117.
40. World Game Lecture Series by R. Buckminster Fuller, University of Pennsylvania Museum, Philadelphia, 1975.
41. Lecture Series by R. Buckminster Fuller, University of Oregon, Eugene, 1–12 July 1962.
42. World Game Lecture Series by R. Buckminster Fuller, Drexel University, Philadelphia, 1977.
43. Lecture Series by R. Buckminster Fuller, University of Oregon, Eugene, 1–12 July 1962.
44. Ibid.
45. R. Buckminster Fuller *Critical Path* (New York: St. Martin's Press, 1981), p. 147.
46. Lecture Series by R. Buckminster Fuller, University of Oregon, Eugene, 1–12 July 1962.
47. World Game Lecture Series by R. Buckminster Fuller, Drexel University, Philadelphia, 1977.
48. J. Baldwin, "Dymaxion Transports," *Automobile Magazine* (July 1988), p. 111.
49. Athena V. Lord, *Pilot for Spaceship Earth* (New York: Macmillan Publishing Co., Inc., 1978), p. 120.

50. Lecture Series by R. Buckminster Fuller, University of Oregon, Eugene, 1–12 July 1962.
51. Sidney Rosen, *Wizard of the Dome* (Boston: Little, Brown, 1969), p. 130.
52. Alden Hatch, *Buckminster Fuller at Home in the Universe* (New York: Crown Publishers, Inc., 1974), p. 199.
53. Lecture Series by R. Buckminster Fuller, *The Future of Business,* Lake Tahoe, CA, 17–21 August 1981.
54. Alden Hatch, *Buckminster Fuller at Home in the Universe* (New York: Crown Publishers, Inc., 1974), p. 200.
55. Lecture Series by R. Buckminster Fuller, University of Oregon, Eugene, 1–12 July 1962.
56. Ibid.
57. J. Baldwin, "Dymaxion Transports," *Automobile Magazine* (July 1988), p. 111.
58. Lecture Series by R. Buckminster Fuller, University of Oregon, Eugene, 1–12 July 1962.
59. Ibid.
60. Interview with Dave Anderson of Ford Motor Company, May 1988.
61. Ibid.
62. World Game Lecture Series by R. Buckminster Fuller, University of Pennsylvania, Philadelphia, 1976.
63. World Game Lecture Series by R. Buckminster Fuller, Amherst, MA, 1978.
64. Ibid.
65. World Game Lecture Series by R. Buckminster Fuller, University of Pennsylvania, Philadelphia, 1976.
66. Lecture Series by R. Buckminster Fuller, *The Future of Business,* Lake Tahoe, CA, 17–21 August 1981.
67. Ibid.
68. R. Buckminster Fuller and Robert Marks, *The Dymaxion World of Buckminster Fuller* (Garden City, NY: Anchor Books, 1973; originally published by Southern Illinois University Press, 1960), p. 203.
69. Ibid., p. 204.
70. World Game Lecture Series by R. Buckminster Fuller, New York University, New York, 14–21 July 1979.
71. Ibid.
72. Ibid.
73. World Game Lecture Series by R. Buckminster Fuller, University of Pennsylvania Museum, Philadelphia, 1975.
74. Ibid.
75. Calvin Tomkins, "Profiles, in the Outlaw Area," *The New Yorker* (6 January 1966), p. 36.
76. World Game Lecture Series by R. Buckminster Fuller, University of Pennsylvania Museum, Philadelphia, 1975.
77. Ibid.
78. Lecture Series by R. Buckminster Fuller, *The Future of Business,* Lake Tahoe, CA, 17–21 August 1981.
79. "The Dymaxion American," *Time* (10 January 1984), p. 50.
80. Lecture Series by R. Buckminster Fuller, *The Future of Business,* Lake Tahoe, CA, 17–21 August 1981.
81. Ibid.
82. Robert Snyder, ed., *Buckminster Fuller: Autobiographical Monologue Scenario* (New York: St. Martin's Press, 1980), p. 93.

83. Alden Hatch, *Buckminster Fuller at Home in the Universe* (New York: Crown Publishers, Inc., 1974), p. 203.
84. Ibid.

CHAPTER 15

1. R. Buckminster Fuller and Robert Marks, *The Dymaxion World of Buckminster Fuller* (Garden City, NY: Anchor Books, 1973; originally published by Southern Illinois University Press, 1960), p. 153.
2. Lecture Series by R. Buckminster Fuller, University of Oregon, Eugene, 1–12 July 1962.
3. World Game Lecture Series by R. Buckminster Fuller, Amherst, MA, 1978.
4. World Game Lecture Series by R. Buckminster Fuller, University of Pennsylvania, Philadelphia, 1976.
5. Ibid.
6. World Game Lecture Series by R. Buckminster Fuller, Drexel University, Philadelphia, 1977.
7. Ibid.
8. Ibid.
9. Ibid.
10. Ibid.
11. World Game Lecture Series by R. Buckminster Fuller, New York University, New York, 14–21 July 1979.
12. Ibid.
13. World Game Lecture Series by R. Buckminster Fuller, University of Pennsylvania Museum, Philadelphia, 1975.
14. R. Buckminster Fuller and Robert Marks, *The Dymaxion World of Buckminster Fuller* (Garden City, NY: Anchor Books, 1973; originally published by Southern Illinois University Press, 1960), p. 131.
15. World Game Lecture Series by R. Buckminster Fuller, University of Pennsylvania Museum, Philadelphia, 1975.
16. Alden Hatch, *Buckminster Fuller at Home in the Universe* (New York: Crown Publishers, Inc., 1974), p. 206.
17. Athena V. Lord, *Pilot for Spaceship Earth* (New York: Macmillan Publishing Co., Inc., 1978), p. 127.
18. World Game Lecture Series by R. Buckminster Fuller, University of Pennsylvania Museum, Philadelphia, 1975.
19. World Game Lecture Series by R. Buckminster Fuller, Washington, DC, 16–18 July 1982.
20. Alden Hatch, *Buckminster Fuller at Home in the Universe* (New York: Crown Publishers, Inc., 1974), p. 208.
21. World Game Lecture Series by R. Buckminster Fuller, University of Pennsylvania Museum, Philadelphia, 1975.
22. World Game Lecture Series by R. Buckminster Fuller, Washington, DC, 16–18 July 1982.
23. "The Dymaxion American," *Time* (10 January 1984), p. 50.
24. Ibid.
25. World Game Lecture Series by R. Buckminster Fuller, Amherst, MA, 1978.

26. Alden Hatch, *Buckminster Fuller at Home in the Universe* (New York: Crown Publishers, Inc., 1974), p. 210.
27. World Game Lecture Series by R. Buckminster Fuller, Amherst, MA, 1978.
28. Ibid.
29. Ibid.
30. Alden Hatch, *Buckminster Fuller at Home in the Universe* (New York: Crown Publishers, Inc., 1974), p. 214.
31. Lecture Series by R. Buckminster Fuller, University of Oregon, Eugene, 1–12 July 1962.
32. Ibid.
33. Lecture Series by R. Buckminster Fuller, *The Future of Business,* Lake Tahoe, CA, 17–21 August 1981.
34. Ibid.
35. Alden Hatch, *Buckminster Fuller at Home in the Universe* (New York: Crown Publishers, Inc., 1974), p. 216.
36. World Game Lecture Series by R. Buckminster Fuller, New York University, New York, 14–21 July 1979.
37. World Game Lecture Series by R. Buckminster Fuller, University of Pennsylvania Museum, Philadelphia, 1975.
38. "The Dymaxion American," *Time* (10 January 1984), p. 50.
39. Ibid.
40. Ibid.
41. R. Buckminster Fuller and Robert Marks, *The Dymaxion World of Buckminster Fuller* (Garden City, NY: Anchor Books, 1973; originally published by Southern Illinois University Press, 1960), p. 8.
42. "The Dymaxion American," *Time* (10 January 1984), p. 50.
43. Lecture Series by R. Buckminster Fuller, University of Oregon, Eugene, 1–12 July 1962.
44. Robert Snyder, ed., *Buckminster Fuller: Autobiographical Monologue Scenario* (New York: St. Martin's Press, 1980), p. 158.
45. Sidney Rosen, *Wizard of the Dome* (Boston: Little, Brown, 1969), p. 127.
46. Lecture Series by R. Buckminster Fuller, *The Future of Business,* Lake Tahoe, CA, 17–21 August 1981.
47. Ibid.
48. Ibid.
49. Calvin Tomkins, "Profiles, in the Outlaw Area," *The New Yorker* (6 January 1966), p. 40.
50. Ibid., p. 46.
51. R. Buckminster Fuller *Ideas and Integrities* (New York: Prentice-Hall, Inc., 1963), p. 9.
52. Lecture Series by R. Buckminster Fuller, University of Oregon, Eugene, 1–12 July 1962.
53. Ibid.
54. Ibid.
55. Lecture Series by R. Buckminster Fuller, *The Future of Business,* Lake Tahoe, CA, 17–21 August 1981.
56. Ibid.
57. Interview with R. Buckminster Fuller, Andrew D. Basiago, Pacific Palisades, CA, 20–21 October 1981.
58. Lecture, R. Buckminster Fuller *Children in the Universe,* University of California, Santa Barbara, 7 February 1973.

59. Ibid.

60. R. Buckminster Fuller, *Inventions: The Patented Works of R. Buckminster Fuller* (New York: St. Martin's Press, 1983), p. xvi.

61. World Game Lecture Series by R. Buckminster Fuller, New York University, New York, 14–21 July 1979.

62. Ibid.

63. Ibid.

64. Ibid.

65. Ibid.

66. "Reprint of talk given by R. Buckminster Fuller on June 5, 1958," *Journal of the Royal Institute of British Architects* (October, 1958).

67. Ibid.

68. Ibid.

69. Ibid.

70. World Game Lecture Series by R. Buckminster Fuller, New York University, New York, 14–21 July 1979.

71. Ibid.

72. Ibid.

73. World Game Lecture Series by R. Buckminster Fuller, Drexel University, Philadelphia, 1977.

74. Ibid.

75. World Game Lecture Series by R. Buckminster Fuller, University of Pennsylvania, Philadelphia, 1976.

76. Calvin Tompkins, "Profiles, in the Outlaw Area," *The New Yorker* (6 January 1966), p. 79.

77. Lecture Series by R. Buckminster Fuller, *The Future of Business,* Lake Tahoe, CA, 17–21 August 1981.

78. Calvin Tomkins, "Profiles, in the Outlaw Area," *The New Yorker* (6 January 1966), p. 78.

79. Lecture Series by R. Buckminster Fuller, *The Future of Business,* Lake Tahoe, CA, 17–21 August 1981.

80. Interview with R. Buckminster Fuller, Andrew D. Basiago, Pacific Palisades, CA, 20–21 October 1981.

81. Ibid.

82. Ibid.

83. Ibid.

84. Ibid.

85. Ibid.

86. R. Buckminster Fuller *Inventions: The Patented Works of R. Buckminster Fuller* (New York: St. Martin's Press, 1983), p. xviii.

87. Interview with R. Buckminster Fuller, Andrew D. Basiago, Pacific Palisades, CA, 20–21 October 1981.

88. Calvin Tomkins, "Profiles, in the Outlaw Area," *The New Yorker* (6 January 1966), p. 79.

89. William Yarnall, *Dome Builders Handbook No. 2* (Philadelphia: Running Press, 1978), p. 16.

90. Ibid.

91. Ibid.

92. World Game Lecture Series by R. Buckminster Fuller, Amherst, MA, 1978.

93. Alden Hatch, *Buckminster Fuller at Home in the Universe* (New York: Crown Publishers, Inc., 1974), p. 241.
94. World Game Lecture Series by R. Buckminster Fuller, Amherst, MA, 1978.
95. Ibid.
96. Lecture by R. Buckminster Fuller, Burklyn Business School, Kirkwood Meadows, CA, 1 August 1982.
97. Ibid.
98. Sidney Rosen, *Wizard of the Dome* (Boston: Little, Brown, 1969), p. 164.
99. Athena V. Lord, *Pilot for Spaceship Earth* (New York: Macmillan Publishing Co., Inc., 1978), p. 142.
100. World Game Lecture Series by R. Buckminster Fuller, New York University, New York, 14–21 July 1979.
101. R. Buckminster Fuller *Critical Path* (New York: St. Martin's Press, 1981), p. xxiv.
102. Lecture by R. Buckminster Fuller, Burklyn Business School, Kirkwood Meadows, CA, 1 August 1982.
103. Ibid.
104. Ibid.
105. World Game Lecture Series by R. Buckminster Fuller, New York University, New York, 14–21 July 1979.
106. Alden Hatch, *Buckminster Fuller at Home in the Universe* (New York: Crown Publishers, Inc., 1974), p. 243.

CHAPTER 16

1. Calvin Tomkins, "Profiles, in the Outlaw Area," *The New Yorker* (6 January 1966), p. 79.
2. Ibid.
3. World Game Lecture Series by R. Buckminster Fuller, University of Pennsylvania, Philadelphia, 1976.
4. Ibid.
5. Alden Hatch, *Buckminster Fuller at Home in the Universe* (New York: Crown Publishers, Inc., 1974), p. 240.
6. Ibid.
7. Lecture Series by R. Buckminster Fuller, *The Future of Business,* Lake Tahoe, CA, 17–21 August 1981.
8. World Game Lecture Series by R. Buckminster Fuller, University of Pennsylvania Museum, Philadelphia, 1975.
9. Ibid.
10. Ibid.
11. Ibid.
12. R. Buckminster Fuller, *Critical Path* (New York: St. Martin's Press, 1981), p. 50.
13. Ibid., p. 51.
14. World Game Lecture Series by R. Buckminster Fuller, University of Pennsylvania Museum, Philadelphia, 1975.
15. R. Buckminster Fuller, *Critical Path* (New York: St. Martin's Press, 1981), p. 51.
16. World Game Lecture Series by R. Buckminster Fuller, Amherst, MA, 1978.

17. Ibid.

18. Ibid.

19. Ibid.

20. R. Buckminster Fuller, *Critical Path* (New York: St. Martin's Press, 1981), p. 53.

21. Ibid.

22. Lecture Series by R. Buckminster Fuller, University of Oregon, Eugene, 1–12 July 1962.

23. "Reprint of Talk Given by R. Buckminster Fuller on June 5, 1958," *Journal of the Royal Institute of British Architects* (October 1958).

24. World Game Lecture Series by R. Buckminster Fuller, Drexel University, Philadelphia, 1977.

25. Interview with R. Buckminster Fuller, Andrew D. Basiago, Pacific Palisades, CA, 20–21 October 1981.

26. Calvin Tomkins, "Profiles, in the Outlaw Area," *The New Yorker* (6 January 1966), p. 95.

27. Ibid.

28. World Game Lecture Series by R. Buckminster Fuller, Drexel University, Philadelphia, 1977.

29. World Game Lecture Series by R. Buckminster Fuller, New York University, New York, 14–21 July 1979.

30. World Game Lecture Series by R. Buckminster Fuller, New York University, New York, 14–21 July 1979.

31. World Game Lecture Series by R. Buckminster Fuller, University of Pennsylvania Museum, Philadelphia, 1975.

32. Sidney Rosen, *Wizard of the Dome* (Boston: Little, Brown, 1969), p. 88.

33. Lecture Series by R. Buckminster Fuller, *The Future of Business,* Lake Tahoe, CA, 17–21 August 1981.

34. Ibid.

35. Ibid.

36. Lecture Series by R. Buckminster Fuller, University of Oregon, Eugene, 1–12 July 1962.

37. R. Buckminster Fuller, *Critical Path* (New York: St. Martin's Press, 1981), p. 226.

38. Ibid.

39. Ibid.

40. Lecture Series by R. Buckminster Fuller, University of Oregon, Eugene, 1–12 July 1962.

41. Lecture Series by R. Buckminster Fuller, *The Future of Business,* Lake Tahoe, CA, 17–21 August 1981.

42. Ibid.

43. Ibid.

44. World Game Lecture Series by R. Buckminster Fuller, University of Pennsylvania Museum, Philadelphia, 1975.

45. Interview with R. Buckminster Fuller, Andrew D. Basiago, Pacific Palisades, CA, 20–21 October 1981.

46. World Game Lecture Series by R. Buckminster Fuller, University of Pennsylvania Museum, Philadelphia, 1975.

47. World Game Lecture Series by R. Buckminster Fuller, Drexel University, Philadelphia, 1977.

48. Ibid.

49. Alden Hatch, *Buckminster Fuller at Home in the Universe* (New York: Crown Publishers, Inc., 1974), p. 250.

50. World Game Lecture Series by R. Buckminster Fuller, Drexel University, Philadelphia, 1977.

51. World Game Lecture Series by R. Buckminster Fuller, New York University, New York, 14–21 July 1979.

52. World Game Lecture Series by R. Buckminster Fuller, Washington, DC, 16–18 July 1982.

53. Ibid.

54. Ibid.

55. Hugh Kenner, *Bucky: A Guided Tour of Buckminster Fuller* (New York: William Morrow & Company, Inc., 1973), p. 35.

56. Ibid., p. 45.

57. R. Buckminster Fuller and Robert Marks, *The Dymaxion World of Buckminster Fuller* (Garden City, NY: Anchor Books, 1973; originally published by Southern Illinois University Press, 1960), p. 52.

58. Athena V. Lord, *Pilot for Spaceship Earth* (New York: Macmillan Publishing Co., Inc., 1978), p. 93.

59. Interview with R. Buckminster Fuller, Andrew D. Basiago, Pacific Palisades, CA, 20–21 October 1981.

60. Ibid.

61. R. Buckminster Fuller and Robert Marks, *The Dymaxion World of Buckminster Fuller* (Garden City, NY: Anchor Books, 1973; originally published by Southern Illinois University Press, 1960), p. 52.

62. Calvin Tomkins, "Profiles, in the Outlaw Area," *The New Yorker* (6 January 1966), p. 36.

63. Interview with R. Buckminster Fuller, Andrew D. Basiago, Pacific Palisades, CA, 20–21 October 1981.

64. R. Buckminster Fuller and Robert Marks, *The Dymaxion World of Buckminster Fuller* (Garden City, NY: Anchor Books, 1973; originally published by Southern Illinois University Press, 1960), p. 52.

65. Ibid.

66. World Game Lecture Series by R. Buckminster Fuller, Amherst, MA, 1978.

67. Ibid.

68. World Game Lecture Series by R. Buckminster Fuller, University of Pennsylvania, Philadelphia, 1976.

69. Ibid.

70. Ibid.

71. Ibid.

72. World Game Lecture Series by R. Buckminster Fuller, University of Pennsylvania Museum, Philadelphia, 1975.

73. Ibid.

74. Ibid.

75. Ibid.

76. World Game Lecture Series by R. Buckminster Fuller, University of Pennsylvania, Philadelphia, 1976.

77. Ibid.

78. Robert Snyder, ed., *Buckminster Fuller: Autobiographical Monologue Scenario* (New York: St. Martin's Press, 1980), p. 211.

79. Ibid.

80. Peter Meisen, "International Report," *Global Energy Network, International* (1987).

81. Ibid.
82. R. Buckminster Fuller, *Critical Path* (New York: St. Martin's Press, 1981), p. 221.
83. Interview with R. Buckminster Fuller, Andrew D. Basiago, Pacific Palisades, CA, 20–21 October 1981.
84. Ibid.
85. Ibid.
86. Calvin Tomkins, "Profiles, in the Outlaw Area," *The New Yorker* (6 January 1966), p. 40.
87. World Game Lecture Series by R. Buckminster Fuller, Washington, DC, 16–18 July 1982.
88. World Game Lecture Series by R. Buckminster Fuller, Amherst, MA, 1978.
89. Ibid.

CHAPTER 17

1. World Game Lecture Series by R. Buckminster Fuller, University of Pennsylvania Museum, Philadelphia, 1975.
2. Alden Hatch, *Buckminster Fuller at Home in the Universe* (New York: Crown Publishers, Inc., 1974), p. 253.
3. Lecture Series by R. Buckminster Fuller, *The Future of Business,* Lake Tahoe, CA, 17–21 August 1981.
4. Ibid.
5. Ibid.
6. Ibid.
7. Alden Hatch, *Buckminster Fuller at Home in the Universe* (New York: Crown Publishers, Inc., 1974), p. 254.
8. Lecture by R. Buckminster Fuller, Burklyn Business School, Kirkwood Meadows, CA, 1 August 1982.
9. Alden Hatch, *Buckminster Fuller at Home in the Universe* (New York: Crown Publishers, Inc., 1974), p. 225.
10. Lecture by R. Buckminster Fuller, Burklyn Business School, Kirkwood Meadows, CA, 1 August 1982.
11. Ibid.
12. Lecture Series by R. Buckminster Fuller, *The Future of Business,* Lake Tahoe, CA, 17–21 August 1981.
13. World Game Lecture Series by R. Buckminster Fuller, University of Pennsylvania Museum, Philadelphia, 1975.
14. Lecture Series by R. Buckminster Fuller, *The Future of Business,* Lake Tahoe, CA, 17–21 August 1981.
15. Ibid.
16. Calvin Tomkins, "Profiles, in the Outlaw Area," *The New Yorker* (6 January 1966), p. 36.
17. E. J. Applewhite, *Cosmic Fishing* (New York: Macmillan Publishing Co., Inc., 1977), p. 23.
18. Ibid., p. 28.
19. Ibid., p. 54.
20. Ibid., p. 1.
21. Ibid., p. 9.

22. Lecture by R. Buckminster Fuller, *Buckminster Fuller's Personal Odyssey*, New Dimensions Foundation, San Francisco.
23. Lecture Series by R. Buckminster Fuller, *The Future of Business*, Lake Tahoe, CA, 17–21 August 1981.
24. E. J. Applewhite, *Cosmic Fishing* (New York: Macmillan Publishing Co., Inc., 1977), p. 44.
25. Ibid., p. 30.
26. Ibid., p. 33.
27. World Game Lecture Series by R. Buckminster Fuller, University of Pennsylvania Museum, Philadelphia, 1975.
28. E. J. Applewhite, *Cosmic Fishing* (New York: Macmillan Publishing Co., Inc., 1977), p. 67.
29. Ibid., p. 70.
30. Ibid., p. 71.
31. Ibid., p. 73.
32. Ibid., p. 123.
33. Ibid., p. 125.
34. Ibid.
35. Ibid., p. 127.
36. Ibid., p. 128.
37. Ibid., p. 138.
38. Ibid., p. 139.
39. Ibid.
40. World Game Lecture Series by R. Buckminster Fuller, University of Pennsylvania Museum, Philadelphia, 1975.
41. R. Buckminster Fuller, *Synergetics* (New York: Macmillan Publishing Co., Inc., 1975), p. 109.
42. E. J. Applewhite, *Cosmic Fishing* (New York: Macmillan Publishing Co., Inc., 1977), p. 144.
43. Ibid., p. 145.
44. Ibid., p. 146.
45. World Game Lecture Series by R. Buckminster Fuller, University of Pennsylvania Museum, Philadelphia, 1975.
46. World Game Lecture Series by R. Buckminster Fuller, Washington, DC, 16–18 July 1982.
47. Calvin Tomkins, "Profiles, in the Outlaw Area," *The New Yorker* (6 January 1966), p. 38.
48. David Jacobs, "An Expo Named Buckminster Fuller," *New York Times Magazine* (23 April 1967).
49. Ibid.
50. World Game Lecture Series by R. Buckminster Fuller, New York University, New York, 14–21 July 1979.
51. Ibid.
52. Ibid.
53. World Game Lecture Series by R. Buckminster Fuller, Washington, DC, 16–18 July 1982.
54. Ibid.
55. Ibid.
56. R. Buckminster Fuller and Robert Marks, *The Dymaxion World of Buckminster Fuller* (Garden City, NY: Anchor Books, 1973; originally published by Southern Illinois University Press, 1960), p. 235.

57. Ibid.
58. R. Buckminster Fuller, *Critical Path* (New York: St. Martin's Press, 1981), p. 336.
59. Ibid.
60. World Game Lecture Series by R. Buckminster Fuller, Amherst, MA, 1978.
61. Ibid.
62. Ibid.
63. World Game Lecture Series by R. Buckminster Fuller, New York University, New York, 14–21 July 1979.
64. World Game Lecture Series by R. Buckminster Fuller, Amherst, MA, 1978.
65. R. Buckminster Fuller, *Critical Path* (New York: St. Martin's Press, 1981), p. 317.
66. Ibid., p. 316.
67. World Game Lecture Series by R. Buckminster Fuller, New York University, New York, 14–21 July 1979.
68. Ibid.
69. Interview with Jaime Snyder, Lloyd Sieden, February 1988.
70. R. Buckminster Fuller, *And It Came to Pass—Not to Stay* (New York: Macmillan Publishing Co., Inc., 1976), p. 56.

Index